Papermaking Science and Technology

a series of 19 books covering the latest technology and future trends

Book 14

Process Control

Series editors
Johan Gullichsen, Helsinki University of Technology
Hannu Paulapuro, Helsinki University of Technology

Book editor
Kauko Leiviskä

Series reviewer
Brian Attwood, St. Anne's Paper and Paperboard Developments Ltd

Book reviewers
Donald F. Swing, Jr., Mead Coated Board, Inc.
Henry X. Swanson, Mead Corporation

Published in cooperation with the Finnish Paper Engineers' Association and TAPPI

ISBN 952-5216-00-4 (the series)
ISBN 952-5216-14-4 (book 14)

Published by Fapet Oy
(Fapet Oy, PO BOX 146, FIN-00171 HELSINKI, FINLAND)

Copyright © 1999 by Fapet Oy. All rights reserved.

Printed by Gummerus Printing, Jyväskylä, Finland 1999

Printed on LumiMatt 100 g/m^2, Enso Fine Papers Oy, Imatra Mills

Certain figures in this publication have been reprinted by permission of TAPPI.

Foreword

Johan Gullichsen and Hannu Paulapuro

PAPERMAKING SCIENCE AND TECHNOLOGY

Papermaking is a vast, multidisciplinary technology that has expanded tremendously in recent years. Significant advances have been made in all areas of papermaking, including raw materials, production technology, process control and end products. The complexity of the processes, the scale of operation and production speeds leave little room for error or malfunction. Modern papermaking would not be possible without a proper command of a great variety of technologies, in particular advanced process control and diagnostic methods. Not only has the technology progressed and new technology emerged, but our understanding of the fundamentals of unit processes, raw materials and product properties has also deepened considerably. The variations in the industry's heterogeneous raw materials, and the sophistication of pulping and papermaking processes require a profound understanding of the mechanisms involved. Paper and board products are complex in structure and contain many different components. The requirements placed on the way these products perform are wide, varied and often conflicting. Those involved in product development will continue to need a profound understanding of the chemistry and physics of both raw materials and product structures.

Paper has played a vital role in the cultural development of mankind. It still has a key role in communication and is needed in many other areas of our society. There is no doubt that it will continue to have an important place in the future. Paper must, however, maintain its competitiveness through continuous product development in order to meet the ever-increasing demands on its performance. It must also be produced economically by environment-friendly processes with the minimum use of resources. To meet these challenges, everyone working in this field must seek solutions by applying the basic sciences of engineering and economics in an integrated, multidisciplinary way.

The Finnish Paper Engineers' Association has previously published textbooks and handbooks on pulping and papermaking. The last edition appeared in the early 80's. There is now a clear need for a new series of books. It was felt that the new series should provide more comprehensive coverage of all aspects of papermaking science and technology. Also, that it should meet the need for an academic-level textbook and at the same time serve as a handbook for production and management people working in this field. The result is this series of 19 volumes, which is also available as a CD-ROM.

When the decision was made to publish the series in English, it was natural to seek the assistance of an international organization in this field. TAPPI was the obvious partner as it is very active in publishing books and other educational material on pulping and papermaking. TAPPI immediately understood the significance of the suggested new series, and readily agreed to assist. As most of the contributors to the series are Finnish, TAPPI provided North American reviewers for each volume in the series. Mr. Brian Attwood was appointed overall reviewer for the series as a whole. His input is gratefully acknowledged. We thank TAPPI and its representatives for their valuable contribution throughout the project. Thanks are also due to all TAPPI-appointed reviewers, whose work has been invaluable in finalizing the text and in maintaining a high standard throughout the series.

A project like this could never have succeeded without contributors of the very highest standard. Their motivation, enthusiasm and the ability to produce the necessary material in a reasonable time has made our work both easy and enjoyable. We have also learnt a lot in our "own field" by reading the excellent manuscripts for these books.

We also wish to thank FAPET (Finnish American Paper Engineers' Textbook), which is handling the entire project. We are especially obliged to Ms. Mari Barck, the project coordinator. Her devotion, patience and hard work have been instrumental in getting the project completed on schedule.

Finally, we wish to thank the following companies for their financial support:

A. Ahlstrom Corporation

Enso Oyj

Kemira Oy

Metsä-Serla Corporation

Rauma Corporation

Raisio Chemicals Ltd

Tamfelt Corporation

UPM-Kymmene Corporation

We are confident that this series of books will find its way into the hands of numerous students, paper engineers, production and mill managers and even professors. For those who prefer the use of electronic media, the CD-ROM form will provide all that is contained in the printed version. We anticipate they will soon make paper copies of most of the material.

List of Contributors

Leiviskä Kauko, Professor, Department of Process Engineering, University of Oulu

Nyberg Timo, Professor, Automation Department, Tampere University of Technology

Tornberg Jouni, Chief Research Scientist, VTT Electronics, Oulu

Preface
Kauko Leiviskä

Automation is a necessary part in mill investments and its role is increasing when technology is advancing. It is needed to complete operatorsí activities in difficult and demanding situations and provide mill staff and other people with timely and precise information on raw materials, production, products and other things effecting mill and its environment.

Automation is a multidisciplinary area. It requires knowledge from electronics, information technology, mathematics, work sciences, etc. Good process knowledge is crucial in actual installations. In writing a book on process automation, dealing with all areas in impossible. In this book the viewpoint has been on process control. It is seen already in the name. The aim has been to study the status and future views of control methods, control systems and process analysis methods that are becoming essential in guiding process operations and control design.

Excellent books on instrumentation level applications are available. The emphasis in this book is supervisory level systems and millwide control. Special attention has been given to intelligent systems that is an evolving new area of applications.

As editor of this book, I would like to thank the series editors, professor Johan Gullichsen and professor Hannu Paulapuro, for setting guidelines for this work and for their valuable comments during the preparation of this book. I also thank my co-authors for their contributions and the coordinator of this book series project, Ms. Mari Barck for patience during all the phases of text preparation and editing. I would also like to thank all my colleagues and partners whom I had the possibility to work with during so many years. You see many of their names in the reference lists of this book.

Finally, I address my gratitude to my wife, Eija, and my children, Minja and Marko, for their endless patience, and for taking our common time during the preparation of this book.

In Oulu, December 12, 1998

Kauko Leiviskä

Table of Contents

1. Introduction 10
2. Control systems 13
3. Control methods 21
4. Special measurements in pulp and paper processes 49
5. Process control in fiber line 73
6. Process control in chemical recovery 101
7. Process control in mechanical pulping 129
8. Process control in paper mills 147
9. Millwide control 197
10. Process analysis, modeling, and simulation 249
    Conversion factors 294
    Index 296

CHAPTER 1

Introduction

CHAPTER 1

Kauko Leiviskä

Introduction

The amount of automation in total mill investments varies from a few percentage points to 30%–40%. Small investments in automation are typical for mills that produce one or few products with almost constant throughput. Larger investments in automation are common in mills that produce a variety of products from varying raw material sources and that use batch and continuous processes. Increasing requirements for constant, high quality have led to investments in quality analyzers, quality controllers, and inspection and monitoring systems.

What kind of operations and functions require automation? How should the monitoring and control divide between operators and process automation? These questions are difficult to answer, and no general advice or preferences are possible. Continuously controlling complicated processes optimally is very difficult for an operator. He can manage it for a time but usually loses attention during a long run. Differences between operators cause variations in product quality, process operations, etc. Rapid variations and changes in processes are also difficult for humans. Their reactions may not be sufficiently fast. Physically heavy and dangerous operations also require automation. Difficult operations and design and planning problems that repeat after long periods may require computerized support. Otherwise the human operator may not recall the best solutions available.

The degree of automation describes the share of work between the human and the automation system. Some functions can be manual, automatic, or something between these two variants. The reader should note that the degree of automation is not a numerical measure. It only has qualitative meaning.

If direct operation and safety aspects do not determine the degree of automation investment, the costs of automation and the expected savings potential of the investment will determine the investment. The costs of automation are reasonably easy to define, but the estimation of the savings potential requires process knowledge and experience.

The key to successful control is accurate and reliable measurement. In the pulp and paper industry, special measurements and analyzers measuring the quality variables of processes have received primary attention. Several areas remain for future development.

Control problems usually fall into three categories:

- Controlling process variables (temperatures, pressures, flow rates, etc.) to maintain them at the required constant values (setpoints) despite any disturbances acting on them. Terms such as "stabilizing controls" or "regulatory controls" are common. Examples occur everywhere in the process industries.

Introduction

- Performing controlled changes in the states of process variables. This is "servo controls" or "transition controls." For example, the changing of batch reactor temperature according to predetermined programs belongs to this category. More difficult problems occur in managing production rate or quality changes in paper machines or in pulp mill fiber lines. In these cases, several variables require manipulation using a predetermined schedule.
- Defining the optimum values for process variables according to quality or production. These values are usually the setpoints for stabilizing or regulatory controls. This optimization can be steady state or dynamic.

Millwide control is an important area in the pulp and paper industry. It concentrates on coordinating and controlling production, energy, and quality in the entire mill and at the company level. Considerable advantages are possible, but modern data transmission technology still offers additional potential for millwide control.

The following chapters consider the state-of-the-art and future developments in process control in the pulp and paper industry. Chapters 2 and 3 survey the systems and methodological development. Chapter 4 discusses special measurements, and Chapters 5 and 6 cover chemical pulp mill systems. Control of mechanical pulping is the topic of Chapter 7. Chapter 8 discusses paper mill systems. Chapter 9 introduces millwide applications, and Chapter 10 covers the methods and systems applied in process analysis, modeling, and simulation.

CHAPTER 2

Control systems

1	**Development of process automation**	15
2	**Development of digital automation systems**	16
3	**Hierarchical system structure**	17
	References	19

CHAPTER 2

Kauko Leiviskä

Control systems

1 Development of process automation

The development of automation systems started in the 1950s with centralized control rooms and standardized systems. Developments in electrical instrumentation systems found use in all automation functions. Their development resulted in the functionally oriented automation systems of the 1960s. These had separate functional parts for measurements, controls, monitoring, and interlocking.

Computerized automation dates from the early 1960s. The first paper machine control was at Potlach Forests, Inc. in Idaho in 1961. The first digester control system was at Gulf States Paper in Alabama in 1962[1]. The initial application in Europe was at Enso-Gutzeit Oy Kaukopää Mills in Finland in 1963[2]. Large business computers of that time were the first computer control systems. Centralized systems meant that one large computer controlled the entire paper mill. The results were not necessarily good. Technology development was minimal, and the applicability of computers for real time control problems was insufficient. A lack of experience, special measurements, and trained staff also existed.

The 1970s saw the development of dedicated process computers and their application to the control of different subprocesses of pulp and paper mills. Package systems for paper machines, digester houses, etc., with specialized vendors for these systems came into the systems market. This development was made possible because of new instrumentation for paper machines. The computers were reliable and could handle real time systems. Organizations also changed. One branch in an organization handled business computers, and another area was responsible for process computers. The first digital automation systems came to the market at the end of the 1970s.

Modern process automation uses the distribution of control functions in the best possible way by dividing the overall automation problem into smaller independent parts. Simultaneously, collecting information from local control units and displaying it in centralized control rooms centralizes process monitoring. This requires efficient data transfer capabilities. Modern automation systems are undoubtedly the result of versatile knowledge including digital technology, data transmission technology, software engineering, ergonomics, etc. Actual installations rely on process knowledge. Knowing what to do and why is imperative with this new technology.

Development continued during the 1990s. Systems integration has continued on mill and corporate levels, the importance of data networks has increased, and mill networks have undergone integration with office networks and general data highways. Mill-wide automation, office automation, and personal data processing have also become integrated. All information system functions are becoming increasingly common.

CHAPTER 2

Besides automatic, real time control functions, this also requires advisory systems that support the user in decision making and also consult the user concerning the reasons for proposed decisions. Continuous development of user interfaces and the computer tools available to the user are necessary. Such development seems to be leading toward intelligent, knowledge-based systems. Figure 1 shows this development[3].

Standard instrumentation systems	Functionally oriented systems First process computers	Minicomputers Packages First digital automation systems	Distributed systems Hierarchical systems Integration	Networking Personal computing Knowledge systems
1950s	1960s	1970s	1980s	1990s

Figure 1. Development of process automation[3].

2 Development of digital automation systems

The extended use of distributed systems has meant an increased need for data transfer. This was evident in the data transfer inside the automation systems such as the transfer of data from process stations to operator stations. This data transfer becomes even more important when connecting different automation systems. Data transfer is especially important in data acquisition to millwide systems. In these cases, the mill computer must connect to process automation systems and process computers. Today and in the predictable future, we must live with multi vendor systems with only limited standardization. Adding new equipment and programs to existing systems is very difficult especially with older hardware and software.

Process automation is passing through three stages toward an open systems architecture that can connect process automation with any data processing activities in the mill and its surroundings[4].

Until the 1980s, process control used stand-alone computers or automation systems essentially tailored according to the application as Fig. 2 shows. Applications had their own databases, and the naming

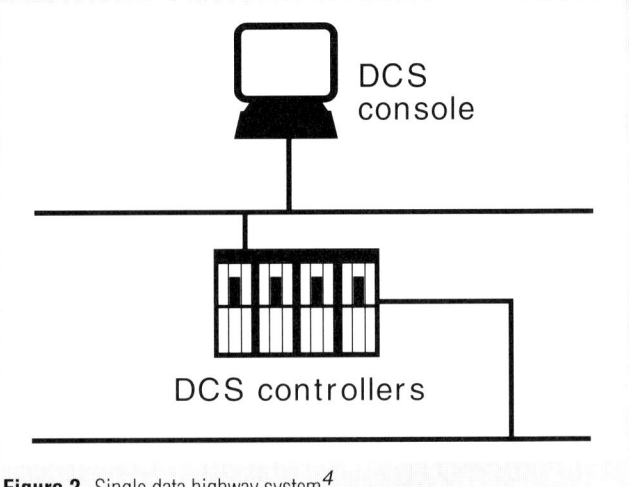

Figure 2. Single data highway system[4].

of variables occurred in applications. User interfaces were also tailored separately for each application. The applications were single-vendor based and interoperability of the systems was essentially impossible. The disadvantages of these closed systems became apparent especially for millwide systems.

Figure 3 shows that existing systems are multiple network systems. Data transfer between the systems of different vendors uses different data buses and networks. The solutions still need tailoring in many cases, and the interoperations have limits. This is important in systems renewal, since renovating one part could influence the entire automation system.

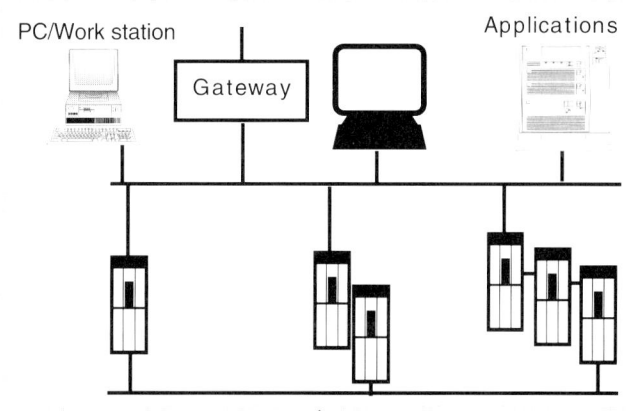

Figure 3. Multiple network solution[4].

The trend today is toward open systems using a single standard network as Fig. 4 shows. All subsystems connect to this network. In addition to databases, they can also share applications and user interfaces. User interfaces will become increasingly dependent on personal computers and work stations. These two technologies are approaching each other. Interfaces will also facilitate the use of general program environments for spreadsheet, text processing, and various analyses including process, control, etc. The target is to integrate information to have it readily available everywhere for all applications. This means an open system environment.

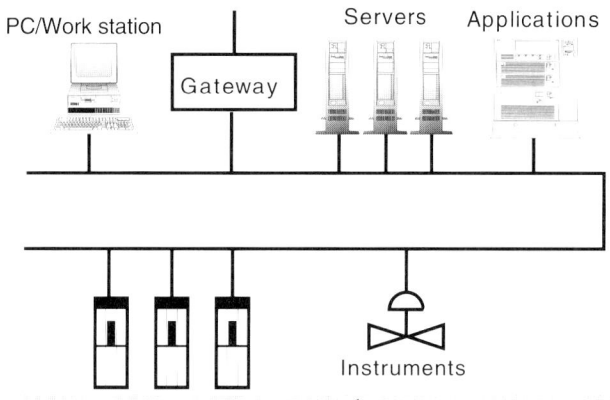

Figure 4. Standard open network solution[4].

3 Hierarchical system structure

Hierarchical presentation is a common way to describe automation systems especially for millwide automation. Figure 5 shows one possible way to use this approach. It is a modification of earlier work[1].

CHAPTER 2

The direct control level and the supervisory control level are near the actual process. The direct control level is responsible for controlling single process variables (flows, temperatures, pressures, etc.) using the setpoints given by operators or the supervisory control level. The supervisory control level calculates the setpoints so that the process or a set of processes operate in the best possible way. Area planning and coordination take care of monitoring and control of production areas such as fiber lines and recovery cycles in a pulp mill. Mill planning and coordination are responsible for the same operations at the mill level. They must define production rate and quality requirements for single processes so that the targets set for the entire mill are met. The upper planning and coordination levels are millwide control levels. They emphasize the real time, on-line requirements for control.

The hierarchical presentation usually refers to operations at each level. Risks are associated with the presentation of hardware and organizational questions using this formal method. Figure 5 shows one such trial. The system realization is mill dependent, and the figure is an oversimplification.

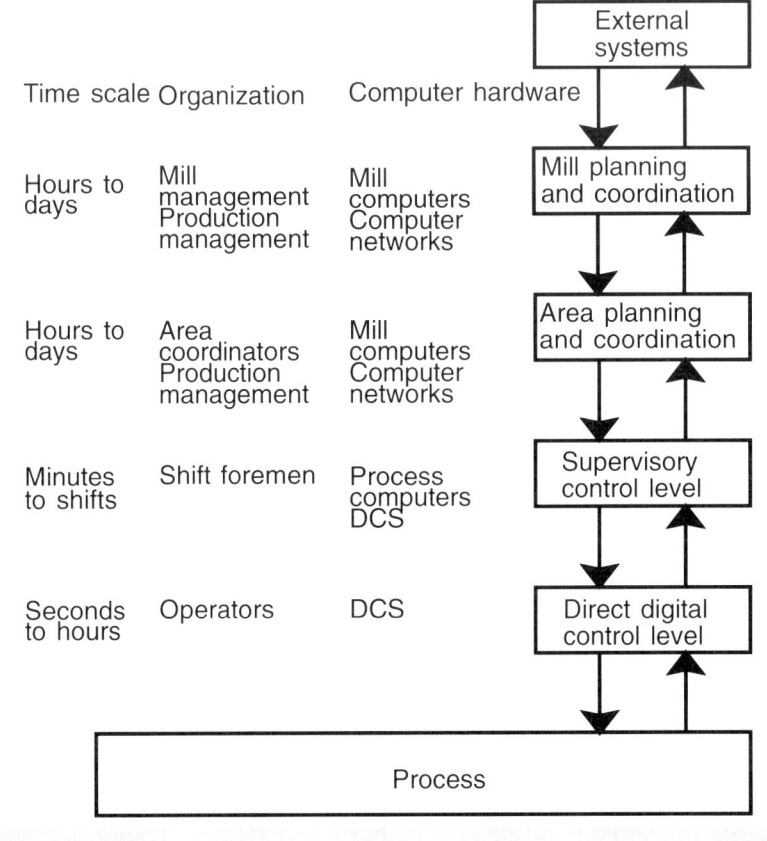

Time scale	Organization	Computer hardware	
			External systems
Hours to days	Mill management Production management	Mill computers Computer networks	Mill planning and coordination
Hours to days	Area coordinators Production management	Mill computers Computer networks	Area planning and coordination
Minutes to shifts	Shift foremen	Process computers DCS	Supervisory control level
Seconds to hours	Operators	DCS	Direct digital control level
			Process

Figure 5. Control hierarchy.

References

1. Uronen P. and Williams T. J., Hierarchical computer control in the pulp and paper industry, Report Number 111, Purdue Laboratory for Applied Industrial Control, Purdue University, 1978, pp.7-10.
2. Korhonen J., in Paper Manufacturing (A. Arjas, Ed.), part 2, Teknillisten Tieteiden Akatemia, Turku, 1983, pp.877-910.
3. Leiviskä K., "Development of process automation from control loops to millwide control," Inaugural Speech, May 9, 1990. University of Oulu. (In Finnish).
4. Furness H., Control Engineering 7:63(1992).

CHAPTER 2

CHAPTER 3

Control methods

1	**Open- vs. closed-loop control**	**21**
2	**Digital control algorithms**	**21**
3	**Automatic tuning of digital PID-algorithms**	**23**
4	**Adaptive control**	**24**
4.1	Gain scheduling	25
4.2	Model reference control	25
4.3	Self-tuning control	26
5	**Multi variable control**	**26**
5.1	Basic concepts	27
5.2	Decoupling	28
5.3	Optimal state controller	28
6	**Expert systems**	**29**
6.1	Operating principles	29
6.2	Expert systems in process control	29
	6.2.1 Stabilizing control	30
	6.2.2 Supervisory control	31
6.3	Other applications	31
	6.3.1 Diagnostics	31
	6.3.2 Expert systems in scheduling	32
	6.3.3 DSS applications	32
6.4	Advantages	32
6.5	Problems	33
7	**Fuzzy logic control**	**34**
7.1	What does fuzziness mean?	34
7.2	Fuzzy logic controller	35
	7.2.1 Fuzzification	35
	7.2.2 Fuzzy reasoning	36
	7.2.3 Defuzzification	36
8	**Neural networks**	**36**
8.1	Basic definitions	37
8.2	How do the networks learn?	37
8.3	Network structures	38
	8.3.1 Perceptron network	38
	8.3.2 Back propagation networks	39
	8.3.3 Competitive learning	39

CHAPTER 3

9	**Statistical process control**	**40**
9.1	Presentation of quality data	40
9.2	Cause-and-effect analysis	41
9.3	Pareto analysis	41
9.4	Control charts	42
9.5	Cusum charts	43
9.6	Capability indices	45
	References	46

CHAPTER 3

Kauko Leiviskä

Control methods

1 Open- vs. closed-loop control

Open-loop control does not use measurement of the process variable to be controlled. The effect goes directly from the input or control action to the output as Fig. 1 shows. Closed-loop control uses the measure-ment of the output variable to form a closed loop. The corrections use the deviation of the measured output from its setpoint or target value. This is a control loop, and the approach is the feedback control as Fig. 2 shows. Table 1 shows a compari-son between open-loop and closed-loop control.

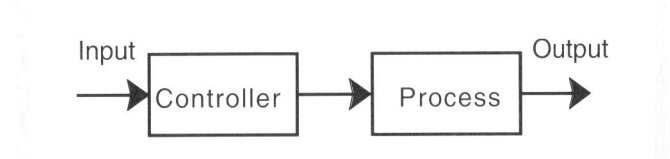

Figure 1. Open-loop control.

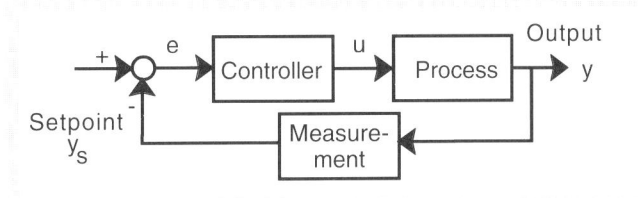

Figure 2. Closed-loop control.

Table 1. Comparison between open-loop and closed-loop control.

	Open-loop	Closed-loop
Measurement	(Disturbance)	Output
Correction	Based on disturbance	Based on deviation
Hardware	Simpler, cheaper	
Tuning requirements	Strong	Easier
Stability problems	Not usual	Sometimes difficult
Efficiency	Takes only one disturbance into account	Takes all disturbances into account

2 Digital control algorithms

PID-algorithms or the corresponding three-term algorithms have wide use in digital con-trollers. This is due to their extensive application range and ease of tuning. Experiences from conventional analog controllers also contribute to their usefulness.

CHAPTER 3

The continuous PID-algorithm has the following form:

$$u(t) = K_C \left[e(t) + \frac{1}{T_R} \int_0^t e(t)dt + T_D \frac{de(t)}{dt} \right] \quad (1)$$

where $u(t)$ is the control
 $e(t)$ the deviation
 K_C the controller gain
 T_R the time of integration
 T_D the time of derivation.

This algorithm must be discretized for digital control. Usually, the simplest forms of discretization are sufficient for control purposes. The following approximation replaces the integral part of Eq. 1:

$$\int_0^t e(t)dt = \sum_{i=0}^n e_i T = T \sum_{i=0}^n e_i \quad (2)$$

where T is the sampling time.

The sum shown above is usually not calculated for each sampling period but saved in memory. The following simpler form has use:

$$S_n = S_{n-1} + e_n \quad (3)$$

$$S_n = \sum_{i=0}^n e_i \quad (4)$$

A digital PID-algorithm is a combination of the equations shown above:

$$u_n = u_o + K_C \left[e_n + \frac{T}{T_R} S_n + \frac{T_D}{T}(e_n - e_{n-1}) \right] \quad (5)$$

where u_o is the base value for the control variable.

Setting the value for the sum (Eq. 4) to zero and the value for the base value to correspond to the existing value of the control variable guarantees an *effortless* change from manual to digital control. Setting the value for the deviation to zero temporarily eliminates the effects of other terms.

One weakness of this algorithm occurs in situations where the control variable meets its limits, but the deviation is not zero. In these cases, the sum continues to increase or decrease until the deviation becomes zero and changes from positive to negative or vice versa. Even after this, a longer time can occur until the value of the control variable changes. This results in a considerable overshoot in the response.

Limiting the value of sum S, i.e., by freezing it when u reaches its upper or lower limit, improves this situation. This leads to an erroneous value of S when the deviation changes its sign later.

Replacing Eq. 5 by the following avoids this difficulty:

$$u_n = u_{n-1} + \Delta u_n \tag{6}$$

Calculating the correction to the control variable then uses the following:

$$\Delta u_n = K_C\left[e_n - e_{n-1} + \frac{T}{T_R}e_n + \frac{T_D}{T}(e_n - 2e_{n-1} + e_{n-2})\right] \tag{7}$$

This equation does not have the base value for the control variable or the sum S. The control now uses one more error or deviation value than before. When switching to automatic control, the control variable obtains its momentary value. Simply setting limits to the control variable avoids wind-up.

A major difference between analog and digital PID-controllers concentrates on the use of the sampling time in digital algorithms. With short sampling times (half of the time constant at maximum), one can apply the same tuning principles as in analog control. Otherwise, special methods developed for digital control are necessary. Earlier publications discussed controller tuning[1,2].

3 Automatic tuning of digital PID-algorithms

Automatic controller tuning tries to use the increased data processing capacity of digital automation systems and single-loop controllers to facilitate controller tuning during start-ups and normal operation. This usually requires generation of the

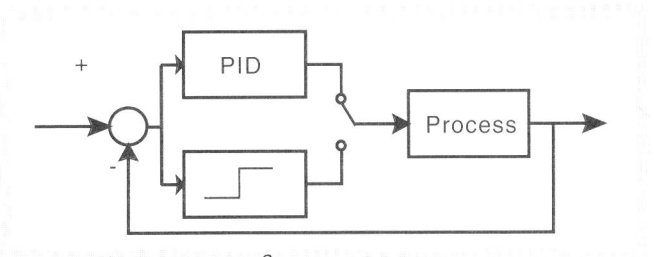

Figure 3. Åström's autotuner[3].

necessary test signal, data collection, calculation of process parameters using the process data, and the calculation of the tuning parameters with some given criteria. The user starts the tuning procedure that is not necessarily automated. In this sense, the automatic tuning methods differ from the adaptive self-tuning control discussed later.

An original method was the autotuner proposed by Åström in 1984. It uses the well-known Ziegler-Nichols method for defining critical frequency and critical gain[3]. In the original method, the controller gain increased until the control loop started to oscillate. Since this is difficult to automate, Åström's method connected a relay or other nonlinear component in the control loop to generate the oscillation as Fig. 3 shows.

Autotuning is a common feature in commercial controllers and automation systems today. Tuning of PID-controllers usually consists of the following stages[4]:

- Data collection from the control loop. The amount, quality, and correctness of data also define the quality of tuning.

- Process test during the data collection to determine the process dynamics. Usually, data collected during normal operation does not contain sufficient information for controller tuning. This then requires artificial changes to process operation. Several test signals such as step, pulse, random, and pseudo random have found use. The comparison of amplitude and duration of signals with process characteristics also requires consideration.

- Modeling of the process. The model type depends on the process. Linear fixed parameter models in time or frequency domain are often common. Test signals also support this fact. Tests occur in a narrow area around the fixed operation point. They do not reveal any nonlinear dynamics.

- Computation of tuning parameters. Usually, some rules of thumb such as the well-known Ziegler-Nichols rules find use.

- Commissioning and testing of results. This stage reveals the correctness of tuning parameters. Models describe the situation during the modeling period, and varying process conditions may cause a need for fine-tuning.

Pennanen[4] introduced a tuning software package based on commercial tools and user interface possibilities. It uses random binary signals (RBS) for testing and auto regressive moving average process with exogenous input (ARMAX) models or a simple first order model with dead time. PC software tuning packages are also available that gather data either via a link to the controller or to an externally attached I/O package for high speed data acquisition.

4 Adaptive control

An adaptive controller can change its tuning or structure when necessary. This facilitates controller tuning and guarantees the optimum performance of the controller when the behavior of the process is changing. Adaptive actions can occur continuously or batch-wise.

Adaptive control is necessary when performance of the conventional fixed parameter feedback controller is insufficient. This is due to a nonlinear process or actuator or to the simple fact that the behavior of the process is time dependent or unknown. In addition, dead time processes and processes with strong interactions can require adaptive features from controllers.

Adaptive control is not new technology. Although the first patents date from the 1950s, the main development work occurred during the 1970s. Recent hardware and software development has made this technology feasible for process automation. In the pulp and paper industry, the literature reports applications of adaptive control in thermomechanical pulp plants[5, 6] and in continuous digester control[7, 8].

Three main approaches of adaptive control exist: gain scheduling, model reference adaptive control, and self-tuning control. The following text presents a short review of each approach. Detailed descriptions and algorithmic solutions are available in any textbook on adaptive control[9].

4.1 Gain scheduling

In practice, a certain process variable such as the production rate usually explains most changes in process dynamics. Production rate changes also change the time constants and dead time of processes. In such cases, tabulation of controller parameters can be a function of production rate or any corresponding variable.

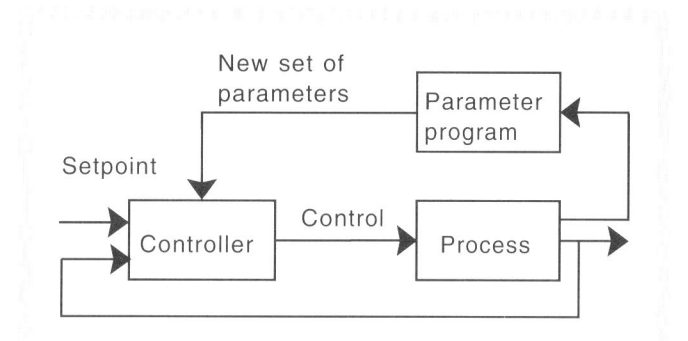

Figure 4. Gain scheduling.

When changes occur, these tables are useful as a source of controller tuning parameters. One should use functional descriptions instead of tables when possible. Figure 4 shows a block diagram for the gain scheduling approach.

4.2 Model reference control

Model reference control uses a model that describes the ideal behavior of the system when the setpoint is changing. The tuning mechanism then forces the response of the real system to follow the dynamics of the reference model by changing the controller tuning.

Figure 5 shows a block diagram of the model

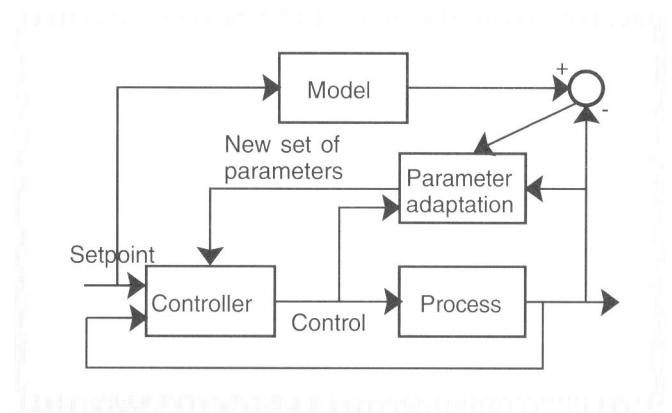

Figure 5. Block diagram of a model reference controller.

reference controller. The control loop consists of two separate loops. One is the conventional feedback loop, and the other includes the controller tuning mechanism. One could also draw the corresponding diagram to describe the situation where tuning changes according to some measurable load disturbance.

The biggest problem in constructing model reference controllers is finding the tuning mechanism that guarantees the controller stability and simultaneously minimizes the control error. Also, lack in model accuracy can cause a problem with controller and tuning stability.

4.3 Self-tuning control

A self-tuning controller offers some advantages in situations where the controller requires continuous tuning because of the disturbance behavior of the process as one example. According to Fig. 6, the self-tuning controller has three parts: the controller with tunable parameters, parameter estimator, and tuning algorithm.

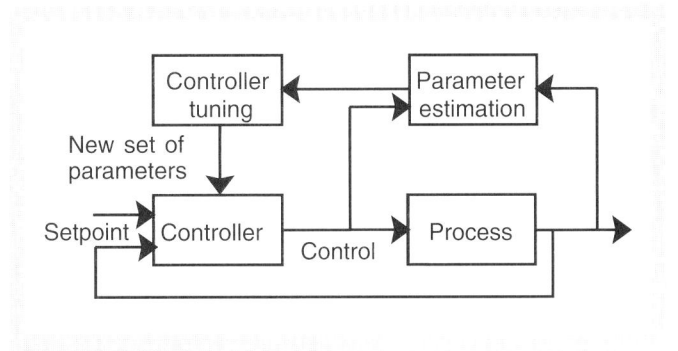

Figure 6. Self-tuning controller.

Parameter estimation continuously calculates parameters in a process model using measurements of control and output variables. The tuning part calculates new tuning parameters for the controller. Several variations concerning the algorithmic realization of the functional blocks exist. The controller tuning can use required gain and phase margins, pole placement, minimum variance control, etc.

In practice, self-tuning control faces several problems. The most difficult is the drifting of controller parameters when the process is stable and no changes occur. One possibility is to use the self-tuning feature only in tuning the controller and forget it when no changes occur. This means loss of some advantages of the self-tuning approach.

5 Multi variable control

Multi variable control means the replacement of conventional compensation technology by computerized methods. The aim is to compensate the effect of process interactions and make the entire process behave optimally.

Many possibilities are available to handle multi variable systems. One is to use analog compensators that try to consider interactions. The computer control uses the upper level supervisory actions based on process models for the same purpose. This requires process models and the conversion of process knowledge to the actual control strategy. These alternatives seem tempting, because the actual control loop can be realized by conventional PID-controllers with all their advantages.

Control methods

Multi variable control methods are common, and many research reports and papers are available. Practical applications seem rare. Optimum state controllers require considerable computer resources in the design and tuning stages. This has impeded their progress.

The advantages and applicability of multi variable methods require consideration on a case-by-case basis. No general advice is available. One must always compare the advantages with the increased maintenance and operability requirements. The following discussion shows the principles of some multi variable strategies. Algorithms and design methods are available in textbooks[10, 11].

5.1 Basic concepts

A multi variable process has several inputs and outputs according to Fig. 7. The inter-actions between variables are often strong. This means that changes in inputs (control variables) influence several output variables. The strength of these interactions defines how well the conven-tional feedback control strategy controls the multi variable process.

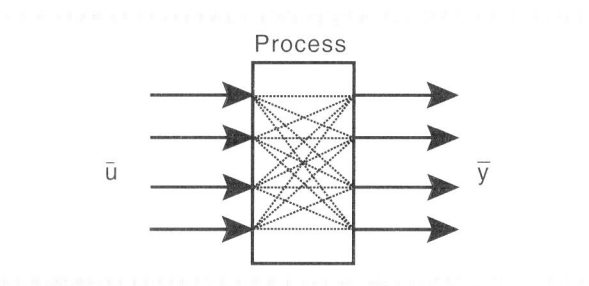

Figure 7. A multi variable process.

The simplest way to handle multi variable processes is to ignore the interactions and design simple, separate control loops as Fig. 8 shows. This approach works sufficiently for most cases. If severe interactions exist, an oscillating or even unstable system can result. Careful selection of input and output pairs and realistic tuning of control loops improves this situation.

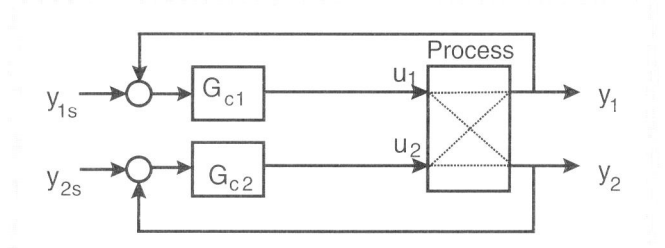

Figure 8. Control of multi variable system using diagonal control strategy.

Multi variable control is necessary in cases where separate control loops influence each other. This results in decreased control performance of the previously mentioned strategy when interacting control loops respond to setpoint changes or load disturbances. To assure stability, the separate loops must have excessive tuning. This decreases their performance.

5.2 Decoupling

One possibility to decrease the effect of interactions is to add a compensator into a diagonal control structure as Fig. 9 shows. The design of this compensator starts by using the situation where the selection of input and output pairs is as good as possible. The compensator and the process should form a system that is as diagonal as possible. Then the control using the separate controllers is possible. Computerized design methods for these compensators exist.

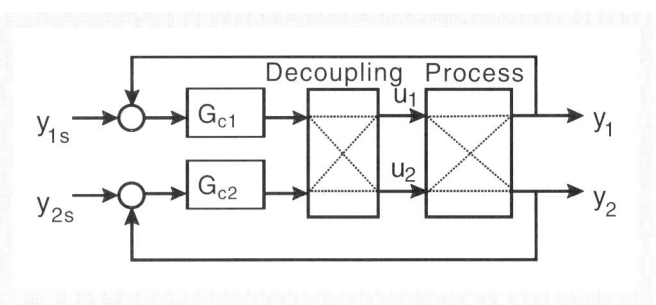

Figure 9. Decoupling controller for 2x2-process.

5.3 Optimal state controller

An optimal state controller minimizes some performance index or objective function. This can include the minimization of deviation between the existing state and its setpoint or minimization of controls or their changes. In optimum state control, the process model is a state equation instead of the transfer function used in previous methods. These presentations convert easily to each other. All state variables should be measurable, or they require estimation. Figure 10 shows the structure of the optimal state controller in the case of a system with two inputs and two outputs.

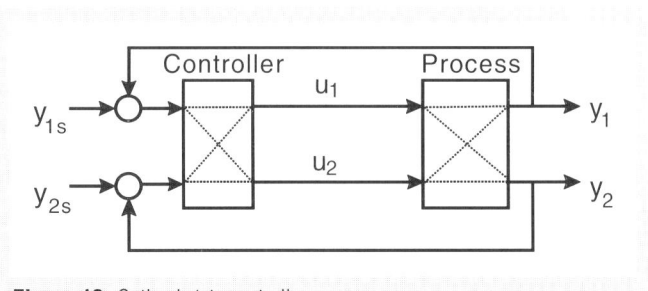

Figure 10. Optimal state controller.

The following features are essential in optimal state control:

- A suitable model must describe the process. If no reliable and robust model exists, no reason exists to apply state control.

- All states should be measurable. If fewer measurements exist than states or if noise corrupts the measurements, state estimators are necessary.

- Changing the weight factors in the objective function tunes the controllers. They usually lack any physical meaning.

- The actual controller can be realized without any computer support. It is necessary in the controller design.

6 Expert systems

The number of reported applications of expert systems has increased rapidly in all fields of engineering. Previous estimates indicated that 300–400 systems are in use in the pulp and paper industry[12, 13]. Applications include processes and the entire field. Typical applications are for quality and productivity management. Research work is very active, and the development rate is high. The technology seems to be maturing. Risks connected with industrial applications are therefore becoming lower.

The role of expert systems depends on the level of control where they are applied. At the lowest control level, the requirements of the actual, time critical control functions must be considered. The integration with existing conventional control systems and with control methods and algorithms is also essential.

On the upper levels in diagnosis and decision support systems (DSS), man with machine interface and explanation mechanisms play an important role. Integration of expert systems with existing mathematical models and methods is important. The use of expert systems also emphasizes storing the primary expert knowledge systematically. On the upper levels, the acquisition, validation, and updating of knowledge are essential. These are problem areas. The following sections come from earlier literature[14].

6.1 Operating principles

Figure 11 shows the main parts and operation of expert systems and any kind of rule-based system.

Dynamic working memory includes starting information, reasoning targets, and the prevailing state of reasoning. This receives updating during the problem solving process. In rule-based systems, an inference engine controls problem solving, interprets rules, and initiates necessary actions. An explanation mechanism shows how reasoning has proceeded from starting information to decisions. It aims to prove decisions correct and teach the user. The knowledge base includes domain knowledge such as facts and rules stored in the computer memory using some presentation methods. Knowledge acquisition means expertise transfer from experts or another existing source into the knowledge system using its presentation methods.

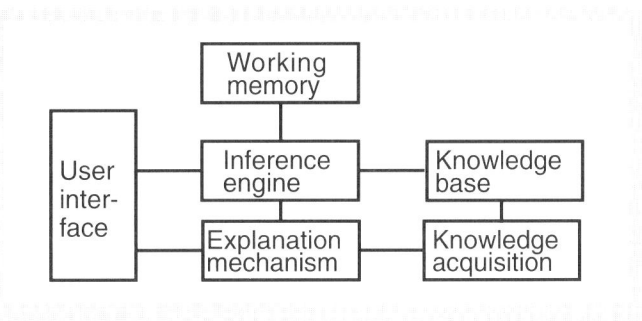

Figure 11. Main parts and operation of expert systems.

6.2 Expert systems in process control

Expert system applications exist for nearly all types of processes whether simple or complex. The most beneficial are complex processes where process performance and behavior are difficult to model with conventional methods. The complexity often relates to the large amount of measurements with uncertainties and high noise levels. The operators have difficulty in system monitoring and following the process performance especially for changes. Slow or varying dynamics and long time delays also make the situation more complex. These processes need the inclusion of human knowledge and qualitative reasoning in control strategies. Uncertainties in systems, environments, and measurements can also require the application of expert system or fuzzy logic tools as the next section shows.

An earlier work reviews the specific features of expert control[15]. Non-monotonic reasoning is necessary because the validity of data changes with time and the state of the system. This requires automatic consideration. Interrupting mechanisms must be available to prioritize between asynchronous events of different importance. A real-time system must also include features like temporal reasoning capabilities. This means it must be possible to connect time and reasons and their sequences. It also requires reasoning under limited time or with erroneous or missing sensor data.

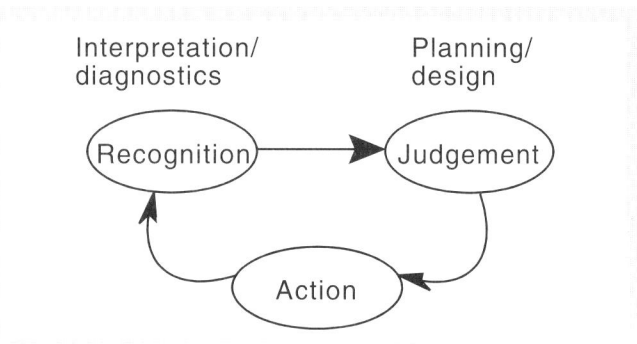

Figure 12. Expert control as a recognition-judgement-action cycle[16].

According to Saito[16], the recognition-judgement-action cycle occurs repeatedly in expert control as Fig. 12 shows.

6.2.1 Stabilizing control

In stabilizing controls, Fig. 13 shows the use of open-loop and closed-loop applications.

Advisory systems use process measurements to give advice to process operators. In this case, only the operator closes the control loop. Two variations of closed-loop control exist. In the first, an expert system can handle the entire control using fuzzy logic control for instance. For the second variation, the expert system tunes conventional controllers in changing control strategies and adapting to varying operational conditions. Using expert knowledge and rule-based systems does this.

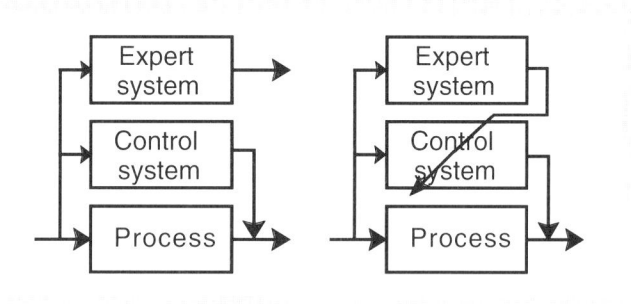

Figure 13. Principles of open-loop and closed-loop expert controllers[14].

Another way of dividing expert system applications in process control would be to examine the state of the process. This means to look for systems for start and stop sequences, normal operation, and state changes. Very few expert systems for start and stop sequences and process or production line state changes currently are in use.

6.2.2 Supervisory control

For supervisory control, expert system applications already exist using process control computers with Fortran and other languages. Customized systems are available for different processes based on process knowledge but without any formal expert system framework.

Expert system tools offer several advantages especially for systems building. They make it possible to use linguistic terms instead of programming approach. System building is faster with minimum programming skills and computer knowledge, and system modification is easier. Supervisory applications suffer from the same limitations as other expert systems as indicated later.

The greatest challenge on the supervisory control level would be building systems that use expert knowledge in the system tuning and implementation stage. This is usually the most time consuming stage in supervisory system projects. Success here depends primarily on the skills of the system engineer. No reports of applications are available.

6.3 Other applications

Needs for other applications are essentially the same as for control applications; Expert systems have use when it is necessary to handle much complicated, varying, and uncertain information and when the decision making uses experience and qualitative reasoning. One important aspect in upper level expert system solutions is that they make it possible to retain the key knowledge in the system and facilitate diminishing the negative effects of changing personnel. They are also useful in training new staff members. Wisely used, they give users a chance to keep their own knowledge current.

Upper level expert systems are primarily advisory systems allowing reduction of work loads.

6.3.1 Diagnostics

Diagnostics and fault detection systems are classical applications of expert systems. Compared with algorithmic-based fault detection the biggest advantage of the expert system approach is that it provides possibilities to follow fault diagnosing in a human way.

Applications vary from troubleshooting of hydraulic and electronic systems to vibration analysis. The essential features are combining the expert systems with mathematical models and other conventional tools and integrating them into the total process monitoring systems.

6.3.2 Expert systems in scheduling

Expert systems in scheduling target higher throughput, meeting delivery times, and satisfying product quality. The functions vary from the allocation of customer orders to the selection of the routing in the mill according to product requirements.

Two main possibilities exist. In the first one, expert systems are used with conventional algorithmic scheduling in the problem formulation stage and in combining qualitative or rule-based information in the system. In this way, the strict mathematical problem formulation adapts to changing requirements. In the second approach, the entire scheduling problem is formulated as a rule-based system, and the solution itself can include some algorithmic parts as Fig. 14 shows.

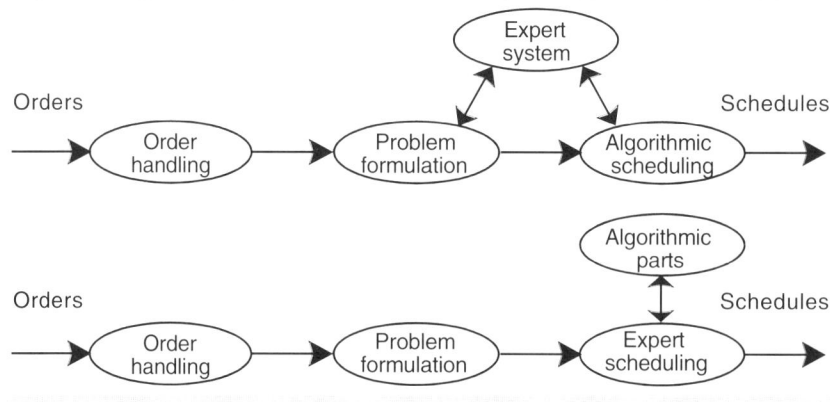

Figure 14. Different ways of applying expert systems in scheduling [14].

The use of expert system approaches in scheduling allows inclusion of qualitative information about limitations, bottlenecks, quality variables, strategy changes, etc., into the solution. This is not necessarily possible in algorithmic approaches. Using linguistic variables and fuzzy presentations further enhances this by considering the vagueness that is common in production systems such as soft constraint.

6.3.3 DSS applications

At a higher level in the control hierarchy, connections with conventional data processing systems and data bases become more important. DSS systems are by nature advisory systems. This means that operator interface questions are more important compared with earlier mentioned systems. Questions connected with maintenance and updating are also very important. Moving from rule-based systems to object-based ones makes modifications easier.

6.4 Advantages

A regularly mentioned advantage of expert systems is the improved and more consistent decision making by nonexperts and the saving of expert time. Processing uncertain knowledge like the expert would do is also possible. By storing expert knowledge, the expertise remains within the company, although experts may leave the company or move to other positions inside the company. A similar effect also results by using expert systems in training new staff members or keeping staff member skills at an existing level.

Expert systems make fast system development possible using prototyping with certain tools by users without any need for system programmers (primarily concerning the knowledge base). First results are available very rapidly. Prototyping also has some disadvantages especially with large-scale systems. It does not give a clear picture of the job in the beginning and may even be misleading. The need for more efficient tools for systems design is clear.

6.5 Problems

Expert systems also suffer from some limitations and disadvantages. Some of these are the following:

- Lack of systematic methods for knowledge acquisition and validity testing. This problem occurs in all applications. The user expects some automatic or at least systematic methods. In control applications, this is a necessity to check rule-based control with model based controls to assure system optimality and stability.

- Lack of learning capabilities. This occurs especially with deeper knowledge and large scale systems. In this case, some automatic methods that help the system to adapt its operations according to what has happened in the past are needed.

- Lack of proper design methods. For large scale systems especially, prototyping is an inefficient approach. Small prototypes do not give sufficient information about the requirements of the large-scale applications. Full prototypes are very expensive. Some systematic structural design methodology and corresponding supporting functions for system design are necessary.

- Problems of integrating into existing systems. Nearly all applications include interfacing and integration needs, interfacing with program languages, conventional simulation and optimization programs, and outside hardware and software is necessary. In process control applications, one must consider the degree of integration on a case-by-case basis.

- Need for training. As a new technology, expert systems applications require considerable training especially for the staff responsible for development, updating, and maintenance. This applies to hardware and software. The need for training increases in some instances due to the lack of maintenance support and knowledge from some vendors and by the need to keep company knowledge confidential.

- Problems in uncertainty processing. Uncertainty is an unavoidable part of complex environments. Information is partly uncertain or subjective, range of applicability of models is usually limited, and many interactions with other application areas require consideration.

CHAPTER 3

7 Fuzzy logic control

Fuzzy logic control means using a control mechanism to convert the control rules devised by the operator into the control system. This kind of control provides an excellent opportunity to consider uncertainties and inaccuracies in process control.

Fuzzy logic is not a new approach. Zadeh first introduced the theory of fuzzy sets in 1965[17]. The introduction gave a general theoretic background for different applications. Fuzzy logic control is a component of artificial intelligence. First control applications developed primarily during the 1980s. Many industrial applications have originated in Japan.

Commercial products based on fuzzy sets theory are also commercially available. These products include microprocessors with special designs for fuzzy reasoning. Applications vary from different artificial intelligence applications to sensors and home electronics. Commercial industrial controllers for fuzzy logic control are also available today. Various computer programs and program packages for the development and installation of fuzzy logic control exist. Universities and research institutes are conducting considerable research work. The number of actual industrial applications is low but growing. In the pulp and paper industry, applications occur in lime kiln, digester, bleaching, and thermomechanical pulp control[18–22].

7.1 What does fuzziness mean?

Fuzziness is a concept that tries to describe inaccuracies and uncertainties that are common in daily discussions. It represents a set of methods that makes the use of linguistic variables and descriptors in process control possible. For example, concepts such as young, middle-aged, and old are common when speaking about someone's age. These linguistic variables do not have any exact value but

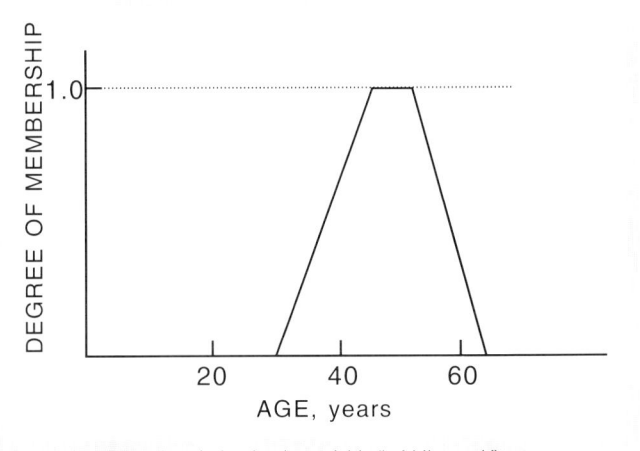

Figure 15. Fuzzy description for the variable "middle aged."

rather a vague range of values. According to a common textbook example, the fuzzy set of Fig. 15 can describe middle-aged.

This picture means that no exact single-valued limits apply to the fuzzy variable "middle aged." The term can have some flexibility. This difference is obvious when comparing the fuzzy description in Fig. 15 with the crisp mathematical description in Fig. 16. One can interpret these figures so that the crisp valued mathematics is only a subset of more versatile fuzzy mathematics. This is because common mathematics cannot handle expressions such as "almost middle aged" or "little over middle aged" that are possible in fuzzy logic.

7.2 Fuzzy logic controller

Fuzzy logic control is a tool that can transfer an operator's knowledge in a linguistic form into a controller. This means that operator experience becomes a part of control algorithms in the form used in human reasoning. Instead of exact numerical values, fuzzy logic controllers operate with values like low, high, medium, sufficient, etc. This form of presentation also helps collect knowledge from people who speak in qualitative terms rather than in numerical values.

According to Fig. 17, a fuzzy logic controller has three main parts: fuzzification, fuzzy reasoning, and defuzzification. In its operation, a fuzzy logic controller uses membership functions and rule sets.

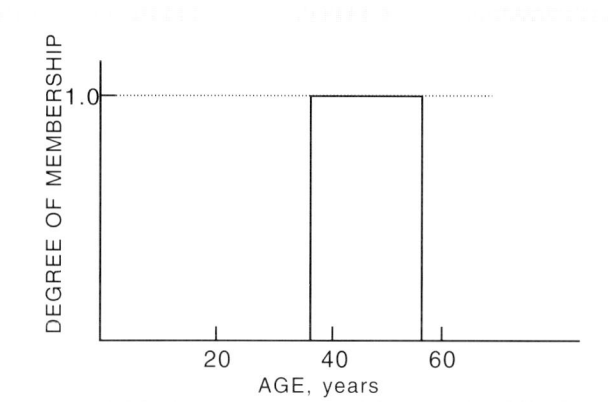

Figure 16. Crisp description for the variable "middle aged."

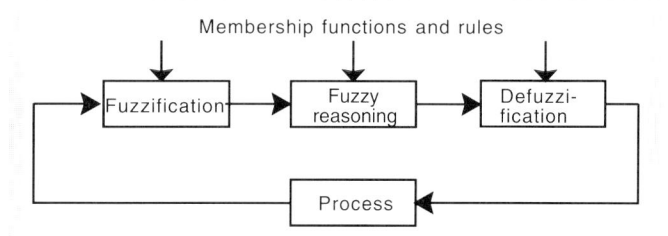

Figure 17. The structure of a fuzzy logic controller.

7.2.1 Fuzzification

Fuzzification means the methods and principles used when converting exact measurement values to linguistic variables. In principle, this happens by dividing the measurement range into as many subareas (labels) as seems reasonable. For instance, the level of a tank can have three areas:

- Low, when height is < 50%
- Good, when height is 20%–80%
- High, when height is > 50%.

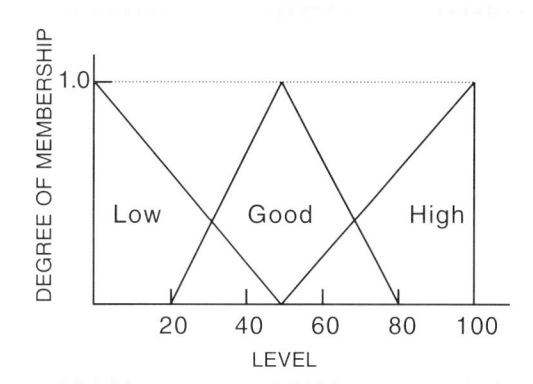

Figure 18. Fuzzy sets for "low," "good," and "high" tank levels.

This selection is arbitrary and case dependent. Developing the fuzzy sets that model this case uses the membership functions that Fig. 18 shows.

CHAPTER 3

In the conventional theory of sets, an element either belongs or does not belong to some set. In the theory of fuzzy sets, there is no exact limit for an element to belong or not to belong to a fuzzy set. Instead, the degree of membership shows a "partial" inclusion into a fuzzy set.

7.2.2 Fuzzy reasoning

Fuzzy reasoning uses rules for the fuzzified measurement values. In the tank level case, the rule set could consist of the following rules (Note that this is not a complete rule base. This would consist of nine rules):

1. IF level is "low" and "decreasing," THEN input flow is "increased considerably."

2. IF level is "low" and "stable," THEN input flow is "increased slightly."

3. IF level is "good," THEN input flow is "unchanged."

4. IF level is "high," THEN input flow is "decreased slightly."

5. IF level is "high" and "increasing," THEN input flow is "decreased considerably."

Note that there are now new fuzzy variables. "Level change" has three values (decreasing, stable, and increasing), and "input flow" with five values (increased considerably, increased slightly, unchanged, decreased slightly, and decreased considerably). These should also be defined by giving suitable membership functions that describe the level trend and the amount of input flow. The rules are handled by using the membership functions and logical operations.

7.2.3 Defuzzification

Defuzzification is necessary in converting the fuzzy output values to the crisp values used in the real world. In the tank level example, linguistic output values are converted to changes in the input flow. Several methods are available for this[23–25].

8 Neural networks

Biological nervous systems and mathematical theories for learning have inspired neural networks[26]. Their learning ability and distributed parallel structure characterize them. One can consider neural networks as black box modeling methods. Various methods are available for neural computing[27]. Back propagation is probably the most popular method in supervised learning. Self-organizing maps are common networks in unsupervised learning[28]. Learning algorithms have a close relation to optimization methods, e.g. the back propagation method corresponds to gradient descent. A huge number of programming and development tools is available commercially and in public domain on the Internet.

Neural networks have use in control in the pulp and paper industry[29], in digester quality analysis[30], and in thermomechanical pulp control[31].

Control methods

8.1 Basic definitions

A neural network consists of a large number of simple parallel processing units connected with each other. Each connection between the units has a weight that describes the strength of the connection. The weights are tuned, i.e., the neural network is learning a certain application. Figure 19 shows this kind of simple processing unit and its connections with the environment.

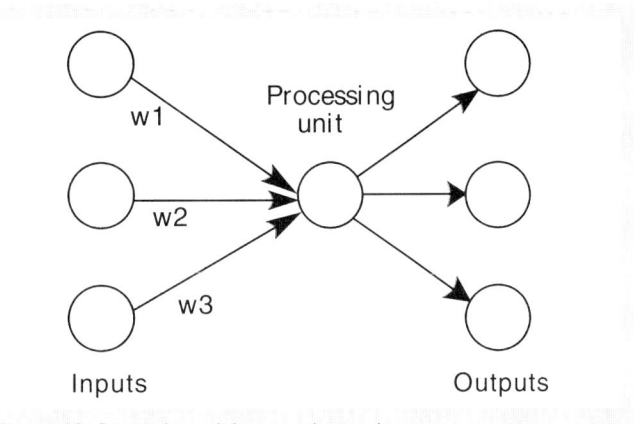

Figure 19. Processing unit in a neural network.

Figure 20 shows a multi layer network having three layers. The lowest level is the input layer, and the highest level is the output layer. The intermediate level is the hidden layer. The hidden layer does not have outside connections, but it has importance in network learning.

8.2 How do the networks learn?

The short-term knowledge in neural networks is stored in the states as neurons, and their connections correspond to long-term knowledge. In this connection, learning means the adaptation of weights or adding or removing connections between neurons. This divides networks into static and dynamic neural networks. In static networks, the weights are changing, but the connections remain unchanged. In dynamic networks, the connections also change.

Some networks learn from examples. A group of examples (the training set) is given to a network, and the network tries to learn (generalize) them. The contents of a learning set vary from one application to another: pictures, hand-written signals, speech, measurement data, etc. During learning, optimum values for the weights in the neural network are determined. Learning continues until the desired performance occurs. In practice, this means that the network produces required outputs. The training set can be introduced to the network several times so the learning proceeds stepwise.

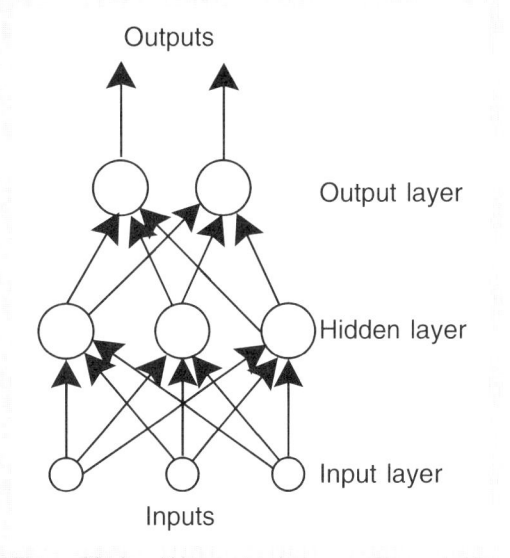

Figure 20. A multi layer network.

CHAPTER 3

Learning algorithms have two main groups:

- Supervised learning
- Unsupervised learning.

In supervised learning, the required outputs corresponding to certain input set are known in advance. Initial weights are set randomly. In unsupervised learning, no external reference signals (training set) are used.

8.3 Network structures

8.3.1 Perceptron network

Perceptron learns classification using supervised learning. The material to be classified is usually binary coded (0 or 1), and the results from classification are also given as binary values. Perceptron networks are simple two-layered networks consisting of an input layer and a processing layer with adaptive weights between them. Additional layers can be added, but they cannot be trained. Layers are completely connected with each other, and the weights are adjusted in the learning phase.

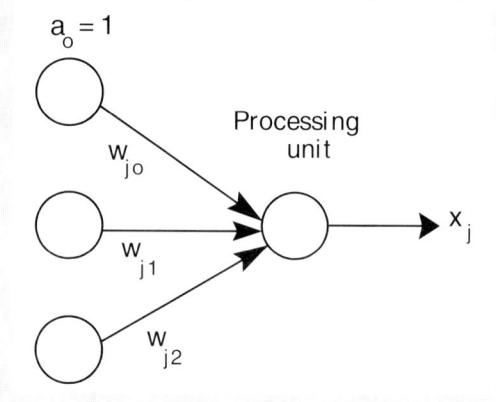

Figure 21. Processing unit in perceptron network.

Figure 21 shows the processing unit of the perceptron network. It calculates the weighted sum of inputs as follows:

$$S_j = \sum_{i=1}^{n} a_i w_{ji} \qquad (7)$$

Figure 21 also shows the bias input that always equals 1. Its weights are adjusted during learning like all other weights. The processing unit checks if the above calculated weighted sum is smaller or greater than some preset threshold according to the following rules:

$$\text{If } S_j > 0 \text{ then } x_j = 1 \qquad (8)$$

$$\text{If } S_j \leq 0 \text{ then } x_j = 0 \qquad (9)$$

The simplest training algorithm starts by calculating the difference between the required output and the output calculated by the network:

Control methods

$$e_{jp} = t_{jp} - x_{jp} \tag{10}$$

where e is the error
           t the required output
           x the value calculated by the network.

Index j refers to the output unit and p to an element in the training set. Because the perceptron uses only values of 0 or 1, the error can take values 0, 1, and -1. Each weighting factor is corrected according to the following equation:

$$w_{ji,new} = w_{ji,old} + Ce_j a_i \tag{11}$$

where C is the training speed and a is 0 or 1.

This means that the weighting factor is corrected only if the input connected to it is active. C is usually smaller than 1. Learning is slower with small values of C than with larger values.

Examples are calculated singly during training. After completion of the entire training set, values can be calculated several times. The performance of the network improves quickly in the beginning, but the improvement decreases with time. The network converges. This means that the network has either learned or not learned. The square sum of the error measures the performance of the network:

$$RMS = \sqrt{\frac{\sum_p \sum_j (t_{jp} - x_{jp})^2}{n_p n_o}} \tag{12}$$

where n_p is the number of elements in the training set
            n_o the number of units on the output layer.

8.3.2 Back propagation networks

Back propagation is the most usual network solution. Figure 20 above showed an example of the back propagation network. It had at least two completely connected layers. Often the number of layers exceeds two. In the previous figure, the bottom layer presented the input variables. This was the only layer connected to outside inputs. The top layer was the output layer. The layer between them was the hidden layer.

8.3.3 Competitive learning

Competitive learning uses unsupervised learning. Training material does not include reference values or required outputs. In the training phase, the network classifies the material into the given number of classes.

CHAPTER 3

According to Fig. 22, a competitive learning network consists of two layers: input layer and competitive layer. The layers completely connect to each other. The input layer processes inputs, and the competitive layer classifies the data. The input can only receive binary values of 0 or 1. The weights are between 0 and 1, and the sum of weights connected to a certain processing unit equals one.

In the competitive layer, the units compete with each other. The competition follows "the winner takes all" principle. This means that the unit having the biggest sum wins. Its output is set to one, and the outputs from other units are set to zero.

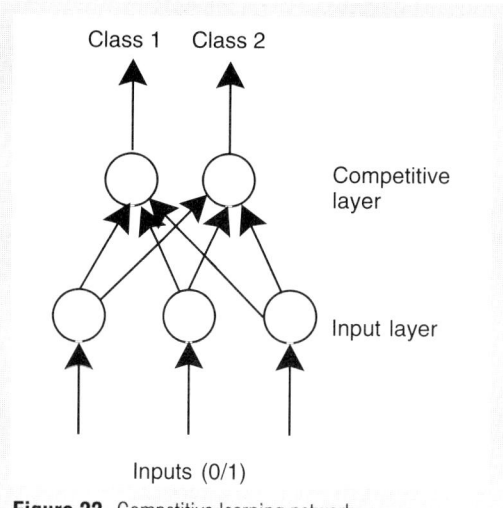

Figure 22. Competitive learning network.

9 Statistical process control

Statistical process control (SPC) strives to continue improvement in product quality and production costs. A definition of quality[32] is "the degree of fitness for purpose or function" emphasizing the fact that quality is a measure of satisfying customer needs. In practice, SPC means a group of tools used in estimating the controllability of the process and the quality of its products while decreasing variations in both. The methods are not new. Shewhart developed them during the 1920s and 1930s. They have been in use for many years. Some applications are in the pulp and paper industry[33–38].

9.1 Presentation of quality data

SPC data originates from several sources. It depends on raw materials, the production processes, and the final or intermediate products. Two types of data are possible: discrete data (attributes) describing the amount of defective products in a sample and continuous data (variables) describing analyzed or measured values. Both data types should be recorded so they are easy to use and process. Common techniques are to use tables, histograms, bar charts, line graphs, pie charts, etc. Figure 23 shows the histogram for some kappa number samples. The histogram tells the average value and the spread of the variable in question, but it does not show its timely development.

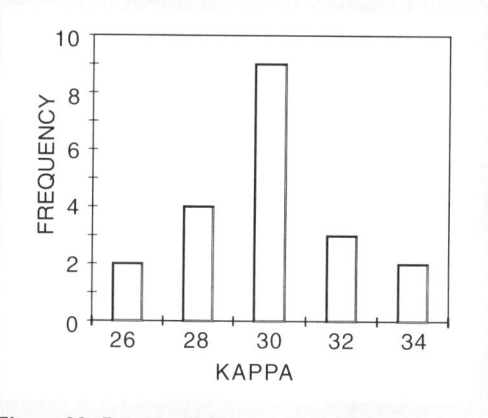

Figure 23. Example of a histogram.

Control methods

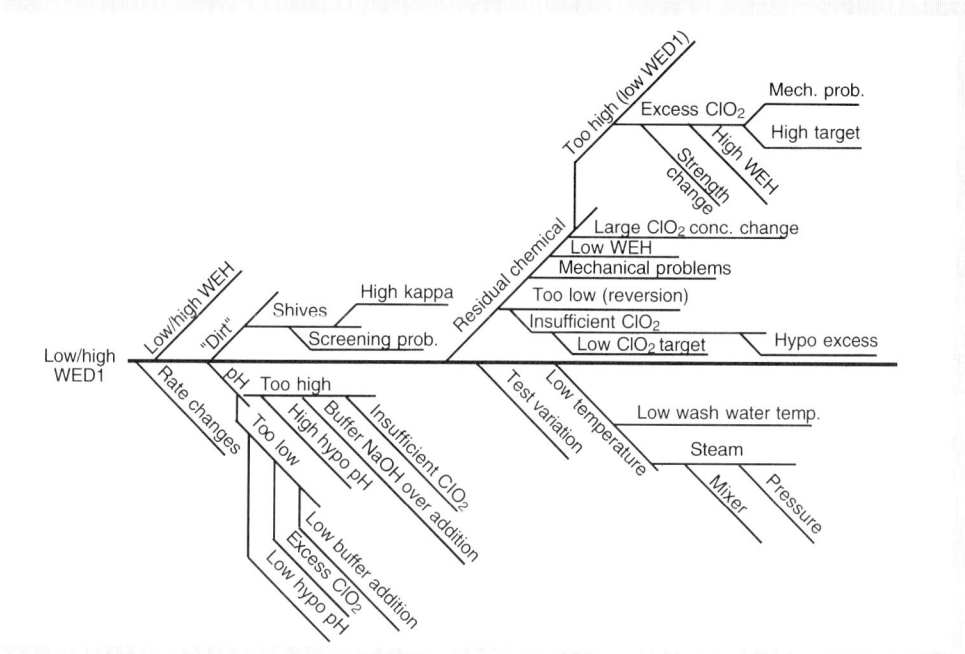

Figure 24. Ishikawa diagram for wet Elrepho hypochlorite brightness (WEH) in the first chlorine dioxide stage (WED$_1$)[35].

9.2 Cause-and-effect analysis

Ishikawa diagrams are useful in identifying reasons and effects of process variables. Figure 24 shows an example of wet Elrepho brightness in the first chlorine dioxide stage[35]. The central axis represents the brightness under investigation. Linked nearest to the axis are the general causes that make the brightness too high or too low. Additional branches show more specific causes. This presentation is "a fishbone chart" because of its usual form. The chart clarifies the relationships between process variables and the mechanism of influence but complicated interconnections are not easy to include.

9.3 Pareto analysis

Pareto analysis is a technique for arranging data according to priority or importance. It uses the common 80/20 rule that 20% of reasons cause 80% of defects. Pareto diagrams are useful in ranking the reasons for studied causes and in directing attention to the most obvious ones and ones that are most

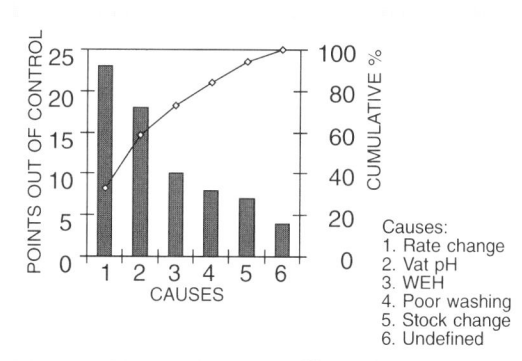

Causes:
1. Rate change
2. Vat pH
3. WEH
4. Poor washing
5. Stock change
6. Undefined

Figure 25. The Pareto diagram for WED$_1$[35].

CHAPTER 3

easily corrected. Figure 25 shows an example using the diagram for causes that have caused the wet Elrepho brightness in the first chlorine stage to go out of control, WED_1[35]. The left-hand scale shows the number of points out of control (interpreted from the original data[35]). Identified causes (See the right-hand lower corner) are bars. The solid line tells the cumulative percentage of points denoted in the right-hand scale.

Pareto diagrams are also useful in showing the results of corrective actions. The vat pH is the most easily corrected parameter in the previous case. Figure 26 shows the situation after installing automatic control of the buffer NaOH addition in response to pH variations[35]. This shows a marked decrease in points out of control caused by vat pH. The total number of points out of control has also decreased.

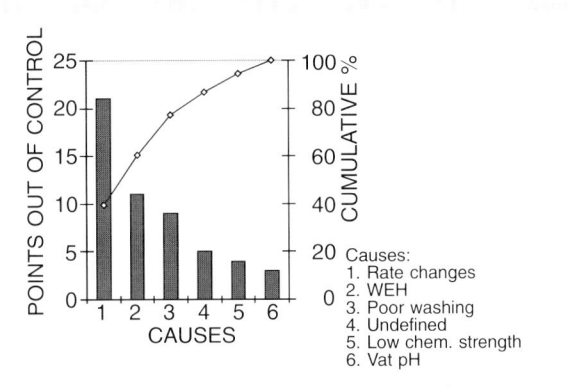

Figure 26. The effect of buffer NaOH control (vat pH) for the Pareto diagram of Fig. 25[35].

Causes:
1. Rate changes
2. WEH
3. Poor washing
4. Undefined
5. Low chem. strength
6. Vat pH

9.4 Control charts

Control charts are useful to follow the timely development of quality variables. Figure 27 shows the principles of drawing one control chart: the mean control chart. If the process is running satisfactorily, one expects all means of successive samples to lie inside the upper and lower action limits. Actually, the probability of a sample to go beyond these lines is 1/1000. In this kind of situation, immediate control actions are necessary to put the process under control again. The other two lines drawn in the picture are upper and lower warning lines located so the chance for a sample to go outside these lines is 1/40. If the sample mean goes outside these limits, a danger exists that the process is going out of control. Two successive samples outside the warning limits usually lead to adjustments in the process. Control charts for range, median, moving average, and standard deviation are in use[32].

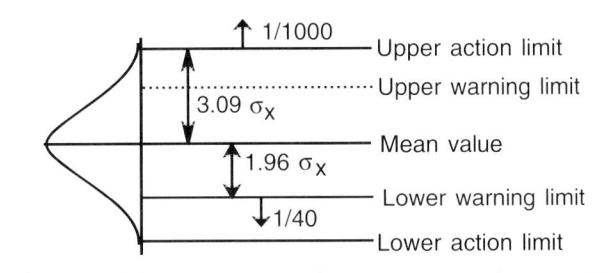

Figure 27. Control chart for the mean value.

Control charts are an illustrative way to describe changes in process operation. Figures 28 and 29 show control charts for the 3-term moving average of Elrepho chlorine brightness before and after a major statistical quality improvement campaign. The upper and lower control limits before were 44.9 and 37.1 (not shown in Fig. 28). After the campaign, they were set to 42.6 and 39.4.

Control methods

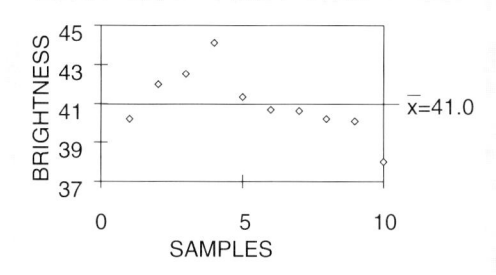

Figure 28. A 3-term moving average chart for wet Elrepho chlorine brightness before a statistical quality improvement campaign[35].

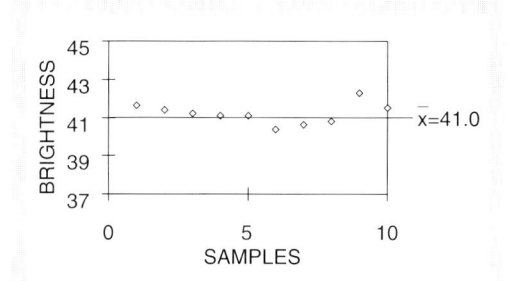

Figure 29. A 3-term moving average chart for wet Elrepho chlorine brightness after a statistical quality improvement campaign[35].

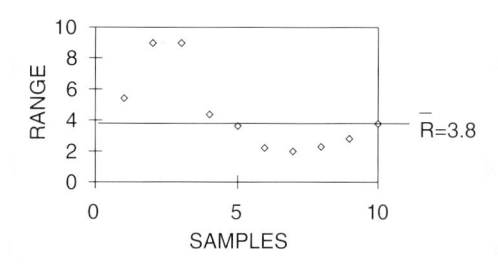

Figure 30. A 3-term moving range chart for wet Elrepho chlorine brightness before a statistical quality improvement campaign[35].

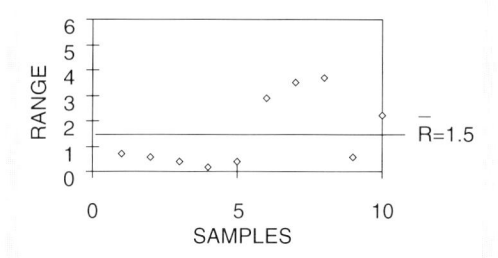

Figure 31. A 3-term moving range chart for wet Elrepho chlorine brightness after a statistical quality improvement campaign [35].

The respective range charts shown in Figs. 30 and 31 show the effects of improved statistical control on process variability. The figures show 3-term moving ranges before and after the campaign. The upper control limit of the range chart decreased from 8.9 to 3.9, and the mid-range changed from 3.8 to 1.5.

9.5 Cusum charts

Cumulative sum (Cusum) charts use all the available information to highlight small changes and trends. Figure 32 illustrates this method using the same pulp kappa samples as the previous mean chart. Cusum charts total all the deviations from the target value. Although kappa value has been on the average close to the target

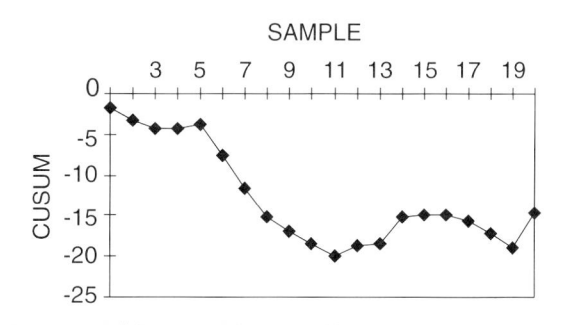

Figure 32. Cusum chart for the kappa samples of Fig. 23.

CHAPTER 3

with a small standard deviation, the value has been under the target and decreasing during the first 11 samples in this case. A corrective action was necessary.

Cusum charts are good for showing trends in measured variables. Figure 33[37] shows a 6-day trend of pulp viscosity. According to the figure, some large variations existed during the first two days after which the viscosity was stable. This is even more evident in Fig. 34 that displays the exponentially weighted moving average chart for viscosity.

The Cusum chart in Fig. 35 shows very clearly the points in time where something special has happened. These are times A, B, and C where the magnitude of the trend is changing. Analyzing what has really happened requires some caution because the actual reason might have occurred in some upstream process several hours previously. Correlation analysis discussed in Chapter 10 continues this example. Analysis showed that point A in Fig. 35 corresponded to an unbleached kappa number peak that occurred about 14 h earlier[37].

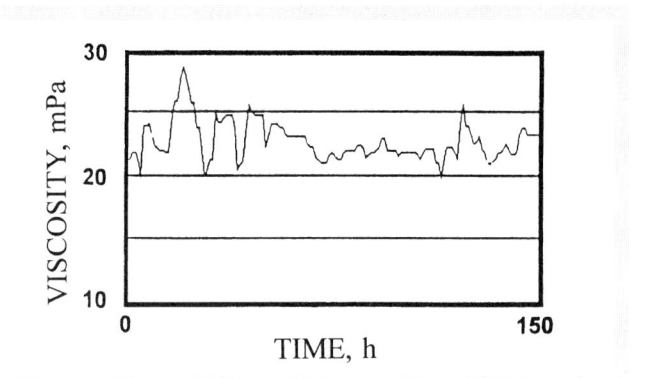

Figure 33. A 6-day trend of pulp viscosity[37].

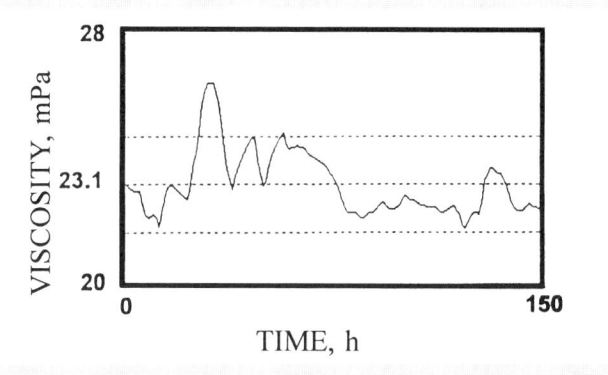

Figure 34. Exponentially weighted moving average trend of pulp viscosity[37].

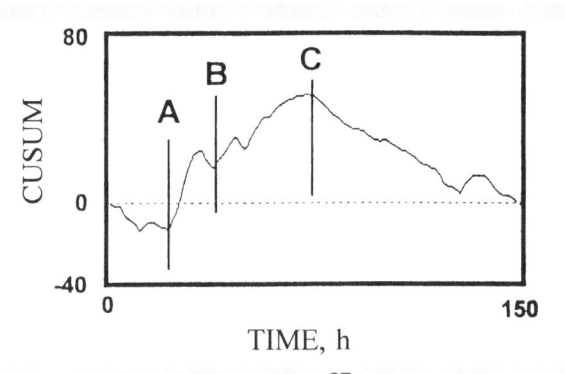

Figure 35. Cusum chart for pulp viscosity[37].

9.6 Capability indices

Process variability and the tolerances set for product quality define the relative precision of the process as Fig. 36 shows. To remain inside specifications, the distance between upper and lower tolerance limits (2T) should be larger than or equal to 6σ[32].

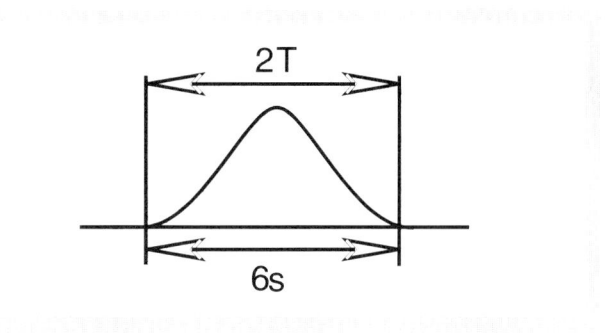

Figure 36. Tolerances and variability.

The relative precision index measures the relationship between tolerances and process variability[32].

$$\frac{2RT}{R_{av}} \geq \frac{6}{d_n} \qquad (13)$$

where R_{av} is the sample range
       d_n Hartley's constant.

Another way is to use process capability:

$$C_p = \frac{2T}{6\sigma} \qquad (14)$$

$$C_{pk} = \left\{ \min\left(\frac{USL - \bar{X}}{3\sigma}, \frac{\bar{X} - LSL}{3\sigma}\right) \right\} \qquad (15)$$

$$\sigma = \frac{R_{av}}{d_n} \qquad (16)$$

where USL is the upper specification (tolerance) limit
       LSL the lower specification (tolerance) limit.

The index value less than 1 shows that the process requires adjustment. The value should approach 2.

CHAPTER 3

References

1. Cluett, W. R. and Wang, L., 1996 Control Systems Preprints, Halifax, Canada, p.75.
2. Corripio, A. B., Digital Control Techniques, AIChE, Washington, Series A, vol. 3, 1982, p. 63.
3. Åström, K. J. and Hägglund, T., Automatica 20(5):645(1984).
4. Pennanen, J., 1994 Control Systems Preprints, SPCI, Stockholm, Sweden, p. 197.
5. Allison, B. J., Ciarniello, J. E., Tessier, P. J.-C., et al., 1994 Control Systems Preprints, SPCI, Stockholm, p. 289.
6. Sutinen, R., Saarinen, K., and Leiviskä, K., ABB Review (9):10(1994).
7. Allison, B. J., Dumont, G. A., and Novak, L. H., 1990 Control Systems Preprints, Finnish Society of Automatic Control, Helsinki, Finland, p. 157.
8. Brattberg, Ö., 1994 Control Systems Preprints, SPCI, Stockholm, p. 298.
9. Åström, K. J. and Wittenmark, B., Adaptive Control, Addison-Wesley, Reading, 1989.
10. Balchen, J. G. and Mummé, K. I., Process Control Structures and Applications, Van Nostrand Reinholt, New York, 1988, pp. 44–78.
11. O'Reilly, J., Multivariable Control for Industrial Applications, Peter Peregrinus Ltd., Exeter, 1987.
12. Fadum, O., Pulp & Paper 67(3):85(1993).
13. Parker, G., Pulp & Paper 66(9):129(1992).
14. Leiviskä, K., 1991 IFAC Workshop Expert Systems in Mineral and Metal Processing Proceedings, Pergamon Press, Oxford, p. 191.
15. Årzen, K.-E., Automatica 25(6):813(1989).
16. Saito, T., 1989 6th IFAC Symposium Automation in Mining, Mineral and Metal Processing Proceedings, Pergamon Press, Buenos Aires, p. 31.
17. Zadeh, L. A., Information and Control 8(3):338(1965).
18. Haataja, K. and Ruotsalainen, J., 1994 MEPP'94 Proceedings, Mariehamn, Finland.
19. Lampela, K., Kuusisto, L., and Leiviskä, K., TAPPI J. 79(4):93(1996).
20. Myllyneva, J., Leiviskä, K., Kortelainen, J., et al., 1997 International Mechanical Pulping Conference Proceedings, SPCI, Stockholm, p. 381.

21. Nilsson, L. and Langkjaer, T. F., 1997 EUFIT Conference Preprints, ELITE Foundation, Aachen, p. 1901.

22. Ostergaard, J.-J., 1993 EUFIT Proceedings, ELITE Foundation, Aachen, p. 552.

23. Driankov, D., Hellendoorn, H., and Reinfrank M., An Introduction to Fuzzy Control, Springer, Berlin, 1993, pp. 103–141.

24. Wang, L.-X., Adaptive Fuzzy Systems and Control, Design and Stability Analysis, Prentice Hall, Englewood Cliffs, New Jersey, 1994, pp. 100–144.

25. Zimmermann, H. J., Fuzzy Set Theory and its Applications, 2nd edn., Kluwer Academic Publishers, Boston, 1991.

26. Huang, S. H. and Zhang, H.-C., Computers in Industry 26(2):107(1995).

27. Dayhoff, J. E., Neural Network Architectures, Van Nostrand Reinhold, New York, 1990.

28. Kohonen, T., Self-Organization and Associative Memory, 2nd edn., Springer, Berlin, 1987.

29. Beaverstock, M. and Wolchina, K., Pulp & Paper 66(9):134(1992).

30. Haataja, K., Leiviskä, K., and Sutinen R., 1997 IMEKO World Congress Proceedings, Finnish Society of Automation, vol. XA, Tampere, p. 1.

31. Kooi, S. B. L. and Khorasani, K., Tappi J. 75(6):156(1992).

32. Oakland, J. S., Statistical Process Control, Heinemann, London, 1986, pp. 70–82.

33. Aldrich, W. D., 1990 24th EUCEPA Conference Proceedings, Control, Maintenance, Environment, SPCI, Stockholm, p. 29.

34. Andrews, I. M. and Barnes, R., Pulp & Paper 59(11):103(1985).

35. Corbi, J.-C., Nay, M. J., and Belt, P. B., Tappi J. 69(2):60(1986).

36. Smith, K. E., Pulp & Paper 59(1):132(1985).

37. White, K. and Roberts, C., 1994 Control Systems Preprints, SPCI, Stockholm, p. 52.

38. Whitley, J. C., III, Pulp & Paper 61(1):99(1987).

CHAPTER 4

Special measurements in pulp and paper processes

1	**Introduction**	**49**
2	**Application oriented special measurements**	**50**
2.1	Pulp consistency	50
	2.1.1 Mechanical sensors	50
	2.1.2 Optical sensors	51
	2.1.3 Absorption and scattering-based principles	52
	2.1.4 Microwave consistency sensor	53
	2.1.5 NIR consistency analyzer	53
2.2	Cooking process measurements	54
	2.2.1 Kappa analyzers	55
	2.2.2 Cooking liquid analyzer	55
2.3	Bleaching control sensors	56
2.4	Pulp quality analyzers	57
	2.4.1 Freeness analyzers	57
	2.4.2 Fiber length analyzers	58
	2.4.3 Pulp quality monitor	59
	2.4.4 PulpExpert	61
2.5	Wet end measurements	61
	2.5.1 On-line cationic demand (anionic trash) measurement	61
	2.5.2 Retention monitoring	62
	2.5.3 Flocculation sensing	63
	2.5.4 Air content measurement	63
3	**On-line measurements of paper quality**	**64**
3.1	Basis weight	64
3.2	Moisture	64
3.3	Filler (ash) content	65
3.4	Formation	65
3.5	Fiber orientation	66
3.6	Caliper (web thickness)	66
3.7	Smoothness (roughness)	67
3.8	Gloss	68
3.9	Opacity, color, and brightness	68
3.10	Porosity	69
3.11	Coating weight	69
3.12	Holes and dirt spots	70
	References	71

CHAPTER 4

Jouni Tornberg

Special measurements in pulp and paper processes

1 Introduction

Properties of pulp and paper are difficult to measure. The requirements vary considerably depending on the application such as pulp type, place of installation in the process, etc. Most measurements except certain basic ones require knowledge regarding the process conditions and special installation knowledge.

The fibers are 1–6 mm long, 20–30 µm wide. Pulp also contains wood-based fines below 1 µm to 50 µm in size. These size distributions have a major effect on the measurements made during the various processes. The fillers and coating pigments have a size distribution like visible light. This is usually the same order as the wavelength used in optical measurement instruments making the measurements difficult to perform and highly dependent on application.

The most important feature of process measurements is repeatability especially in the control loop. Accuracy is not usually so important, but careful attention is necessary for repeatability and long-term stability. Maintenance and calibration costs are obviously causal factors in the long run. An instrument should require minimum service and calibration.

Several typical basic measurements used in various process industries also apply to pulp and paper manufacturing. These naturally include flow, pressure, and temperature measurements at numerous points in the mill and pH and conductivity measurements.

Flow measurements of stock slurries today almost entirely use the electromagnetic measurement principle, but orifice plates are still used widely for the measurement of steam and pure fluids. Not only are the orifice plates inexpensive but they frequently have quicker speeds of response than magnetic flowmeters. Fast time constants allow a control loop to control fast disturbances. Mass flow meters are widely used for chemical mixtures and pulp additives. In some special applications, other methods such as ultrasonic methods have found use, but they are not common.

CHAPTER 4

2 Application oriented special measurements

2.1 Pulp consistency

Consistency measurement is an important measurement in the pulp and paper industry as is flow. The properties of the pulp and paper slurry vary considerably in different processes. From the measurement point of view, consistency measurements are therefore not basic measurements but special measurements that are very application dependent.

Pulp consistency is a parameter that indicates the percentage of dry matter in the pulp usually expressed as [% C_s] or [g/L]. 1%C_s is equivalent to 10 g/L. Laboratory measurements and the calibration of on-line meters rely on standardized measuring methods such as the SCAN or TAPPI test. In this test, a given amount of pulp is filtered, and the resulting fiber cake and its filter paper are oven-dried at 103 ± 2°C. Consistency is then calculated after weighing.

The main factor restricting and specifying consistency meter applications is the measurement range. Developing an instrument that can measure all the pulp consistencies used in the process is difficult. The measurement range would need to extend from almost 0% to 60% or 0 to 600 g/L. The most useful measurement ranges are low consistencies of 0%–2% C_s and intermediate consistencies of 2%–12% C_s. The range above 12% C_s is becoming more common today due to the increasing use of medium-consistency process technology.

Pulp type is also a significant factor for the functioning and accuracy of consistency meters. The flow of pulp at the measurement point can influence selection and properties. Some methods are useful only at a certain flow rate.

On-line consistency measurements are always indirect measurements of rheological or physical properties of the pulp. Conversion of this value yields an indicator of consistency. The measurement principles vary according to the consistency range. The most common methods are mechanical, optical, and microwave approaches.

2.1.1 Mechanical sensors

Mechanical consistency meters use the shear force of the pulp. Two types of sensors have common use for this: rod sensors using an intercepting motion and rotor sensors with a rotating motion as Fig. 1 shows. The former goes directly in the process pipe and requires a continuous, steady flow. This causes a force on the rod equivalent to the fiber consistency that can be measured electrically or pneumatically.

Rotating sensors are usually in a secondary chamber attached to the process pipe into which a propeller on the measurement device diverts the pulp to be measured. The device is suitable for use over a more extensive consistency range than a blade transmitter. It does not require such a steady flow as the rod sensor transmitter. It has an electrically operated propeller and a sensor that measures the shear force exerted by the pulp. This is then equivalent to its consistency. The rotating consistency sensor may also use a simpler method with one sensor that rotates at a constant speed and measures the rotation torque in its axle. Rotating consistency meters may require more maintenance and are more difficult to repair than blade type transmitters.

Special measurements in pulp and paper processes

Mechanical measurement devices are most suitable for use in the intermediate consistency range, although steps have been taken to expand the range by adjusting the form of the sensor. The maximum range is currently over 15%. The degree to which the effect of flow rate can be eliminated depends on the shape of the sensor.

The type of the fiber influences all mechanical consistency measurements. Good sampling and sensor verification methods become as important as good sensor installation.

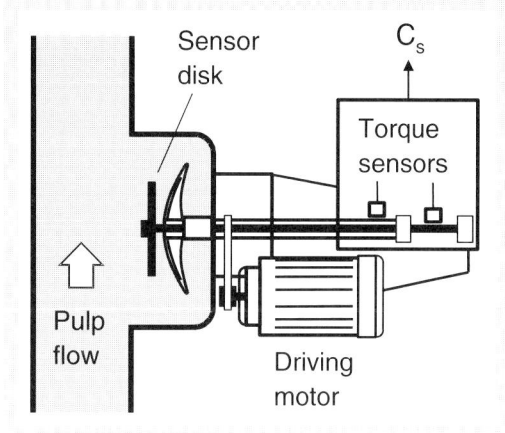

Figure 1. Rotating consistency transmitter (courtesy of BTG).

A blade transmitter operating on the rod sensor principle is the most common type of consistency measurement instrument because of its low price. Numerous manufacturers are available.

2.1.2 Optical sensors

Optical consistency measurement principles have become more common especially in demanding and specialized applications. Optical measurement principles include the absorption, scattering, and polarization of visible and infrared light. Numerous variations of these exist including some instruments designed for higher accuracy and a wider range of applications using combinations of the principles.

Optical methods have produced various specialized applications of consistency measurements. A common one is the measurement of wire retention and adjustment in a papermaking machine using low consistency measurement instruments as the sensor.

Optical methods are sensitive to changes in fillers. For example, the use of titanium dioxide in certain paper grades requires a separate compensation in the distributed control system.

Depolarization principle

Instruments based on optical polarization have the highest performance in low consis-tency applications. A polarization-based instrument may also incorporate other measu-rement principles such as absorption and scattering.

The polarization-based method involves conducting the pulp from the sampler through a pipe to the measurement cuvette and then back to the process. The meters use linear-polarized, narrow-band, or monochromatic light.

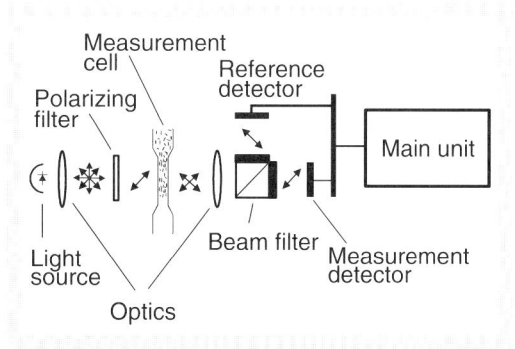

Figure 2. Principle of depolarization consistency measurement (courtesy of Valmet Automation Inc.).

CHAPTER 4

Polarization results from placing a film or prism between the source of light and the measurement cell as Fig. 2
shows. An optical measurement device on the other side of the cuvette measures the depolarized perpendicular beam of the emitted light and the attenuated equipolarization beam. The purpose of measuring the attenuation signal is to compensate for the attenuation of the depolarization signal caused by absorption. The relation between the depolarization signal and the attenuated one indicates consistency. The cuvette itself is only a few millimeters thick, since the use of a thicker device would lead to signal saturation at low consistencies.

More versatile versions of the depolarization principle are also available. These combine various measurement principles within the same device. The retention application mentioned below is one such example.

2.1.3 Absorption and scattering-based principles

Absorption-based consistency measurements use the Lambert-Beer Law:

$$I = I_o e^{-\lambda l c} \tag{1}$$

where λ is the specific absorption constant of the material
l the distance traveled by the beam in the sample
c the concentration to be measured (consistency).

Two possibilities exist. One can measure the attenuation using two or more wavelengths for which the material has different absorption coefficients. One can also use two or more measurement distances. Absorption measurements are traditionally performed by altering the wavelength or by scanning. Infrared analyzers use the second technique. Combining the methods allows measurement of attenuation at different distances with changes in wavelength.

One measurement instrument that uses two detectors and measures attenuation over two distances is the unit of Fig. 3. The detector used to measure the longer distance here actually measures diffuse back reflection. The consistency signal comes from the ratio between the reflected signals. The implementation relies on a narrow-band light emitting diode (LED) source. The method is useful over a wide measurement range from a low consistency of 0.5% C_s to the intermediate 7% C_s region.

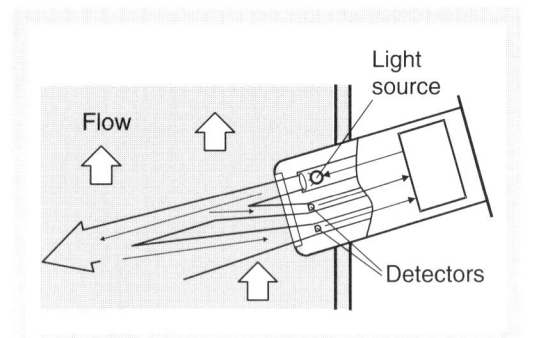

Figure 3. Back-scattering principle for consistency measurement (courtesy BTG).

2.1.4 Microwave consistency sensor

A microwave consistency meter became available in the 1990s. The meter measures changes in the time of flight of a microwave as it passes through the stock as Fig. 4 shows.

The meter sends out a microwave signal from one antenna and times how long it takes to reach a second one. The following formula determines the propagation velocity (v) of microwaves in a material:

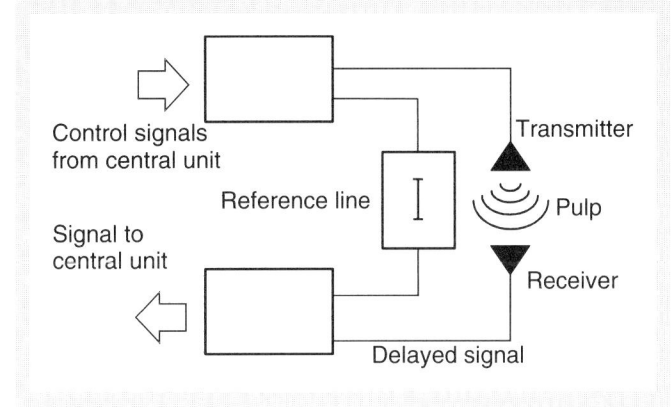

Figure 4. The principle of microwave consistency measurement principle (courtesy Valmet Automation Inc.).

$$v = \frac{c}{\sqrt{E_r}} \tag{2}$$

where c is the speed of light in a vacuum
        E_r the relative permittivity of the material.

The dielectric properties of the suspension influence the time of flight. The relative permittivity for water of approximately 80 at 25°C is over a decade higher than that for fibers (about 3). More water on the microwave path means slower propagation of microwaves. This can determine the water content of a sample and measure consistency.

The manufacturer claims that the measurement is totally independent of the fiber properties of the stock. Items such as freeness, fiber length and its distribution, wood species, brightness, color, pulp grade, or mechanical vs. chemical pulp also do not influence the measurement. Applications lie in control throughout the entire pulp and paper process from the pulping, screening, bleaching, and refining stages to the final basis weight control. The measurement device is expensive, but it is the wave of the future for flows that contain multiple blended fibers and additives. No other measurement principle will be repeatable with these flows.

2.1.5 NIR consistency analyzer

A special consistency analyzer was developed for thermomechanical pulp (TMP) quality measurements in the late 1980s. It uses near-infrared reflectance (NIR) spectroscopy to measure the moisture content of fibers flying in the steam phase in the pipe after the refiners.

The infrared consistency sensor uses the resonance vibration of water that is visible as absorption bands in the infrared region of the spectrum. Pronounced absorption occurs at 1450 nm due to the second harmonic of O-H stretching vibration. The absorption at 1940 nm is a combination frequency of O-H bend and asymmetric O-H stretch vibration that is specific to water. These absorption bands have wide use in other IR moisture analyzers including moisture sensors for paper.

Figure 5 shows the spectra of TMP[1]. The figure shows five spectra of pulp samples representing different consistency levels. Pulp consistency measurement used an on-line sensor using four fixed wavelengths of the spectrum. Two locations were in the absorption bands of water, and the other two were in a region where the effect of water was minimal.

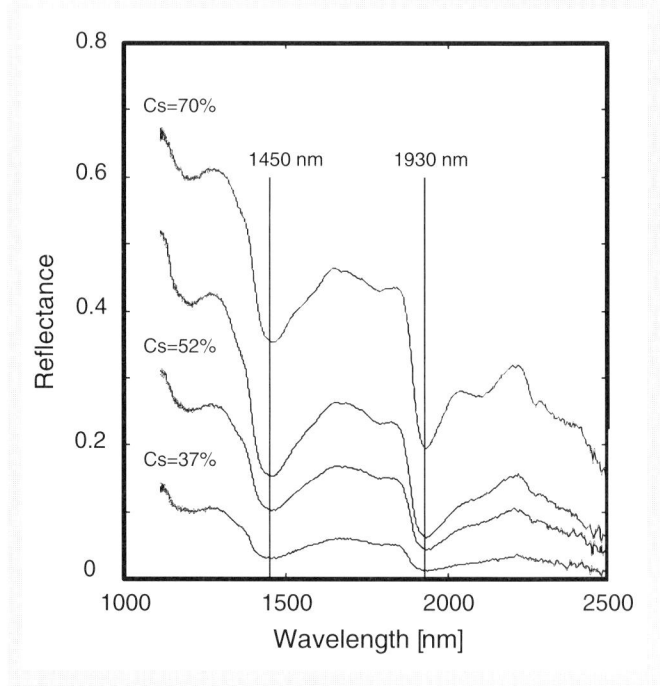

Figure 5. Near-infrared spectra for TMP pulp of different consistency levels.

2.2 Cooking process measurements

A primary parameter for cooking process control is the quality of the wood chips. This essentially depends on water content and chip size. The IR and microwave methods applied to the measurement of water content will not have extensive discussion here. Interesting methods use machine vision and image analysis to measure chip size.

In addition to chip moisture measurements, white liquor strength measurement, black liquor residual measurement, and temperatures all are used for feedforward control of the cook. Estimates or measurements of relief steam flows are needed in batch digesters.

A blow line kappa analyzer traditionally controls the cooking process. The actual analyses usually occur in the laboratory, although the recent development of on-line kappa analyzers has increased the popularity of an on-line approach.

A cooking analyzer capable of performing a number of measurements is also available now for monitoring the cooking process.

2.2.1 Kappa analyzers

The kappa number is normally measured optically as UV absorption in the 280 nm wavelength region at which lignin has a prominent absorption band. A number of arrangements are available from different manufacturers.

Figure 6 shows the kappa measurement principle for one unit. UV light shines on a sample flowing through a measuring cell. Two detectors measure the attenuated and scattered light. One has a position to measure primarily the amount of light passing straight through the cell. The other is mounted at an angle to the optical axis to sense the light scattered in the cell especially at low scattering angles. The analyzer can also measure the lignin dissolved in the sample and the fiber species.

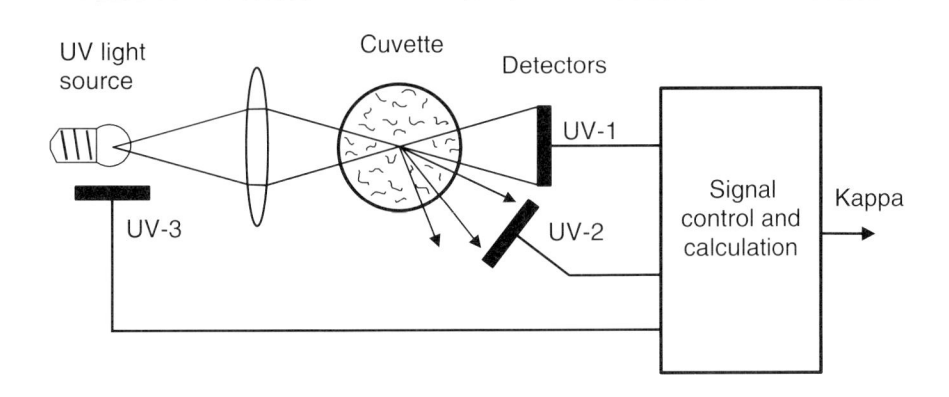

Figure 6. Kappa measurement principle (courtesy STFI).

Another kappa analyzer also uses UV absorption with a xenon light source. The UV light goes through the sample with light scattering from the fibers measured at different wavelengths. The analyzer can also measure the weight-weighted length of the pulp fibers for calculation of the hardwood to softwood ratio. The measurement device includes a washing unit that washes and screens the sample thoroughly and circulates it in the measurement device. A built-in consistency transmitter controls the sample consistency before measurement to eliminate the effect of process fluctuations on the results.

Kappa measurements have been applied to the blow line of the digester and to lignin measurements at the oxygen bleaching stage where control depends on reliable kappa measurements before and after the reactor.

2.2.2 Cooking liquid analyzer

A cooking liquor analyzer allows continuous analysis of alkali concentration and the concentrations of total dissolved solids and dissolved lignin of individual batches in a batch digester house. It can also handle several circulations in a continuous digester as Fig. 7 shows. The device employs continuous sampling directly from a process pipeline.

CHAPTER 4

Figure 7. Cooking liquor analyzer (courtesy of ABB).

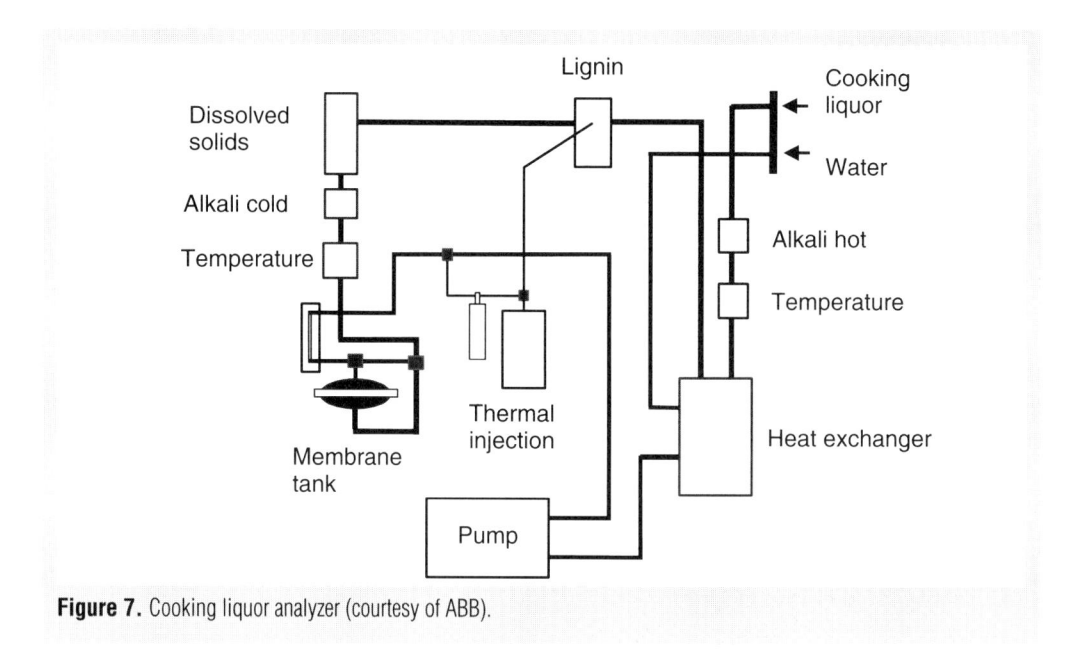

The sample is filtered through an adaptive spiral filter system that prevents any harmful solids from entering the sensor units. Measurements are performed at the actual process concentration, i.e., no dilution is necessary.

In kraft cooking, the unit measures the following concentrations employing different measurement principles:

–Alkali (dual sensor conductivity)

–Dissolved lignin (UV-absorption)

–Total dissolved solids (refractive index).

Alkali measurement operates on a dual sensor conductivity principle that improves the measurement accuracy compared with existing conductivity-based alkali analyzers. Each sensor performs four electrode measurements eliminating contamination effects.

Dissolved lignin measurement uses an UV absorption principle in the same wavelength area as for kappa analyzers. Rather than a large, constant dilution, the sensor uses a very short optical path. This allows UV-absorption measurement at the process concentration. A refractometer measures the total dissolved solids content.

2.3 Bleaching control sensors

The pioneer system for controlling the bleaching stage was the Kajaani dual sensor measurement and control principle developed in the 1970s. Several manufacturers have since introduced similar control systems including sensors. The original dual sensor principle includes brightness and residual chemical sensors as Fig. 8 shows.

Special measurements in pulp and paper processes

The brightness of the paper or paperboard is a measure of the reflectance factor at a wavelength of 457 nm. In-line sensors installed in an on-going process can detect the brightness. The measurement principle involves back scattering at several wavelengths around 457 nm. The exact wavelengths can vary slightly from one manufacturer to another.

Residual chemical measurement uses an electrochemical principle also applied in other process industries such as metal enriching processes. The sensor includes three electrodes. Two electrodes produce a set potential (voltage) in the sample, and the third electrode measures the current produced due to the concentrations of certain chemicals in the slurry.

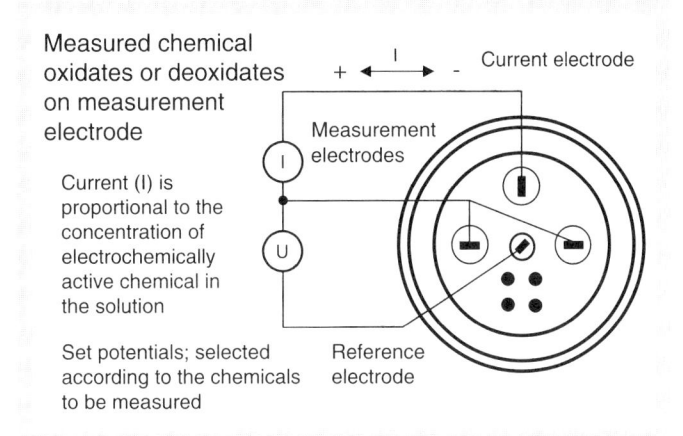

Measured chemical oxidates or deoxidates on measurement electrode

Current (I) is proportional to the concentration of electrochemically active chemical in the solution

Set potentials; selected according to the chemicals to be measured

Figure 8. Principle of residual chemical sensor (courtesy of Valmet Automation Inc.).

Kappa analyzers also have use in bleaching control now due to the conversion to elemental chlorine free (ECF) and totally chlorine free (TCF) processes. These commonly use oxygen stages and require measurement of kappa number, brightness, and residual chemicals for control purposes.

2.4 Pulp quality analyzers

The quality of the pulp is naturally a very important factor in pulp and paper processes. The fiber measurement technologies developed in the 1980s and 1990s have produced a major step toward better control.

Laboratory testing of freeness was the only measurement used initially in mechanical pulping. Several manufacturers later introduced continuous, on-line units. One company then developed the pulp quality analyzer (PQM) for this purpose. Other organizations later introduced fiber length analyzers and an on-line pulp quality control system. These devices also have application for chemical pulp quality measurements.

2.4.1 Freeness analyzers

The laboratory freeness test is an entirely empirical procedure that gives an arbitrary measure of the rate at which a suspension of 3 g of pulp in 1 L of water can drain. The result depends primarily on the quantity of debris (fines) present and to a smaller extent on the degree of fibrillation of the fibers, their flexibility, and their fineness[2].

Freeness over the years has become a general pulp quality factor, although it depends on many of the separate pulp properties listed above. Nevertheless, on-line freeness instruments are useful, and they have wide use for process monitoring.

CHAPTER 4

Several companies manufacture on-line freeness measurement devices. A typical on-line freeness tester measures the time for a given volume of water at a constant pressure head to drain through a fiber mat. The measured drainage time is adjusted for variations in consistency and temperature before conversion to a freeness value (CSF)[3].

One unit differs from the other freeness measurement principles[4]. The sample first flows into a constant volume chamber, and the surplus from the sampling device discharges into a sewer. Pulp temperature is measured at this point. The sample then drops onto a wire screen with a constant initial vacuum connected to the suction chamber under the wire. Water is first drained from the sample that forms a pad on the wire. Then air begins to flow through the pad. The flow of water and air depends on pulp drainage. A more refined pulp has stronger resistance to these flows. After a preset suction time, the vacuum below the pulp pad is measured to calculate drainage.

Besides drainage, the final pressure difference over the pad depends on temperature and the quantity of fibers on the wire. Compensation for pulp consistency changes is made after the suction phase. When the pressure reading is ready, the fibers remaining on the wire screen are weighed. The measured values then allow calculation of the drainage using an equation determined by calibration. Finally, the pulp pad is removed from the wire, and the analyzer is automatically cleaned with water jets[4].

2.4.2 Fiber length analyzers

Fiber length is an important basic variable contributing to the quality and properties of the pulp. It correlates closely with pulp shearing force. Many fiber length measurement applications are available for use in mechanical and chemical pulp and paper processes.

Pulp classifiers have long measured fiber length. The most common is the Bauer McNett classifier that divides the pulp into categories with certain mean length distributions. The amounts of the various fractions are indicated as weight from which their percentages in the pulp can be calculated. The fractions overlap in fiber lengths so these devices do not provide accurate fiber length distributions.

One company has performed pioneering work on fiber length measurements. Their laboratory fiber analyzers and on-line analyzer are pioneers in this field. Other measurement devices have come onto the market more recently.

The operating principle uses the imaging of individual fibers onto a line of detectors using polarized light. The analyzers use a laser-based light source from which the light is conducted to a capillary less than 1 mm in diameter as Fig. 9 shows. An image of the fiber inside the capillary projects onto a row of detec-

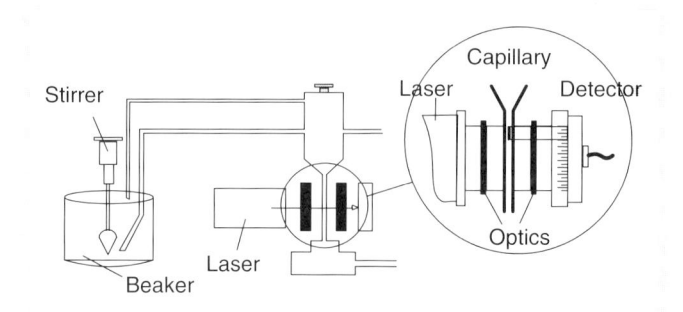

Figure 9. Principle of fiber length analyzer (courtesy of Valmet Automation Inc.).

Special measurements in pulp and paper processes

tor diodes. Another polarizer (analyzer) is between the capillary and the detector and perpendicular to the polarizer on the other side of the capillary This enables a positive image of the fiber flowing inside the capillary to be obtained at the detector. The length results for the fibers are then read into the memory from the detector. This arrangement eliminates air bubbles and the like.

The fiber analyzer can measure coarseness and also provides measurement of the hardwood to softwood ratio on the basis of fiber length distributions and correlation calculations.

An on-line fiber quality analyzer (FQA) measures fiber length, curl, and kink using image analysis of a diluted sample. The entire system consists of the on-line FQA, a sampling module, and a dilution and mixing module capable of receiving material from as many as three sampling lines.

2.4.3 Pulp quality monitor

A pulp quality monitor (PQM) measures three pulp quality variables: freeness, fiber length (three length fractions: fine, medium, and long), and shive content. It has a continuous sampling setup. For a long time, it was the only on-line pulp quality analyzer on the market that used more than one fiber characteristic.

The on-line model measures drainage (freeness) using a built-in drainage tester (DRT) as described above. Fiber length is analyzed in a glass tube of dimensions 10 x 10 mm using three light beams of different diameter as Fig. 10 shows. The transmitted signals are then processed for computational segregation of fine, medium, and long fiber fractions.

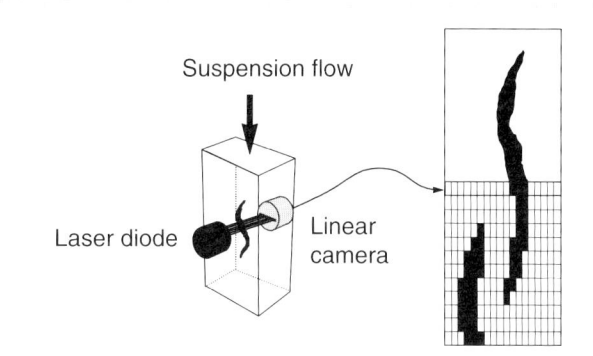

Figure 10. PQM fiber fractionating principle (courtesy Sunds Defibrator AB).

Shive content is analyzed in the same tube using two flat, perpendicular light beams. Shives intercepting the light beams cause disturbances in the transmitted light, and the widths and lengths of the particles are measured using the magnitude and duration of these disturbances. The number of shives detected is displayed in different size categories expressed as shives per gram of pulp.

A PQM laboratory analyzer is a fiber and shive classifier that operates on the image analysis principle giving as its results fiber images, length distributions, five length fractions, average width, coarseness, curl index, softwood to hardwood ratio, shive images, a PQM shive matrix, shive weight, and average length and width.

CHAPTER 4

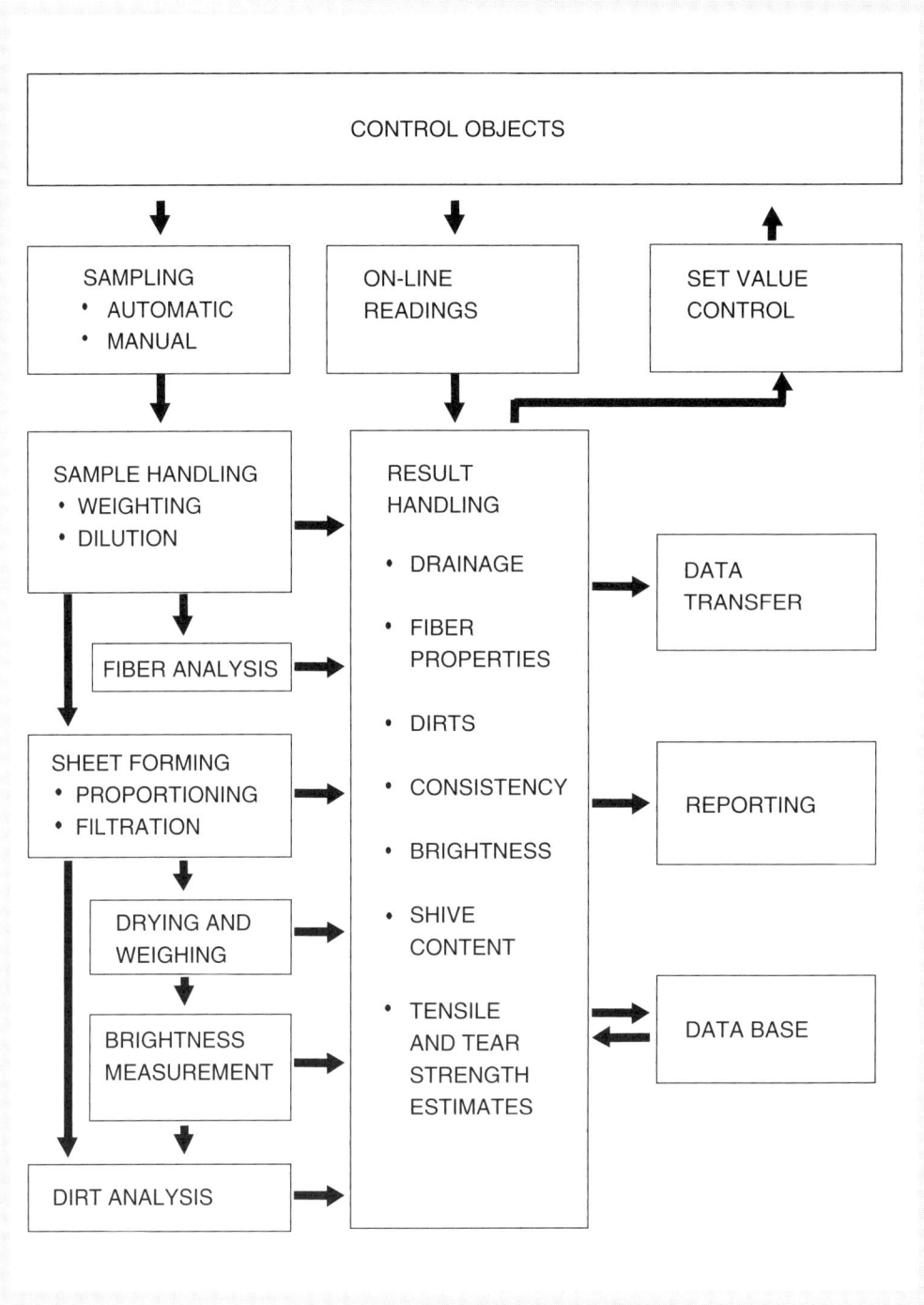

Figure 11. PulpExpert testing procedure (courtesy of Pulp Expert Inc.).

2.4.4 PulpExpert

PulpExpert is an on-line quality control system for paper pulp testing that comprises an automatic sampling system and measurements of various pulp properties. Up to eight sampling lines can be handled with manual introduction of spot samples for analysis.

PulpExpert measures the consistency, drainage (CSF, °SR), brightness, dirt count, fiber length, fiber length distribution, coarseness, curl, light scattering coefficient, and tensile strength of the pulp sample in 10 min as Fig. 11 shows. The individual measurement principles employed in the system are standardized as far as possible (TAPPI, SCAN) and integrated in the unit. The results are computed immediately, presented in the form of graphical reports, and distributed to the mill's data network simultaneously.

PulpExpert has use for quality control in chemical pulping before and after bleaching and between process stages, in groundwood and TMP processes, and for refining, deinking, and the optimization of pulp proportioning to the machine chest in paper mills.

2.5 Wet end measurements

The wet end of the paper machine has evidently been the most difficult area for instrumentation. The list of requirements is still long despite introduction of new measurements in the 1980s and 1990s. Conductivity, pH, and charge (anionic trash) sensors are now used in the process in addition to the flow and basic consistency measurements already in use for a long time. Titration devices have also been used at the wet end for specific chemical analyses.

2.5.1 On-line cationic demand (anionic trash) measurement

The importance of the electrokinetic charge for the wet end chemistry of papermaking systems has long been acknowledged, since the magnitude and total quantity of the charge have a significant effect on drainage, flocculation, and the retention of pigments, dyes, internal size, and wet strength resins.

Many different zeta potential measurement principles have become available over the years including microelectrophoresis, electro-osmosis, the streaming current and streaming potential principles, electrokinetic sonic amplitude, and electrophoretic mass transfer. These methods principally function for zeta potential monitoring. Their use for optimizing the papermaking process has not been very successful because the zeta potential alone does not provide sufficient information on the wet end chemistry under different process conditions.

Other charge measurements such as cationic demand and colloid titration have subsequently been developed to give other parameters related to the total quantity of the charge. Charge titration principle especially has become a standard method on a laboratory scale. Experience gained with laboratory methods has been adapted to on-line charge measurement. The principle is a quantitative polyelectrolyte titration in which the consumption of cationic standard reagent provides a measure of anionic substances in the test solution. This is an indicator of the anionic trash level[5].

CHAPTER 4

Since charge analysis by titration uses the demand for cationic polyelectrolyte, no calibration is necessary. This is a substantial benefit for an on-line instrument.

A commercial on-line particle charge titrator continuously determines quantities of polyelectrolytes, additives, or flocculants by performing on-line charge titration of anionics or cationics such as fibers, fillers, and colloidal matter. The results are in terms of streaming potential (in [mV]), anionic or cationic demand (in [mL]), and, optionally, pH.

2.5.2 Retention monitoring

Early methods for measuring retention in the wet end used total consistency transmitters typically with optical principles. The optical measurement methods were depolarization and attenuation and scattering. These devices enabled monitoring of total wire retention, but control applications were uncommon.

After introduction of filler consistency measurement with total consistency, retention monitoring became an important control tool at the wet end. The control concept using wire water consistency measurement has especially proven effective for stabilizing the short circulation of the paper machine.

The most advanced consistency measurement methods used in retention applications rely on a combination of optical measurement principles including depolarization, absorption (attenuation and extinction), and scattering at several wavelengths from the UV to the near-infrared range (NIR). The optical on-line sensor of Fig. 12 measures light depolarization (NIR laser), extinction and back scattering (NIR laser and UV xenon), and absorption (xenon) at different wavelengths. It then processes signals to monitor total solids and filler consistencies and also flocculation in the sample.

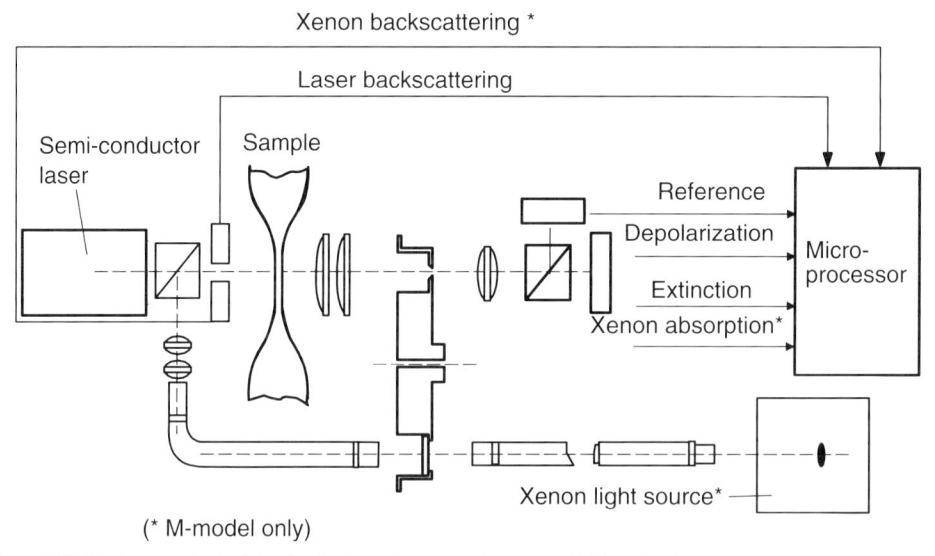

(* M-model only)

Figure 12. Measurement principle of optical on-line sensor (courtesy of Valmet Inc.).

Special measurements in pulp and paper processes

Calibration of the sensors uses reference samples measured in the laboratory. Different furnish compositions use pre-calculated models.

Other commercial products are available for retention monitoring using filler and total consistency sensors. These also use a combination of optical principles to produce consistency measurements.

2.5.3 Flocculation sensing

Flocculation means the tendency of pulp fibers and fines (fillers) to accumulate into clumps or flocs. Classification of the flocs appearing in paper furnish suspensions can use their composition, size, and stability. They may only consist of fibers or fines or a mixture of these. Fiber-fiber flocs are very large and can be a few centimeters in diameter. As the size of the floc increases, the phenomenon becomes local consistency variation in the furnish. Measurement of the size of fines flocs is in micrometers as are filler and fines flocs. Flocculation typically occurs at small size levels in low consistencies and at large size levels in higher consistencies where it may influence web formation in the paper machine.

The paper industry has adopted many methods for measuring or characterizing flocs. These include measurements made on the paper web (formation sensors) or directly on the furnish suspension (flocculation sensors). The latter use variance in consistency measurements or a suitable consistency-dependent signal such as variation in transmitted or back scattered light. As an example, one flocculation meter uses the variance in the back scattering signal.

The causal stages in flocculation measurement are sampling and sample handling. Flow conditions must remain constant, since shear forces may change or disappear. Then the furnish flocculation can change completely in the sample line.

2.5.4 Air content measurement

Entrained air or gas such as carbon dioxide causes many problems in the operation of paper machines. It hampers operation of the pumps, increases energy consumption, interferes with drainage on the wire, and can even cause holes in the paper.

An on-line device for the measurement of entrained air and gas became available in the 1990s. It uses ultrasonic attenuation caused by stabilized and dispersed air bub-

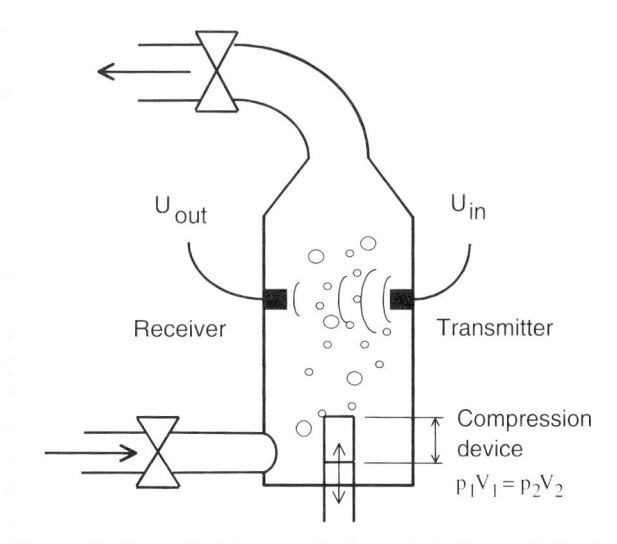

Figure 13. Ultrasonic air content meter (courtesy of Conrex Inc.).

$P_1V_1 = P_2V_2$

bles in the furnish[6]. The stabilized bubbles that are most harmful in the wet end process are particularly effective in attenuating ultrasound.

One company's measurement method is very simple. Ultrasonic energy is transmitted through a cell as Fig. 13 shows of dimension approximately 10 cm. A receiver located on the other side of the cell detects the attenuated signal. The attenuation can be calibrated to the air content reading using a device based on the compression method that is included in the measurement system.

3 On-line measurements of paper quality

3.1 Basis weight

Measurement of basis weight normally uses the beta radiation principle[7]. The paper and fillers including the coating absorb beta radiation (electrons) almost equally. The relationship between transmitted radiation and mass is an exponential one (Beer-Lambert's law, Eq. 1). Calibration of this is very easy to give basis weight.

The meter constructions used in different manufacturers' devices are very similar. Measurement through the paper as Fig. 14 shows is the most common technique. Krypton-85 (Kr^{85}) is the predominant beta source, since it is highly stable, simple, and rugged. Some units have used strontium-90 (Sr^{90}) and promethium-147 (Pr^{147})[7]. Beta radiation measurements can be used from low basis weights of 5 g/m² up to 4900 g/m² grades, covering applications from tissue to heavy board. Prometium is more accurate than Krypton at lower basis weights while Strontium is the choice at higher paper weights.

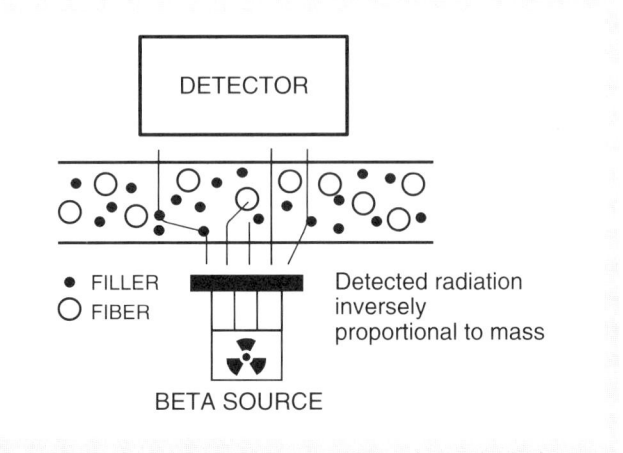

Figure 14. Basis weight sensor (courtesy of Valmet Automation Inc.).

3.2 Moisture

Moisture is the main paper quality variable requiring measurement and control in papermaking. Various approaches to measure this have included IR absorption, microwave attenuation, and radio frequency techniques.

Water absorbs electromagnetic radiation at many wavelength bands in the near infrared region. The 1.94 µm band has primary use as a measurement channel usually with the 1.81 µm band as a reference. Manufacturers typically use two additional channels for measurement and reference purposes in response to increased demands for accuracy and applicability. The 1.45 µm band is used on some board grades as well as microwave or infrared reflection techniques.

Special measurements in pulp and paper processes

One supplier constructs an IR meter using a filter disc device containing an integrating sphere with an embedded, sealed PbS (lead sulfide) IR detector as Fig. 15 shows.

Microwave-based sensors using the attenuation of microwave energy have been on the market since the early 1970s. A typical microwave frequency is 22.2 GHz as in the sensor of Fig. 16 where water has over 100 times higher absorption than other sheet constituents. The specific absorption coefficient of water is nevertheless smaller at microwave frequencies than in the IR range. This provides possibilities for measuring the higher moisture content values of heavy paper grades.

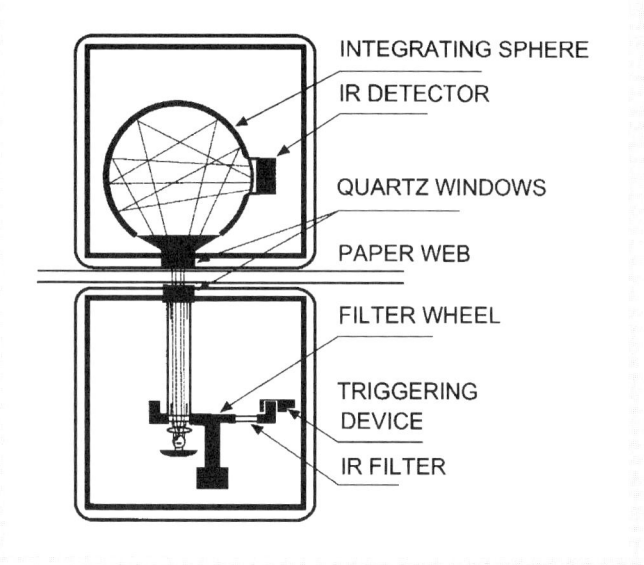

Figure 15. Paper moisture sensor (courtesy of Valmet Automation Inc.).

3.3 Filler (ash) content

Measurement of the filler (ash) content of paper uses the low-energy gamma (X-ray) radiation absorption principle. Gamma radiation is preferable over beta radiation for filler quantity determination since fillers have an attenuation coefficient several times higher than that of fibers at X-ray frequencies. Two suppliers use Iron-55 (Fe^{55}) as the X-ray source and the use of a Tritium/Titanium source has been reported. The sensor is similar in construction to the basis weight sensor of Fig. 14. Two other large suppliers use shaped spectrum machine generated X-rays.

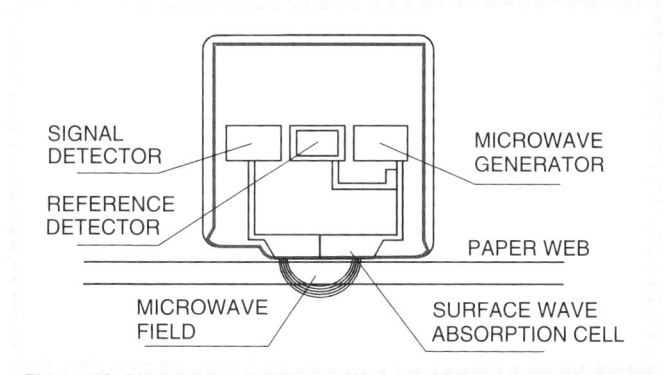

Figure 16. Microwave moisture sensor (courtesy of Valmet Automation Inc.).

3.4 Formation

On-line formation as defined relative to basis weight variation has been a very challenging measurement requirement for years. Since beta radiation is difficult to use with on-line devices, optical sensors have been developed that typically use laser light to provide

CHAPTER 4

information on the formation of the sheet. These attempt to quantify and characterize high frequency basis weight variations in the sheet[8]. The method employs laser light attenuation on a small scale as Fig. 17 shows. Optical systems do not measure basis weight variations but small-scale variations in opacity or cloudiness in the sheet. The signal therefore depends on the filler and coating pigment distributions in the sheet.

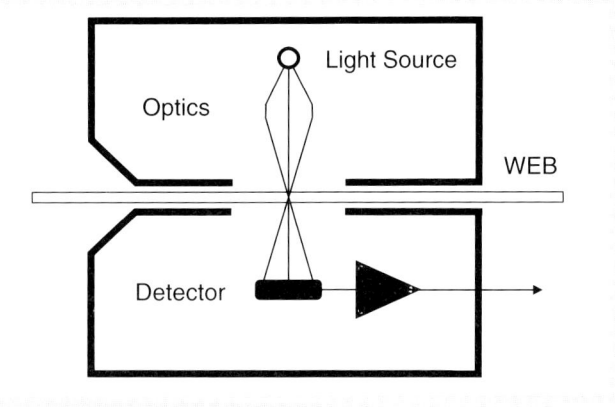

Figure 17. Formation sensor (courtesy of Honeywell Automation Inc.).

3.5 Fiber orientation

A fiber orientation sensor determines the mechanical quality and strength of the paper web by analyzing the orientation of the fibers. A noncontacting laser light beam passes through the web. The orientation of the fibers causes the light to diverge and form an ellipse that is measured by an optical detector. The manufacturers claim that the shape of the ellipse and the orientation of its major and minor axes relate to the tensile strength ratio that then relates to web strength.

3.6 Caliper (web thickness)

Determination of the thickness of the paper defined as a measure of caliper may use the variable magnetic reluctance principle. The thickness of the paper alters the distance between two sensor heads that have a magnetic coupling through the sheet as Fig. 18 shows. Contact pressure is measured and controlled simultaneously. The mag-

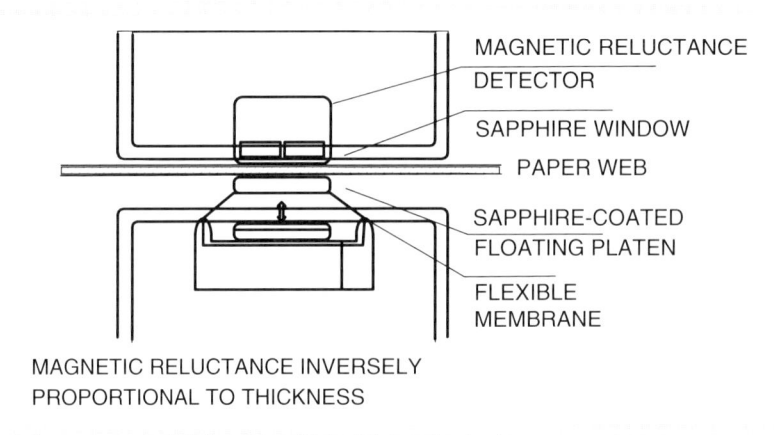

Figure 18. Caliper sensor principle (courtesy of Valmet Automation Inc.).

netic reluctance is inversely proportional to thickness and is converted to an electrical signal that corresponds to the caliper. A typical electronic arrangement is an electromagnetic resonance circuit with a coil wound around a ferrite core and ferrite plate on the opposite side. The two heads produce an electromagnetic oscillation circuit with a resonance frequency dependent on the distance between them. This is then a function of the web thickness.

3.7 Smoothness (roughness)

Printability is the most important printing paper property. Since it is probably impossible to develop a general printability quality sensor, smoothness and gloss sensors usually characterize printability. In the air-leak methods generally used in the laboratory, roughness is a function of the rate at which air passes between a flat surface of specified shape and a sheet of paper. Such methods are difficult to apply to control of a paper machine. In addition, the resulting measures do not correlate unambiguously with printability. Gloss sensors have proven usable in special printing paper applications, but glossy and matte papers cannot be compared by gloss measurements alone when considering printability. Coating porosity measurements can also give valuable information on printability, but their implementation has proven difficult.

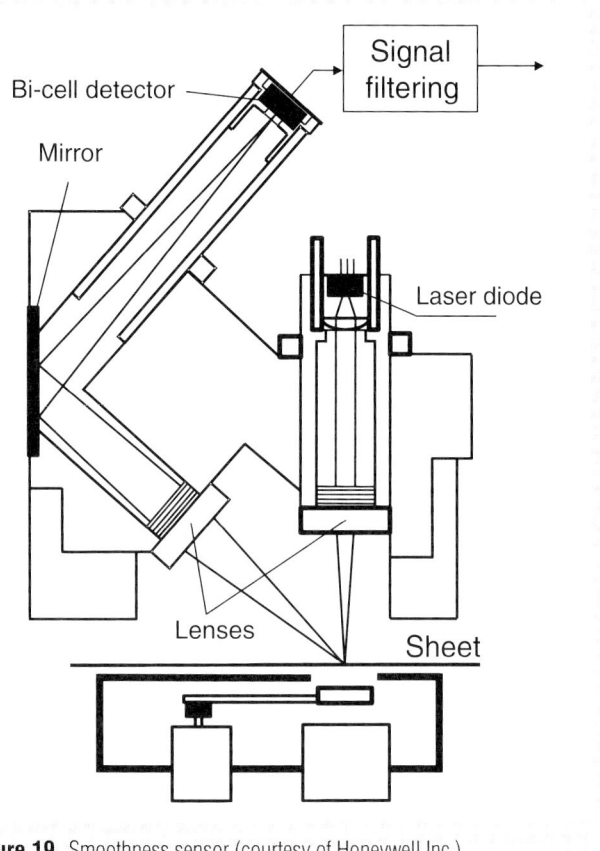

Figure 19. Smoothness sensor (courtesy of Honeywell Inc.).

Smoothness sensors usually use the detection of near-infrared light. This method measures the scattering effect of the surface of the paper on a light beam. The incident angle is 75°. This is the same as with gloss measurements. The measured scatter allows development of physical models to describe the roughness of the paper surface.

A new measurement principle recently introduced uses laser profilometry to measure surface contours as Fig. 19 shows. Laser profilometry has had long use for surface characterization in the laboratory. A laser diode focuses in the normal direction on a 20 mm spot on the sheet, and the receiver optics view the spot from 45°.

The manufacturer reports that the resolution of this triangulation contour measurement is approximately 0.1 mm[9]. This is sufficient for surface analysis.

CHAPTER 4

3.8 Gloss

The arrangement for gloss measurement is very similar to the smoothness sensor principle as Fig. 20 shows. The main difference is the wavelength range that is naturally in the visible area for gloss measurement. The light reaches the web at a standardized 75° angle (TAPPI T 480 om-90) with measurement of the specular reflectance[10].

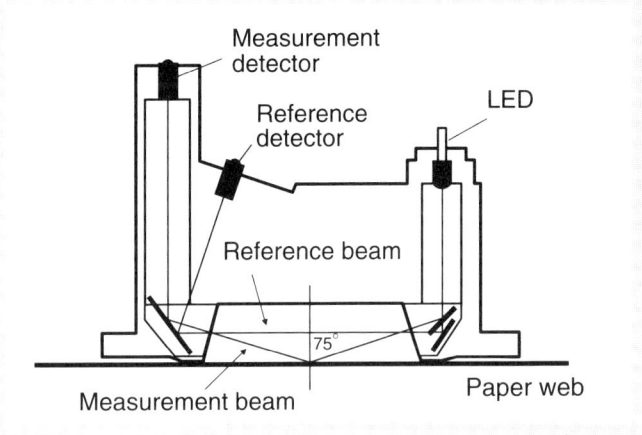

Figure 20. Gloss sensor principle (courtesy of Valmet Automation, Inc.)

3.9 Opacity, color, and brightness

Opacity, color, and brightness are optical variables that can be combined in one device[8] shown in Fig. 21 installed on an on-line scanning sensor platform.

On-line opacity sensors measure the light transmission capacity of the paper. Since the relationship between transmission and opacity is nonlinear, signal processing must be used to calibrate the transmission signal to opacity values.

Color sensing devices use well-defined standard color measurement principles. Color must naturally be measured using reflectance geometry.

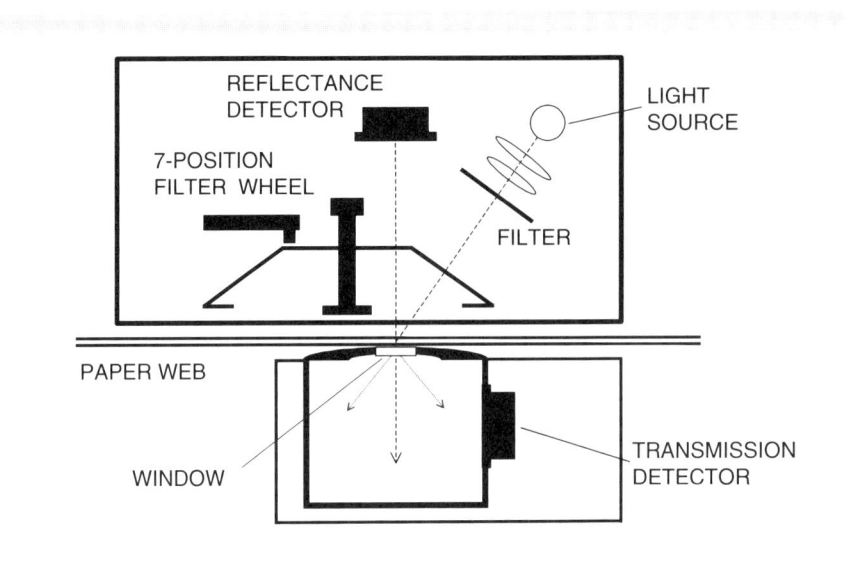

Figure 21. A combination sensor for opacity, color, brightness, and fluorescence measurements in paper manufacturing[8].

Special measurements in pulp and paper processes

3.10 Porosity

Porosity is a very difficult parameter to measure on-line in a paper machine. The principles used are indirect ones such as air permeability (Gurley porosity). Porosity correlates with roughness, but roughness measurements are intended primarily for surface quality characterization. Porosity measurements are aimed at bulk quality characterization.

On-line sensor constructions are cumbersome due to the mechanics necessary for air flow or pressure sensing as Fig. 22 shows[8].

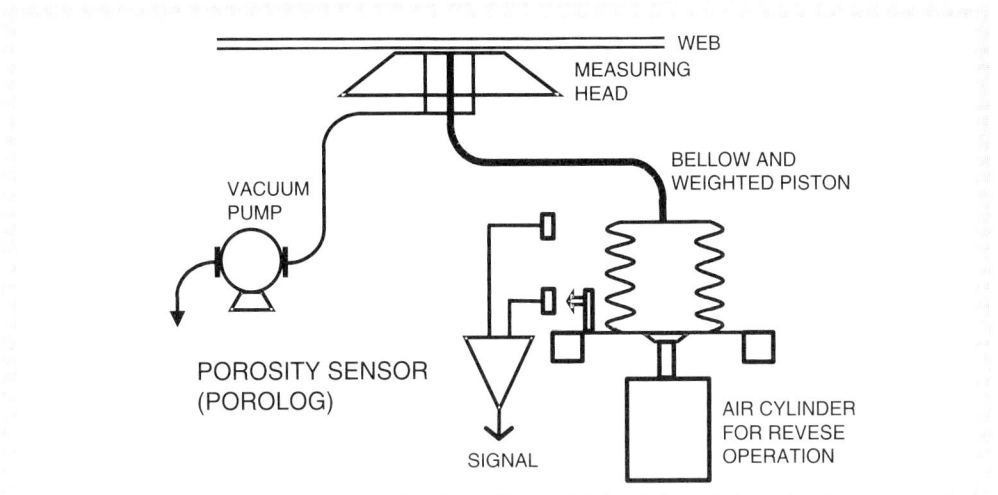

Figure 22. The porosity sensor principle (courtesy of STFI[8]).

3.11 Coating weight

Coating weight measurement can use various basis weight and moisture sensors with beta absorption. The principle of this method is simple. Basis weight sensors before and after the coater provide the difference between the basis weights that equates to the coating weight reading. When applying multiple coatings, one may need as many as six sensors: two for the base paper (basis weight and moisture), two after the first coater, and two after the second coater.

A simpler method developed recently employs near-infrared techniques using absorption of radiation by the coating pigments in the NIR range (800–2500 nm). This method primarily measures clay coating. Unfortunately, NIR techniques cannot measure the calcium carbonate usually applied in the first coating layer. In these cases, clay is usually a trace in the coating, and the total amount of coating can be calculated based on the clay value. The NIR sensor is very similar in construction to that used in NIR moisture sensors, but the wavelength range necessary in the instrument is wider. It extends to 2.2 µm, since the absorption peaks of clay lie at wavelengths of 1.4 µm and 2.2 µm.

CHAPTER 4

Some manufacturers measure a unique calcium carbonate absorption band that occurs in the mid infrared (MIR) near 4.0 μm (2514 cm-1). Other manufacturers use latex as a tracer for $CaCO_3$, that can be sensed in the 2.35 μm (4255 cm^{-1}) range. Both of these techniques have shown to be effective for calcium carbonate measurement especially in board machine applications.

Differential X-ray (and Gamma) techniques are very effective means for measuring inorganic coatings. The coating solids absorb between 3 and 5 times more X-ray energy compared to cellulose and water. This generally provides a high signal to noise ratio with less influence from the MD variability of the base sheet. A moisture compensation is needed because moisture levels before and after coating may be quite different, even though the sensitivity to moisture is small with this technique.

3.12 Holes and dirt spots

Camera-based machine vision techniques inspect for holes and dirt spots. Transmission and reflection techniques are implemented in the same system for different applications. One inspection system uses sodium lamps for illumination and extremely high speed charge-coupled device (CCD) cameras for detection. The most critical aspect of measurement systems intended for detection of holes and other defects is speed. Sophisticated techniques are necessary for camera electronics and signal processing.

CHAPTER 4

References

1. Saarinen, K., "Adaptive control of TMP-plant based on consistency measurement", Ph. Lic. Thesis, Jyväskylä University, Jyväskylä, Finland, 1993.
2. TAPPI T 227 om-85 "Freeness of pulps."
3. Anon., Sunds PQM 400 report 223-E.93.3000, Sunds Defibrator AB, Sundsvall, 1993.
4. Anon., Kajaani Application A47 00.1-E, Kajaani Electronics Ltd., Kajaani, 1995.
5. Bley, L., Pulp & Paper Canada 99 (5): 51(1998).
6. Karras, M., Kemppainen, A., Harkonen, E., et al., U.S. pat. 4,911,013 (March 27, 1990).
7. Puumalainen, P., "On-line measurement of paper quality and the factors effecting on it in the manufacturing process," D. Tech. Thesis, Lappeenranta University of Technology, Lappeenranta, Finland, 1993.
8. Walbaum, H. H. and Lisnyansky, K. Paper Trade J. 167(13):37(1983) and 168(14):34(1983).
9. Chase, L., Belotserkovsky, E., and Goss, J., TAPPI 1996 Finishing and Converting Conference Proceedings, TAPPI PRESS, Atlanta, p. 129.
10. Anon., Valmet IQGloss, A416199, Valmet Automation Inc., 1996.

CHAPTER 5

Process control in fiber line

1	**Continuous flow digester control**	**73**
1.1	Control requirements	73
1.2	Basic control functions	74
1.3	Quality control	79
1.4	Species change control	80
1.5	Benefits	81
2	**Batch digester control**	**82**
2.1	Control requirements	82
2.2	Basic control functions	82
2.3	Cooking plant scheduling and steam leveling	84
2.4	Kappa number control	84
2.5	Tank farm controls	86
2.6	Benefits	87
3	**Brown stock washing control**	**88**
3.1	Control requirements	88
3.2	Control functions	89
4	**Bleach plant control**	**91**
4.1	Control requirements	91
4.2	Oxygen delignification control	92
4.3	Chlorination stage control	93
4.4	Control in the dioxide stages	96
4.5	Control of mechanical pulp bleaching	96
	References	98

CHAPTER 5

Kauko Leiviskä

Process control in fiber line

1 Continuous flow digester control

1.1 Control requirements

Continuous flow digesters have used computerized control since the early 1960s. Dedicated control packages have been commercially available since the 1970s. System platforms have changed with time, but the basic control functions have remained very similar. The control methods and software tools for maintaining the higher level controls have developed considerably. Development of new analyzers has been especially fast.

The pulp digester is a major unit operation in the pulp mill. Its proper control is very important to pulp production in the entire fiber line. Cooking should give the best available quality at minimum costs. Quality variations must simultaneously compensate for the effects of several types of disturbances. The main objective is to guarantee similar cooking history for all chip particles as they pass through the digester. This leads to decreased variation in pulp quality compared with conventional controls.

Continuous digester control must consider three main difficulties[1]:

1. Variations in chip quality. Chip characteristics vary during daily operation. The material itself is very heterogeneous, starting from tree size and age. After chipping and blending with purchased chips, defining the characteristics of the material entering the digester is difficult. Even more difficult is influencing these variations by processing or control means before the digester. Chip quality variations cause two types of problems: out-of-target cooking conditions and operational problems. Operational problems include inconsistent tons per revolution delivered by the chip meter resulting in inaccurate alkali to wood ratios, level control problems, inconsistent plug movement, etc.

2. Measurement problems. A serious problem is that the desired quality control variable, kappa number, is not directly measurable while the cooking occurs. This applies to batch and continuous digesters. Various indirect measurements are necessary. Measurements traditionally used in the kraft cooking process are temperature and alkali measurement using liquor conductivity or automatic titrators. The chip feed and its moisture content provide an estimate of the liquor to wood ratio. The final cooking result (kappa number) is available from the blow line after the digester by using laboratory analysis or an automatic device.

CHAPTER 5

3. Long process delays. Operators sometimes have difficulties understanding process delays especially when changes occurred during the previous shift and the results are not visible until the following shift. The control system can not eliminate delays, but a good control strategy considers them without any human hesitation. A good process management system also provides tools to understand the process delays and other process behavior better. With most existing systems, long delays from kappa number measurement or analysis make using feedback or feedforward control to change process conditions for stabilizing the pulp quality difficult.

1.2 Basic control functions

The basic control functions for the continuous digester according to the process stages are as follows [1]:

Chip feeding

- Chip feeding control
- Production rate control
- Rate change control.

The nominal chip meter speed (or chip screw rpm) is adjusted according to the operator entered production rate target. It also depends on the pocket filling degree of the chip meter, chip meter volume, bulk density of the chips, and the yield. In rapid production rate change, the chip meter speed changes immediately to the new target. Coordination of other controls is such that variations in kappa number are as small as possible. In scheduled production rate change, the cooking factors and the production rate follow a previously determined schedule that considers the actual process time delays and digester volumes. This minimizes deviations from desired pulp quality.

Dosage controls

- Alkali dosage control
- Alkali profile control
- Liquor to wood control.

To produce pulp with a constant quality, alkali dosage must have the correct ratio compared with the amount on moisture-free wood. One must also compensate for variations in alkali concentration of white liquor and chip quality. In some cases, alkali is dosed in several locations (so-called alkali-profile control). In modified cooking, this is a primary technique. Measuring the alkali concentration using a technique such as near infrared range (NIR) measurements in digester circulation allows control of residual alkali. Correcting the alkali to wood ratio or the basic alkali trims in some cases corrects this. Black liquor dosage controls the liquor to wood ratio.

Impregnation vessel controls

- Chip level control
- Bottom dilution control.

Digester vessel controls

- Chip level control
- Liquor level control.
- Temperature profile control
- *H*-factor control.

Stable plug movement requires stabilization of chip level. Changing the outlet device speed or the chip meter speed can accomplish this. In the first case, the outlet device speed is the primary control variable. If this exceeds its limits, blow flow is adjusted. The chip meter is used only during species changes. In the second case, the chip meter speed controls the chip level to provide rapid response to short-term chip level variations with minimum upset to digester operation. The blow flow is adjusted when the chip meter speed reaches its control limits. The nominal blow flow is controlled according to production rate. Previous literature also presents opposite strategies using blow flow.

Some advanced control methods also find use in chip level control. Sastry[2] reports the use of a self-tuning regulator. The biggest advantage of this approach is that blow flow is the only variable necessary for control. More recently, Allison *et al.*[3] studied the use of general predictive control (GPC) in chip level control. They used the chip level and chip meter rpm as inputs to the chip level controller. Chip meter rpm was the second input variable merely to reduce the blow flow manipulation and thereby also reduce disturbances to chip column movement. The chip meter was manipulated to maintain the required production rate. The two-input strategy decreased the blow flow manipulations by more than 50%. Figure 1 shows the block diagram of this approach.

Brattberg[4] has reported the use of an adaptive controller as a cascade loop to control chip level with chip feeder speed. The blow flow here is a feedforward signal. The standard deviation of kappa number with low kappa pulp decreased by more than 40% after the installation of the adaptive controller.

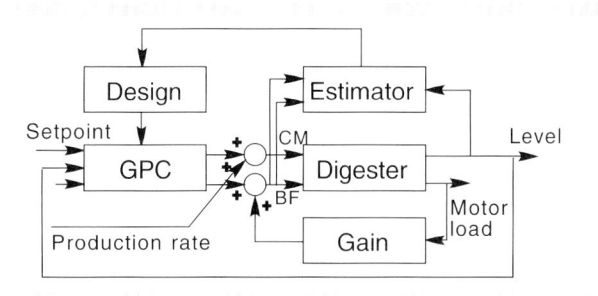

Figure 1. Two-input strategy for chip level control[3]. In the figure, CM denotes the chip meter speed and BF the blow flow.

CHAPTER 5

Temperature profile control tries to maintain the desired delignification degree under varying conditions. The initial temperature targets depend on the production rate and grade and species produced. H-factor and kappa number control correct these targets.

Equation 1 describes the delignification of kraft pulping[5]:

$$\frac{dL}{dt} = -kCL \qquad (1)$$

where L is the lignin content of chips
C the effective alkali concentration of the cooking liquor in the chips
k the temperature dependent reaction rate constant
t the cooking time.

The temperature dependency of the reaction rate constant, k, is usually presented by the Arrhenius equation:

$$k = k_o e^{B - \frac{E}{RT}} \qquad (2)$$

where E is the activation energy
R the universal gas constant
T the absolute temperature
k_o and B are constants.

The parameters depend on wood species, cooking stage, etc. H-factor describes the joint effect of time and temperature usually given as the following integral[6]:

$$H = \int_{t_o}^{t} k_r dt \qquad (3)$$

where t_o is the initial time for calculation
t the actual time
k_r the relative reaction rate coefficient defined by the following equation:

$$k_r = e^{43.20 - \frac{16113}{T}} \qquad (4)$$

where T is the absolute temperature.

The temperature must exceed 100°C for faster increase in the H-factor. The H-factor defined in Eq. 3 does not include the effect of chemical concentrations and liquor to wood ratio. The parameters in Eq. 4 depend on wood species.

According to Jutila[5], most control systems use the cooking model in the following form:

$$k = f(C_1, H) \tag{5}$$

where K is the kappa number
       C_1 the alkali concentration of the cooking liquor in the beginning or during the cook.
       H Vroom's H-factor or its modification.

Jutila[5] also gives a good survey and comparison of models. Other authors present models used in commercial systems[7–9]. Chari[10] and Hatton[11] introduce other modeling approaches.

H-factor control in continuous flow digesters uses the measured/calculated temperature profile and retention times in the different digester zones and the H-factor calculated with this information as Fig. 2 shows. The measured and target H-factor are compared

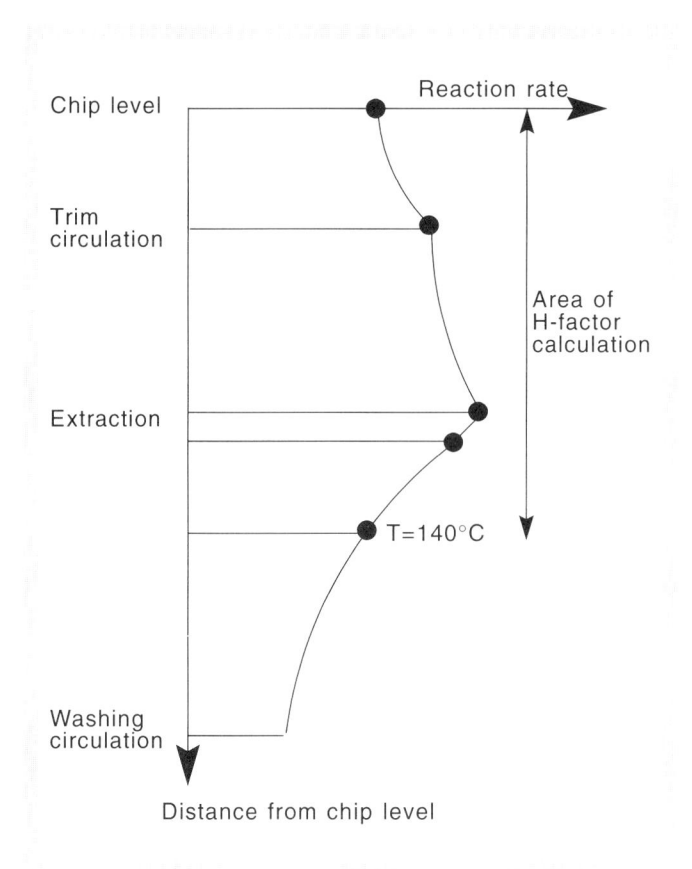

Figure 2. A method for H-factor calculation in continuous digesters[1].

CHAPTER 5

and the temperature profiles corrected when necessary. The temperature profile control then maintains the single temperatures in their targets considering the predetermined interactions between the controlled temperature points as Fig. 3 shows.

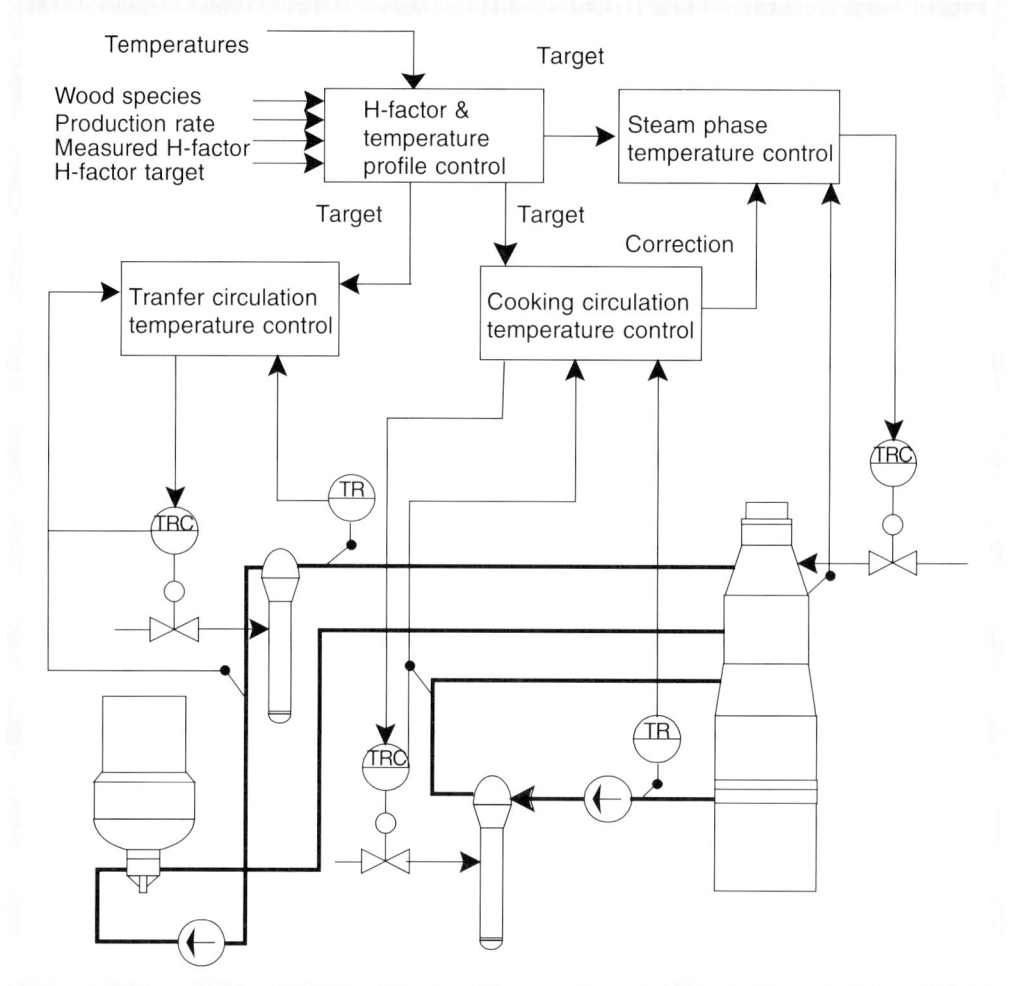

Figure 3. *H*-factor and temperature profile control [1].

Blow line controls

- Blow flow control
- Bottom dilution control.

1.3 Quality control

Control of pulp quality is becoming increasingly more important in cooking plants. Although pulp quality is a large, complicated topic, this discussion only covers kappa number. Traditionally, kappa number is the controllable quality variable of blow line pulp. It is monitored continuously.

Kappa number control uses well maintained alkali and temperature profiles. Retention time is another crucial variable usually determined by the production rate requirements. Kappa number control tries to provide constant conditions for cooking reactions. Because of disturbances, this kind of feedforward procedure cannot assure a stable quality in all conditions without any feedback. In conventional systems, it uses analyzed samples of pulp after the digester. Because of long retention times and analyzing delay, a direct feedback from laboratory kappa number is impractical. Model reference adaptive control strategies, Smith predictors, or other approaches have found use.

The situation improves with use of analyzers. The blow line kappa sensor decreases the analyzing delay and allows control interval decrease.

Liquor analyzers further improve kappa control. By measuring liquor concentrations from the liquor circulation of the impregnation vessel, transfer circulation after impregnation vessel, trim circulation, and extraction black liquor, one can estimate kappa number in different stages of the cooking process. With a good kappa number estimate, control actions are possible before the pulp reaches the blow line of the digester. Using neural networks in kappa number estimations seems to give reliable results even in cases where only transfer circulation measurements have been used[12] as Fig. 4 shows. Another example of using neural networks in modeling of kappa number in continuous digesters is available[13].

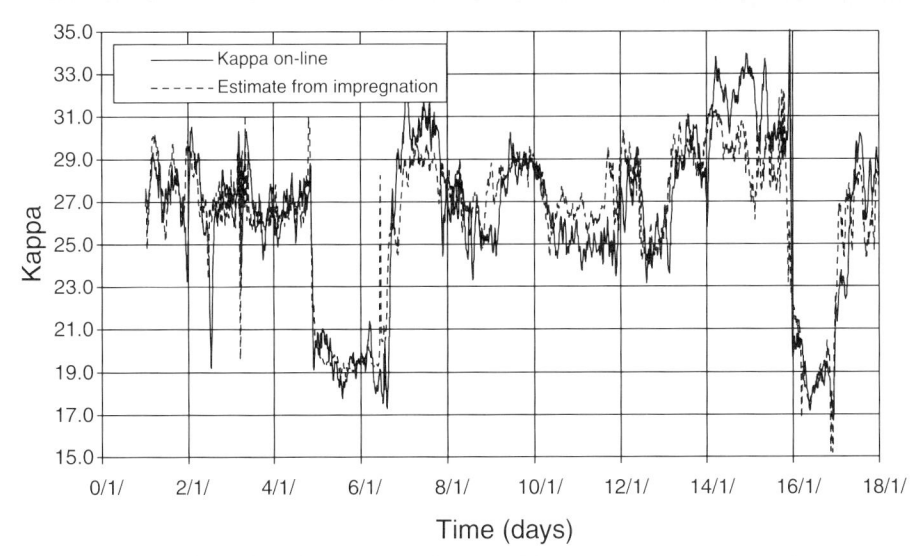

Figure 4. Kappa estimation from transfer circulation liquor[12].

Kappa number control usually uses the H-factor. The difference between the estimated kappa number and kappa number target is useful in correcting the H-factor target and the temperature profile. This model gives estimates of the cooking result inside the cooking zone that are faster than using conventional control with laboratory kappa number. Temporary variations in the cooking phase are observed and compensated for before pulp reaches the washing zone. This minimizes disturbances in pulp quality. The model is updated with blow line kappa or laboratory kappa numbers. Statistical methods filter variations in laboratory kappa number analysis caused by analyzing errors or process fluctuations. Figure 5 shows an example of such quality control strategy[1].

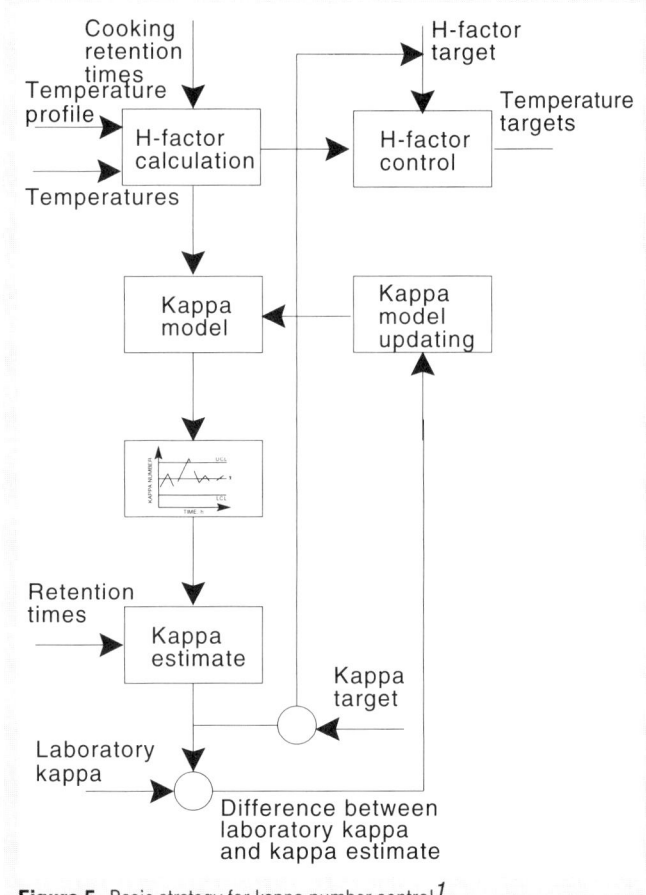

Figure 5. Basic strategy for kappa number control[1].

1.4 Species change control

Species change strategy is a common topic in the Nordic countries, although few written articles about this item exist. This text uses one available publication[1]. Opinions vary on managing species changes that have led to different species change strategies for various mills.

The yield and chip bulk density change with a species change. This means that the strategies from hardwood to softwood and vice versa should be different. It also means that the chip volume necessary to produce the same production in tons of hardwood is about 25% less than with softwood. The chip meter rpm, blow flow, or both will require correction during species changes.

Problems in species changes are especially common when changing from softwood to hardwood. The reason is that when birch (a heavier material) comes into the digester, the chip level starts to decrease rapidly. If this is not a consideration in chip level control strategy, an unstable digester operation can easily result.

Note also that the temperature control strategy differs when changing from pine to birch compared with the opposite case. To guarantee high quality pine pulp at the end of the pine period, the change in the actual cooking temperature of birch must occur gradually. The fact that birch is not so sensitive to overcooking also supports this.

Management of species changes is a challenging task for any control system. During species changes, one must typically tend to timing more than 10 cooking variables correctly. The magnitudes of the steps and ramps must also be correct in varying conditions. This makes it impossible even for experienced operators to manage species changes optimally without an advanced control system.

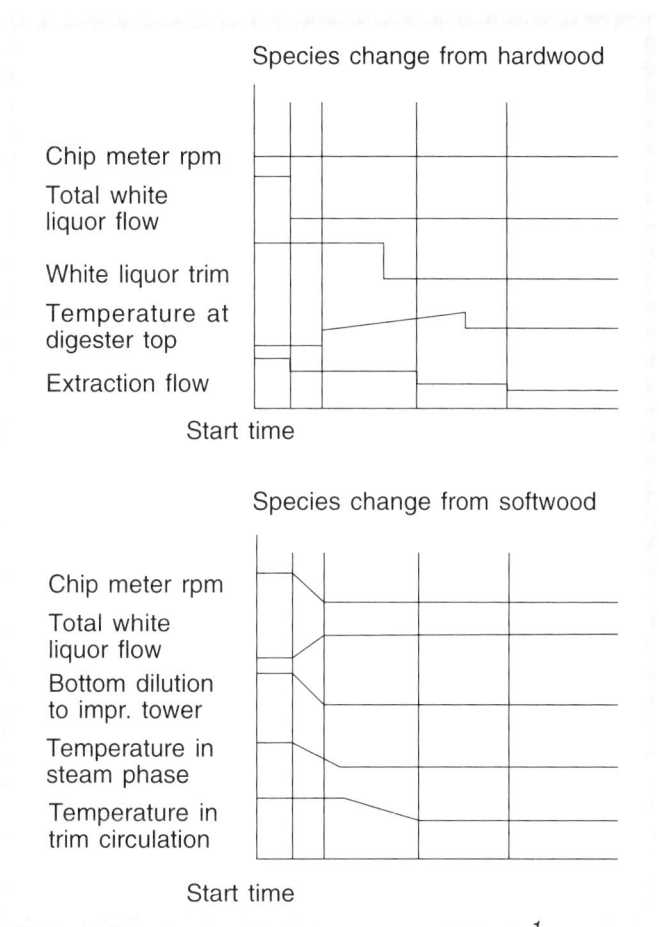

Figure 6. Examples of species change management strategies[1].

Figure 6 shows an example of species change strategy. Using computer controlled species changes makes possible an estimation of the actual species change time with an accuracy of one minute in normal conditions[1]. A typical species change takes 300–350 min.

1.5 Benefits

Another publication[1] discusses the benefits of improved continuous flow digester control in three examples. The benefits seem indisputable. With better control, kappa number standard deviation decreased 40%–45% at kappa levels varying from 26 to 33 for softwood and 30%–50% for hardwood with kappa number varying correspondingly from 16 to 19. Increasing production capacity significantly is also possible because of more stable digester operation due to better total control. Washing losses also decreased by 30%.

In all three mills, the production rate and species changes clearly have better management than with previous control strategy. Operator acceptance of the new management systems has been excellent.

CHAPTER 5

2 Batch digester control

2.1 Control requirements

Computerized conventional batch digester control dates from the end of the 1960s and the beginning of the 1970s. By the middle 1970s, most batch cooking plants had computer control systems.

The control problems in a conventional batch cooking plant resemble those in continuous digesters. The biggest disturbances are due to the measurement and control of moisture content and other characteristics of the chip feed. With a mixture a two chip species, the definition and continuous control of the mass ratio between species presents another problem. Controlling the series of batch reactors instead of one with the continuous digester means that sequencing of the reactor train becomes very important for quality control and steam leveling.

Sequencing controls have become even more important with the advent of new displacement batch cooking methods. In these cases, the availability and suitable quality of liquor is essential in each cooking stage. Tank farm controls have become a critical factor.

2.2 Basic control functions

Early systems had very similar main control functions:

- Production rate control
- Cooking cycle controls
- Scheduling
- Steam leveling
- Quality control.

Figure 7 shows an example of the control hierarchy. Other publications give early applications[14–17].

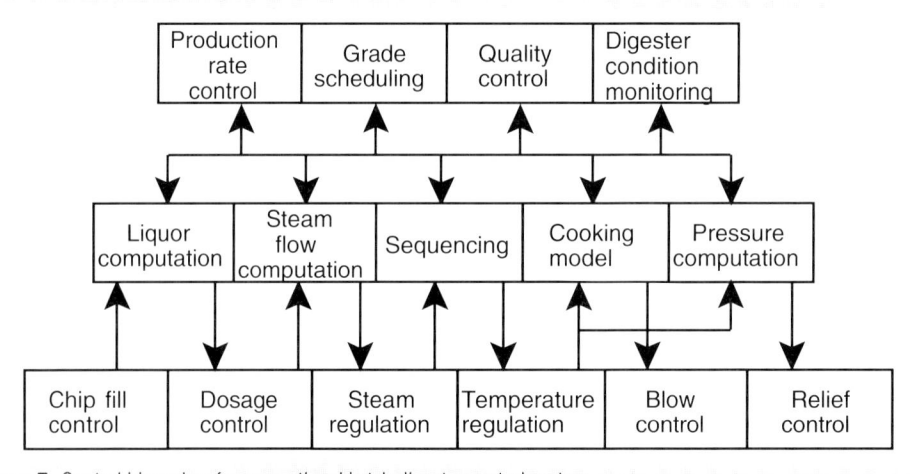

Figure 7. Control hierarchy of a conventional batch digester control system.

Cooking cycle controls include the following:

- Chip feeding
- Liquor filling
- Steam filling
- Heating and cooking
- Blowing.

Liquor filling controls should guarantee the correct alkali to wood and liquor to wood ratios in the digester. They use the measured chip feed, the measured or analyzed chip moisture, and the amount of moisture-free wood calculated from them. The residual effective alkali in black liquor should be included in the alkali to wood ratio calculation. The liquor to wood ratio also includes water from chips. For directly heated digesters, it also includes the steam condensate.

Considerable variations in the black and white liquor concentrations disturb liquor filling controls. White liquor strength and effective alkali require on-line measurement for this purpose. Conductivity measurements and automatic titrators have use[18, 19].

Figure 8 shows a typical ideal temperature profile for standard heating and cooking stages. During the heating stage, the temperature increases according to a linear ramp from initial batch temperature to the actual cooking temperature. The cooking plant scheduling program can change the slope of the ramp, i.e., the time available for heating, to accelerate or decrease the rate of heating.

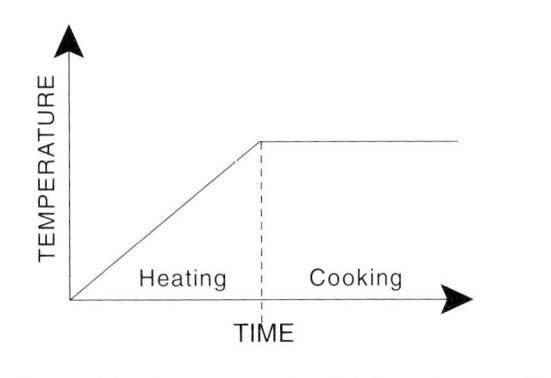

Figure 8. Temperature profile during batch digester heating and cooking stages.

Once the batch reaches the cooking temperature, the temperature controller holds it at that temperature for the time necessary for cooking. H-factor starts accumulating during the heating stage, but the rate of accumulation is much faster at the actual cooking temperature. The target H-factor defines the time of cooking at the cooking temperature. The cooking plant scheduling program can change the cooking time by changing the target cooking temperature. This usually means applying the lowest possible cooking temperature.

CHAPTER 5

2.3 Cooking plant scheduling and steam leveling

Production scheduling in a conventional batch cooking plant tries to keep production rates at their specified target levels or maximized within operational limits[20]. Scheduling programs automatically space the digester blows along each production line to make production as continuous as possible. Scheduling determines the most efficient way to meet the production requirements with the resources available at a given time. This includes the number of digesters in use and the situation on the steam side. The problem depends on the production bottleneck that is chip feeding or the availability of steam.

Another option in scheduling is to apply as low a cooking temperature as available. If desired production rates decrease, the scheduling system takes advantage of the available time by decreasing the setpoint of the cooking temperature[21]. In cooking plants with excess capacity, the cooking occurs at the lowest temperature consistent with quality requirements.

Scheduling programs also calculate the average total steam demand to the cooking plant. Steam leveling uses this value as a target to level the steam flow and prevent instantaneous flows of steam that exceed maximum permissible limits[20]. Good results are possible in leveling medium pressure and low pressure steam and trying to fulfill the steam requirements with low pressure steam as much as possible.

Scheduling should consider different grades and wood species because these influence the parameters of the scheduling algorithms. The system should include an adaptive mechanism to update the parameters in scheduling algorithms using results from the cooking schedules.

2.4 Kappa number control

The use of effective alkali analyzers in the kappa number control uses the fact that effective alkali diminishes very fast in the early part of a cook as Fig. 9 shows[8]. This is due to initial neutralization reactions that occur before bulk delignification takes place. The effective alkali value after the initial drop is more meaningful for the control than its initial value because this tells the effects of chip variations. The kappa batch system used the automatic analysis of a liquor sample just before bulk delignification. This used the Q-factor that includes the carbohydrate concentration, reaction rate constant, liquor strength, and temperature.

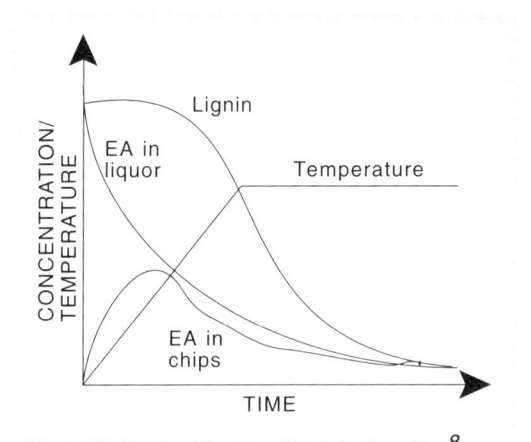

Figure 9. Alkali and lignin profiles in kraft cooking[8].

Using the analyzed value with the target kappa number determines the target H-factor in the feedforward manner. This requires a model such as the following hyperbolic version of Hatton's model[8]:

$$\frac{1}{Y} = A + B[\ln H(EA)^n] \qquad (6)$$

where Y   is the required pulp characteristic (kappa number in this case)
       A and B are parameters
       H   is the H-factor
       EA the effective alkali concentration.

Figure 10 shows the system structure[22].

Wells has given a more recent application of effective alkali sensors for batch digester control[19]. He introduced a simple model that uses the effective alkali measurement to predict kappa number during the cooking process. He also introduced a kappa control strategy that uses the effective alkali measurement to compensate for variations in the cooking conditions. The model has two differential equations:

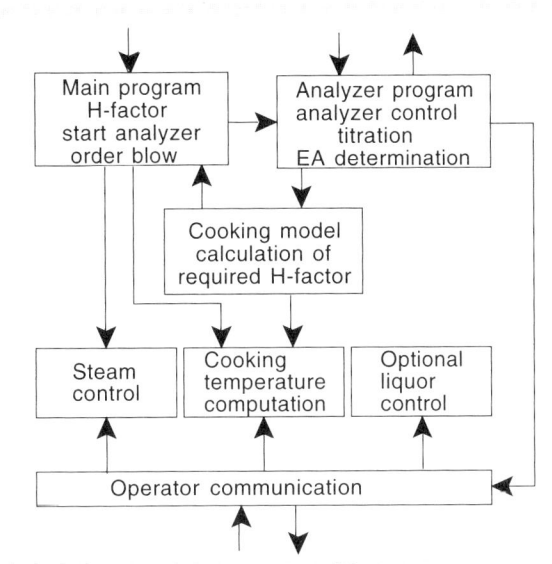

Figure 10. Kappa batch software functions[22].

$$\frac{dL}{dT} = -\alpha_f \exp\left(A_f - \frac{B_f}{T}\right) L^{\upsilon_f} [EA]^{\beta_f} \qquad (7)$$

$$\frac{d[EA]}{dT} = -\alpha_{cf} \exp\left(A_{cf} - \frac{B_{cf}}{T}\right) L^{\upsilon_{cf}} [EA]^{\beta_{cf}} \qquad (8)$$

where L   is the ratio of lignin to cellulose
       EA the concentration of effective alkali
       A and B are constant defined already earlier
       α is the rate constant
       β and υ are the orders of primary and secondary reactions.

The control algorithm proceeds in the following manner. Whenever the estimated and measured effective alkali deviate from each other, solving the model equations finds the initial value for the effective alkali that realizes the measured value and the corresponding value for the cooking temperature. The calculated cooking temperature is then the new target temperature. Solving the equations uses some iterative method such as the Newton-Raphson solution technique.

Another approach is possible[23-25]. This uses a cooking liquor analyzer that can determine from the liquor sample the concentration of the following:

- Dissolved lignin by ultraviolet light detector
- Dry solids concentration by differential refractive index detector
- Residual alkali by conductivity detector.

The original article presented and tested three possible control strategies for the batch digester control:

- Alkali-corrected H-factor model and residual alkali measurements corrected the H-factor model through the entire cook according to the similar approach presented earlier. The best model yielded an approximately 30% better correlation with kappa number than the conventional H-factor model.

- Empirical models for pulping reaction products were constructed from initial charge conditions and lignin and dry solids measurements. Results were slightly better compared with the earlier approach.

- Hunting for kinetic inflection points shown in Fig. 11 gave results comparable with the conventional H-factor model.

The continuous digester case already mentioned the results gained with the new generation cooking liquor analyzer. Another publication reports the laboratory scale testing of the same kind of approach[26].

2.5 Tank farm controls

New displacement batch cooking methods, rapid displacement heating (RDH), and super batch, have initiated the requirement for new control functions in batch digester control systems. In some cooking stages, the methods require the availability of liquors with certain temperatures and concentrations. In certain stages, the free space in liquor tanks is necessary for the continuation of the cooking cycle[27]. Strong interactions also occur between operations, and the possibilities for controlling the next cook depend on the successful control of the previous one. Figure 12 shows this situation. This requires improved coordination and sequencing of the digester and tank farm operations.

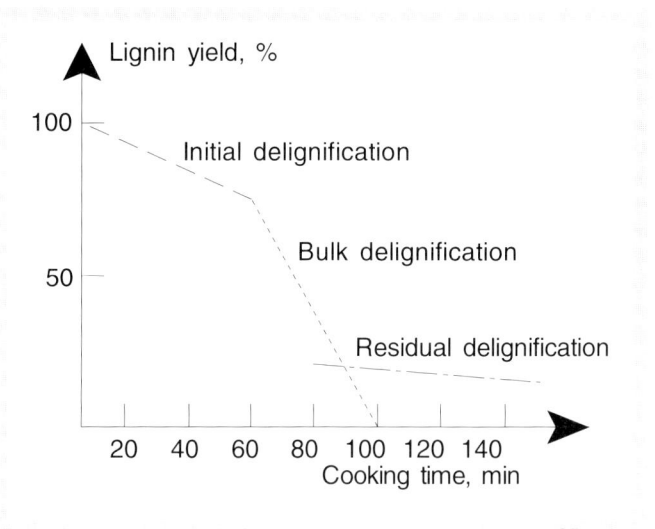

Figure 11. Inflection points occurring after different cooking times[25].

Process control in fiber line

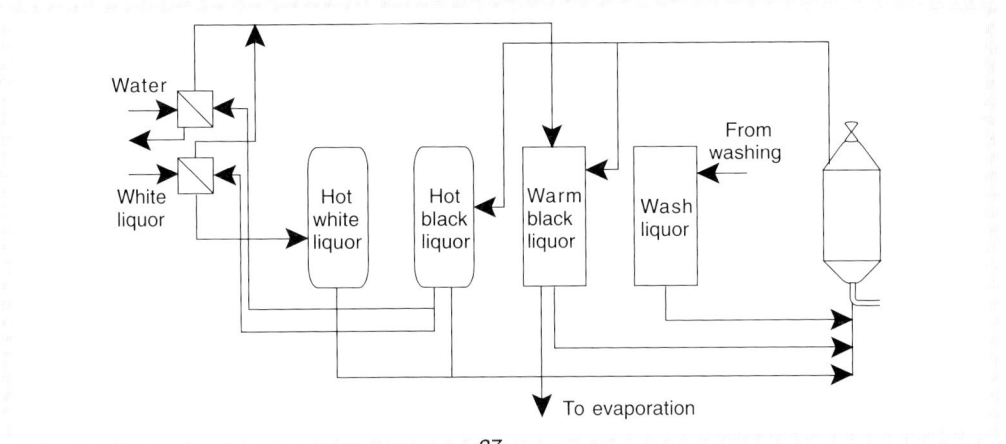

Figure 12. Tank farm system for the RDH cooking plant[27].

Good experiences have resulted from complementing an existing computer control system for batch digesters with tank farm coordination functions. Figure 13 shows that these functions predict changes in the steam balance of the cooking plant and inform the operator of difficulties and possible corrective actions[27]. Another publication reports on studies to check the operation of the RDH control system[28].

2.6 Benefits

The benefits of batch digester control come from two main sources: more constant quality and decreased energy consumption. Figures 14 and 15 show these effects according to

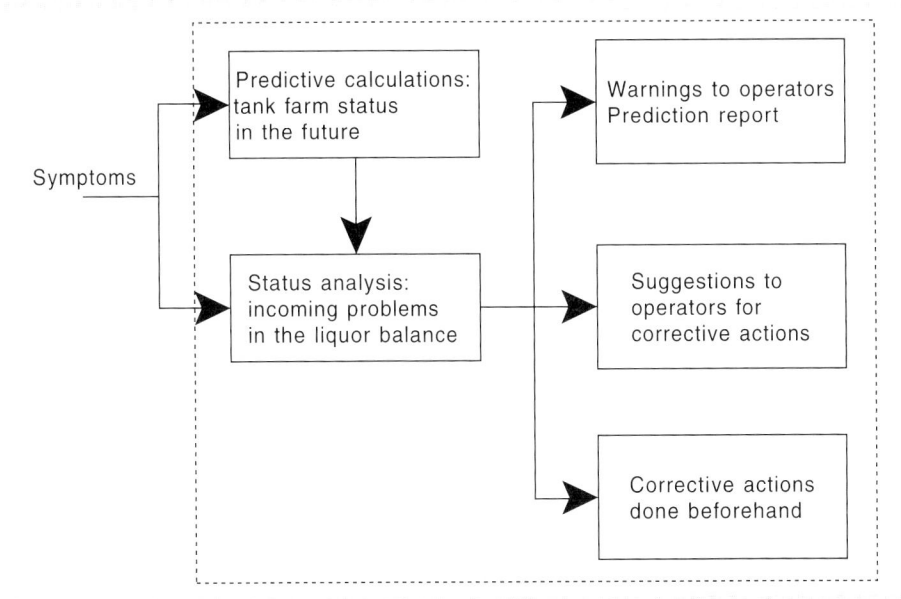

Figure 13. Basic principles of the tank farm coordination system[27].

CHAPTER 5

Finnish research in the middle 1980s[29, 30]. Figure 14 shows that kappa standard deviation has decreased approximately 30%. In Fig. 14a, the benefits result only from the more consistent quality of pulp that simultaneously means savings in raw materials and energy. In Fig. 14b, the decreased variations have increased the kappa number target to give increased yield and a production increase.

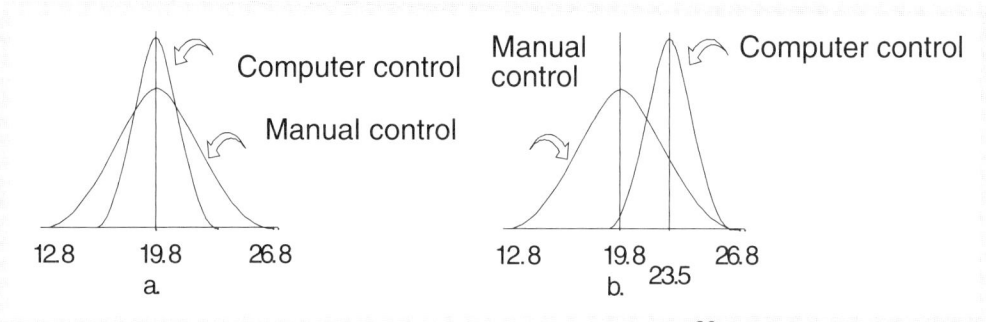

Figure 14. Decreased standard deviation of kappa number due to improved control[29].

Figure 15 shows energy savings from nine batch cooking plants. Figure 15a shows the total savings, and Fig. 15b shows the changing ratio between the high pressure and low pressure steam that is the result of better control and coordination of the cooking plant[29].

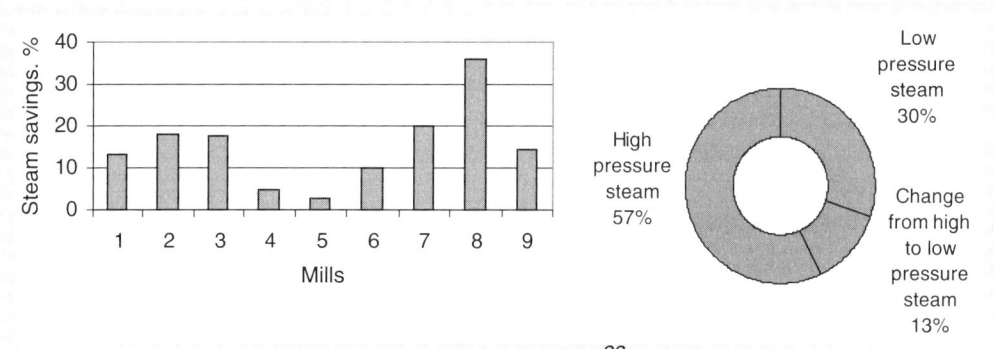

Figure 15. Energy savings achieved by improved batch digester control[29].

3 Brown stock washing control

3.1 Control requirements

The washing plant has a dual function in the pulp mill. It cleans the pulp from the used chemical and other unwanted reaction products that could increase the chemical consumption or harm the environment in the subsequent process stages. It is also the first process in the chemical recovery. This means it must make the best possible recovery of the used chemicals and organic compounds for the energy production possible with the minimum amount of washing water.

Good performance of the washing plant is essential for the entire mill considering both pulp quality and energy consumption. In operation and control, an optimum balance between the energy consumption in the evaporation plant and the costs of lost chemicals and energy is necessary. For control, this essentially occurs with dilution factor control in washing. The selection of washing equipment also has influence. By using more washing water, the washing losses with pulp decrease with a simultaneous increase in the cost of evaporation as Fig. 16 shows[31].

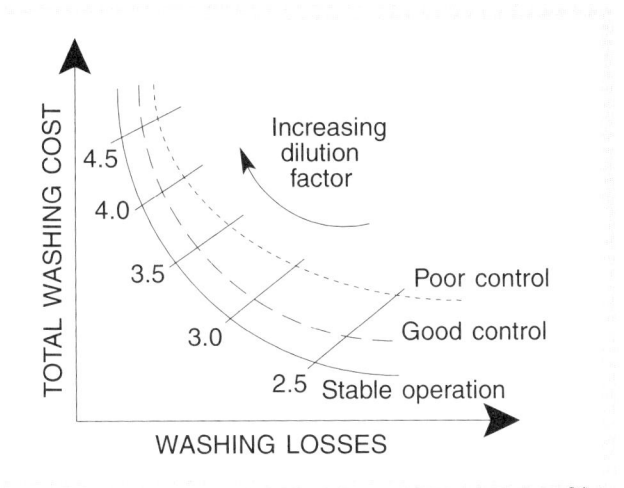

Figure 16. Development of washing costs with poor and good controls[31].

3.2 Control functions

Automatic control functions of brown stock washing usually only consist of low-level control loops[32]. This is primarily due to the lack of sensors to monitor the process continuously. Adjusting drum rotation speed maintains a constant vat level. Flow rate of the shower water stabilizes the liquor level in filtrate tanks.

According to Lundqvist[31], the washing plants are multi stage counter-current processes. For control purposes, the washing plant is a closed system or a series of closed systems. The reference gives the following tasks for the washing control system:

– Maintain dilution factor
– Coordinate flows
– Coordinate buffer contents
– Support decision on dilution factor
– Detect disturbances
– Inform the operator
– Provide support during production rate changes, process upsets, and startups.

The Korsnäs pulp mill installed this system in 1982. Washing losses decreased 50%, and the total reduction in production costs were US$ 500 000 per year[31].

Some approaches try to keep the dilution factor constant on each washer by using feedforward from production rate to the shower water flow[33, 34]. Better results could result by using a conductivity measurement to monitor the concentration of dissolved organic compounds. Figure 17 shows a reproduction of one such approach[32].

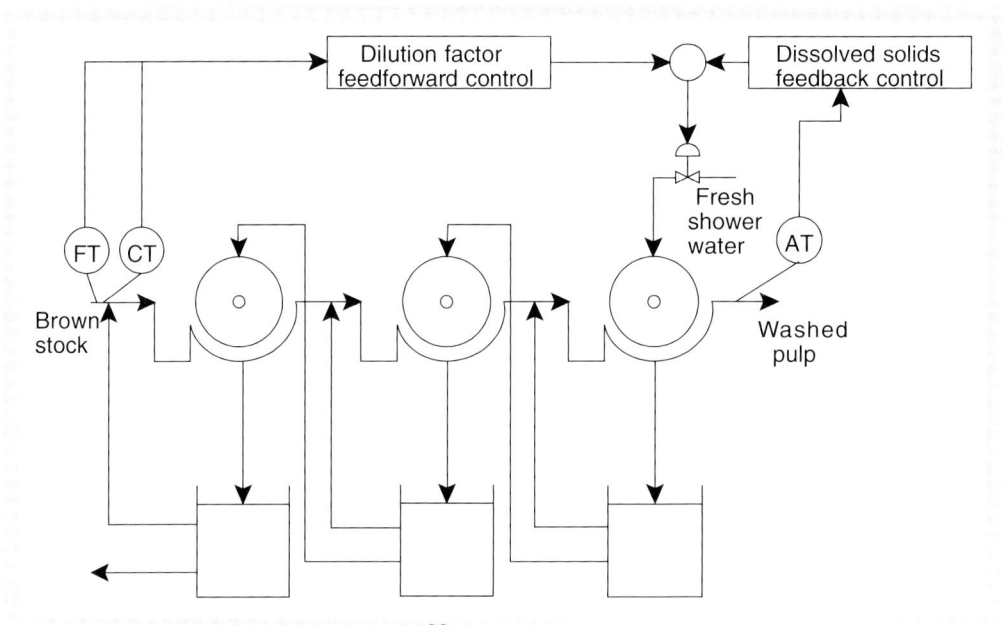

Figure 17. Control strategy for the washing plant[32].

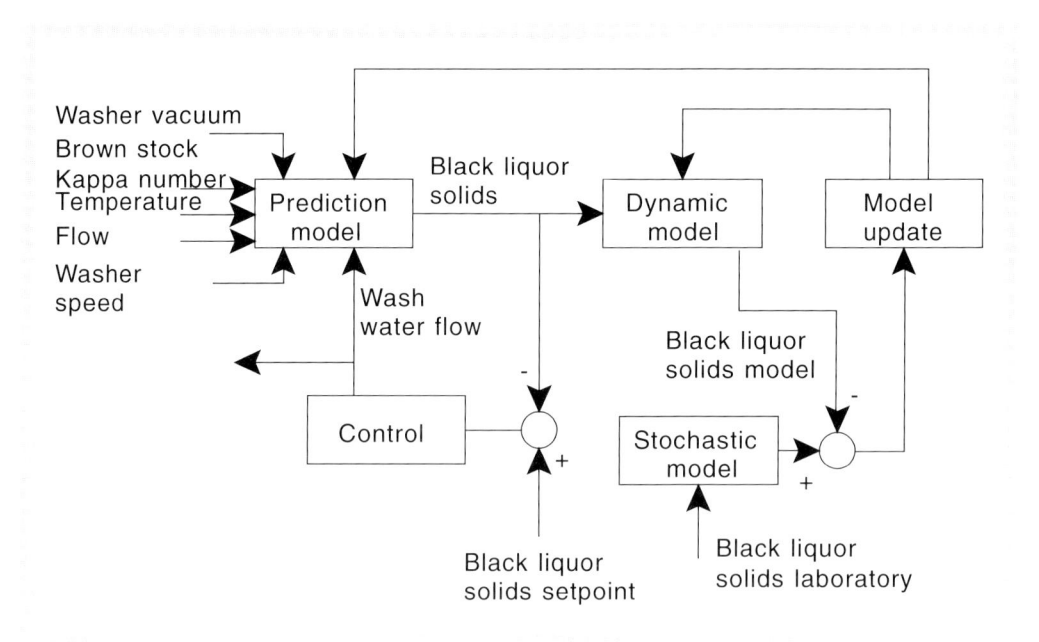

Figure 18. Washing control using the prediction of black liquor solids[35].

The strategy uses feedforward control from the production rate of the washing plant and maintains a constant dilution factor by the shower water. Adjustment of the setpoint of the dilution factor uses the conductivity measurements for the washed pulp leaving the last washer.

Development of this strategy used considerable modeling and simulation work[32].

Advanced methods are also possible in washing control. Freedman[35] reports using a prediction model and Kalman filter updating as early as 1976. In his case, the prediction model calculates the black liquor solids content. The washing water control uses the predicted value with the prediction model updating using laboratory analyses of the black liquor solids as Fig. 18 shows.

Rudd[36] reports using neural networks to control the brown stock washer. Neural networks have use as soft sensors to measure mat consistency, mat density, and soda loss to control the dilution factor as Fig. 19 shows. The target is to stabilize the black liquor solids. Results show a 25% reduction in standard deviation of the black liquor solids using an eight-day trial. The controller also maintained larger disturbances in an automatic mode for the input variables.

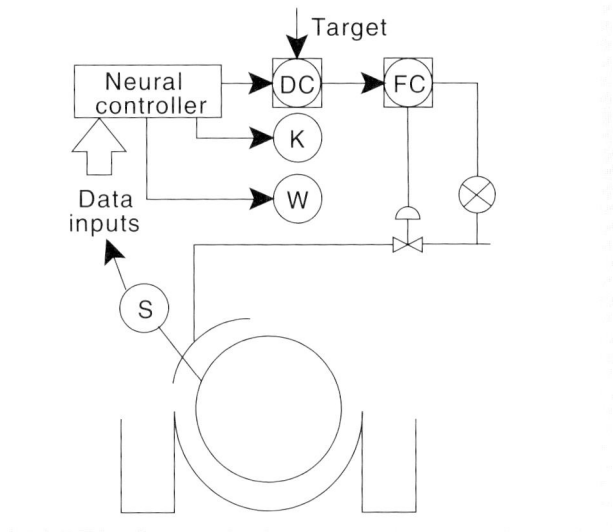

Figure 19. Neural network based control strategy[36].

4 Bleach plant control

4.1 Control requirements

The objective of the bleaching process is to achieve the desired pulp brightness at a desired production rate with minimum expenditure in bleaching chemicals and energy[37]. The process is subject to several disturbances including incoming kappa number variations, washing losses, long delays, channeling, and improper mixing. Manual control therefore commonly uses chemicals dosed in excess of the actual demand. Improved control can diminish the excess, decrease the standard deviation of the quality variables (mostly brightness), and decrease the energy used in bleaching.

The control problems in the bleach plant include random disturbances, measurement noise, dead time, dead time uncertainty, incomplete understanding of process reactions and kinetics, and actuator malfunctions because of the harsh chemical environment[37]. The nonlinear, interconnected nature of the process also makes control difficult.

4.2 Oxygen delignification control

The main disturbance in the oxygen stage is the variation in the incoming kappa number of pulp. This variation occurs so rapidly that control is not possible using conventional laboratory analyses. The situation improves considerably with automatic analysis of incoming kappa number. Existing control strategies use this type of analyzer.

Figure 20 describes the dependencies between variations in the incoming kappa number and possible control actions[38]. The point in the middle shows the reference state that corresponds to the kappa number achieved by a certain alkali charge, temperature, and reaction time. If the incoming kappa number increases, the oxygen kappa also increases if no control action occurs. Simultaneously, the residual alkali increases. If alkali charge increases, the residual alkali increases and the oxygen kappa decreases. The third vector shows the effect of temperature and reaction time. Figure 21 shows the corresponding control strategy.

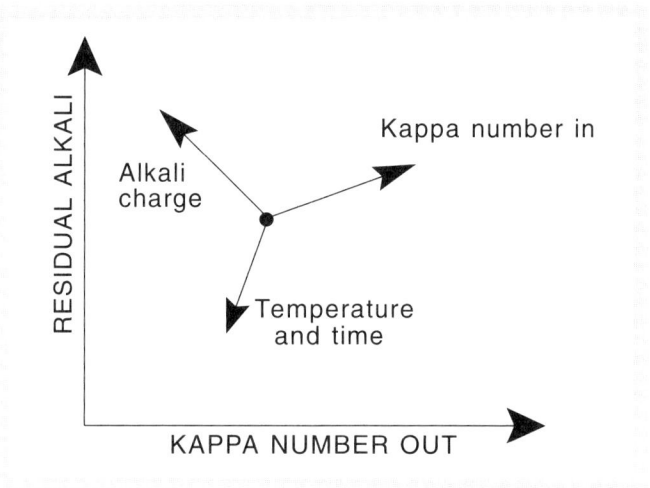

Figure 20. Dependence between operation parameters and control actions in the oxygen stage[38].

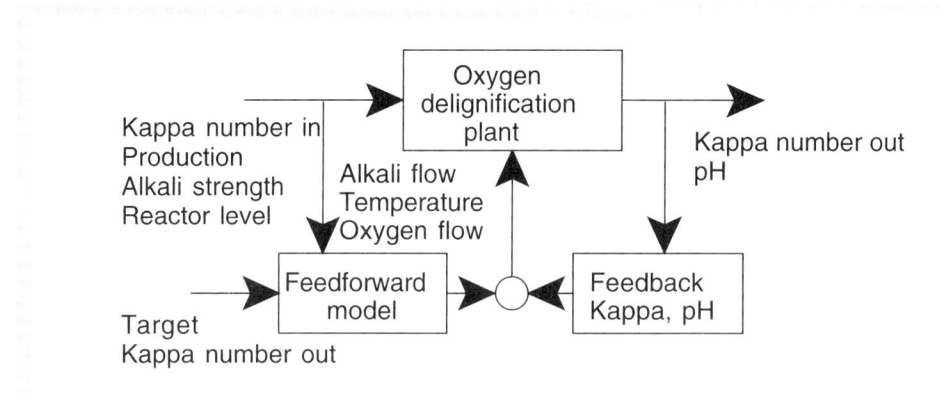

Figure 21. Control strategy for oxygen delignification[38].

Improved control offers the possibility to operate oxygen delignification with a lower final kappa target. This means decreased costs in the final bleaching. Better control also leads to decreased standard deviation of the kappa number after the oxygen stage. One publication reports a 50% decrease[39].

4.3 Chlorination stage control

The chlorination stage control is usually a dual sensor system to measure brightness and residual chemicals a short time after chemical dosing. Figure 22 shows one example[40]. The authors of that work suggest 25 s retention time after chlorine addition to the sensors. The brightness sensor forms a closed loop with the chlorine valve. The residual signal is compared with its targets, and the brightness setpoint adjustment uses the deviation. Chlorine dioxide is added about 5 s downstream from chlorine using a ratio controller to maintain a chlorine dioxide substitution rate of 8% –10% available chlorine in the case previously mentioned. Control limits could decrease by 12%. An 87% reduction occurred in points out of control after tuning the system. Total reduction in the chlorination stage bleaching cost was 13%.

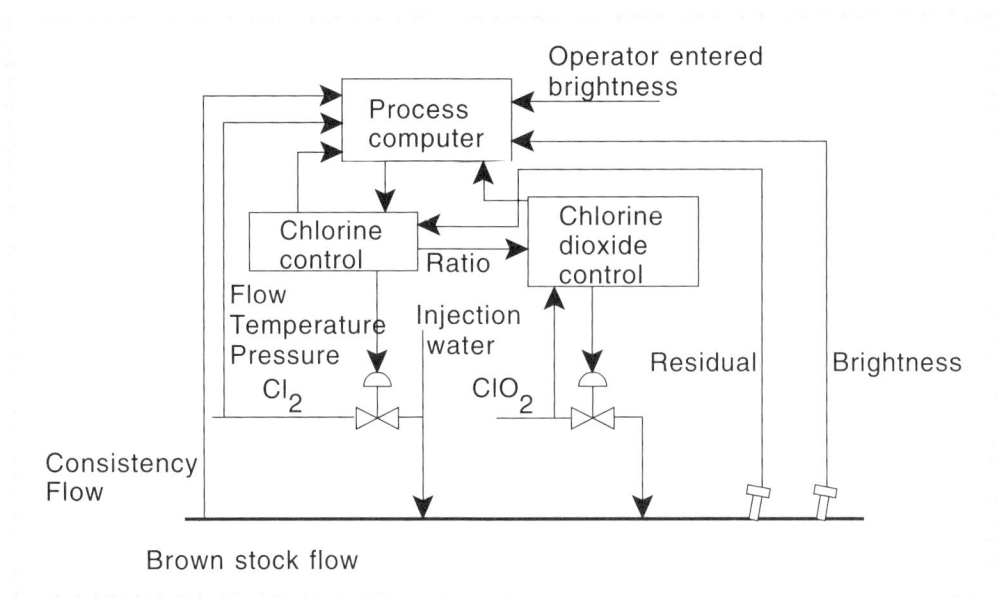

Figure 22. A dual sensor control of the chlorination stage[40].

Earlier application[37] used a sample line that takes a pulp sample after about 1 min retention time from the chlorination tower. The authors found that dead time varies considerably, and they used a Kalman filter-based strategy[41] to compensate for this. Figure 23 shows the brightness control strategy.

CHAPTER 5

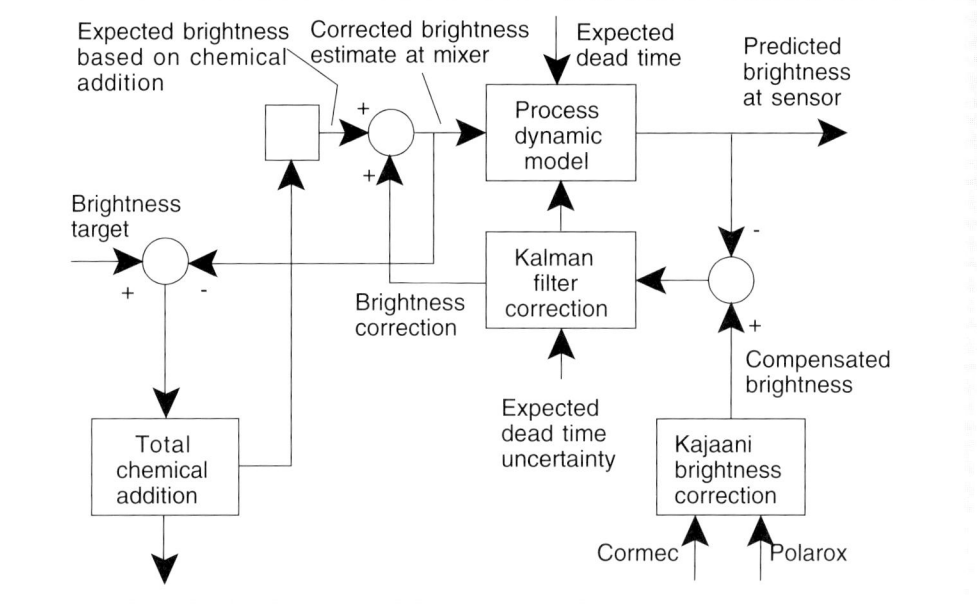

Figure 23. Chlorination stage brightness control[37].

The algorithm works as follows. Compensated brightness is calculated from the brightness and excess chemical sensors. This signal will change in 2–3 minutes after making the change in the chemical charge. The "expected brightness" is calculated from the actual amount of the chemical addition. Kalman filter feedback calculates the brightness correction, and the combination of the expected brightness and the brightness correction produces the "corrected brightness." The total chemical addition depends on the deviation between the brightness target and the "corrected brightness."

Another publication reports fuzzy logic control of D_{100}-stage bleaching[42]. Several reasons make fuzzy logic suitable for this purpose:

– Nonlinear system so that controller tuning depends on the operation point
– Heterogeneous raw material
– Interconnections between process variables
– Several variables influence the optimum ClO_2 dosage
– Continuous measurements and laboratory analyses are easy to include
– Human knowledge is easy to incorporate in the strategy.

Figure 24 shows the main principles of the strategy. The standard deviation of EOP kappa decreased by 60%, and the standard deviation of the final brightness decreased by 40% during a three-week test period after installing the fuzzy logic based system in a conventional five-stage bleaching in Finland.

Process control in fiber line

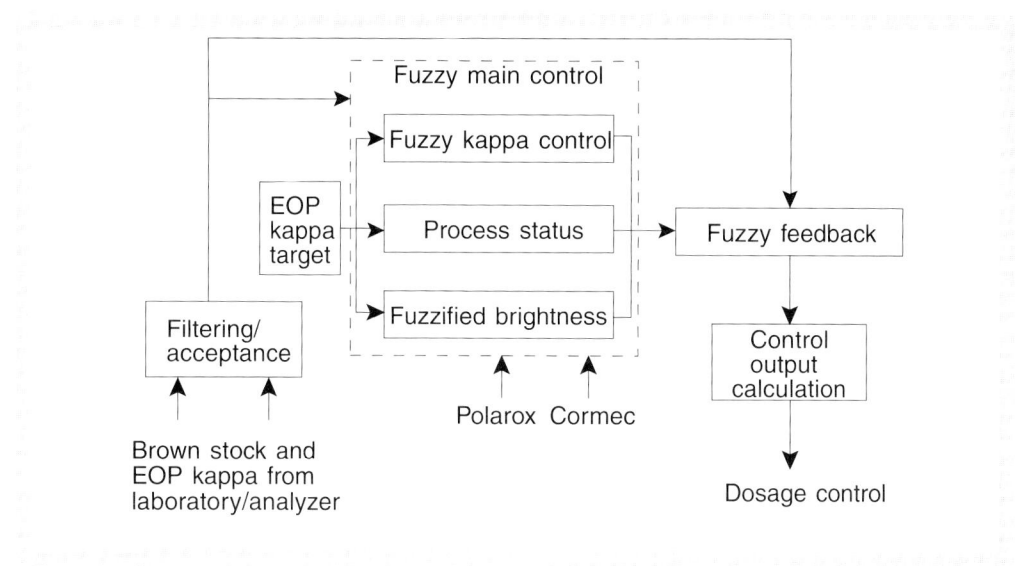

Figure 24. Structure of the fuzzy logic based bleaching control strategy.

The control system observes the following variables:

- Inlet kappa, on-line measurement
- Inlet pulp brightness after the chemical dosage
- Chemical residual after the dosage
- Temperature
- Production rate
- Inlet pulp pH
- EOP kappa using on-line measurement.

Fuzzy logic control uses rules such as the following: "If kappa number is increasing, brightness is decreasing, and pH is increasing, then more chlorine dioxide is necessary." Application is modular and hierarchical and therefore easy to use and maintain. The control output is a setpoint to the dosage controller. The feedforward part uses kappa measurement and brightness and excess chemical sensors before the bleaching stage with inlet pulp pH and COD and conductivity measurements, if available. The feedback controller compares the setpoint and actual value for the EOP kappa and calculates the correction to the feedforward control.

Earlier work reported statistical quality control in the bleach plant[40]. That work used control charts, Pareto analysis, and Ishigawa diagrams to improve chlorination control and developed charts for brightness, viscosity, and dirt count of market pulp. The total reduction in hardwood brightness variation was 60%, and the final hardwood vis-

CHAPTER 5

cosity increased from 8.5 cP to 9.9 cP while its standard deviation decreased from 1.3 to 0.9. Total bleaching chemical costs decreased by 22%.

Diaz et al.[43] used the chlorination stage as an example of using qualitative modeling techniques in process control and diagnostics.

4.4 Control in the dioxide stages

In dioxide stages, a dual sensor strategy resembling that of Fig. 22 has use. Fig. 25 shows an example. The system uses cascade control where brightness control is the master loop. It gives the setpoint for the control loop of the residual chemical. The slave loop also includes the value for pH as the input variable.

The retention time after ClO$_2$ addition is 1–2 minutes. Quality variations have decreased from 40% to 75% depending on the specific case.

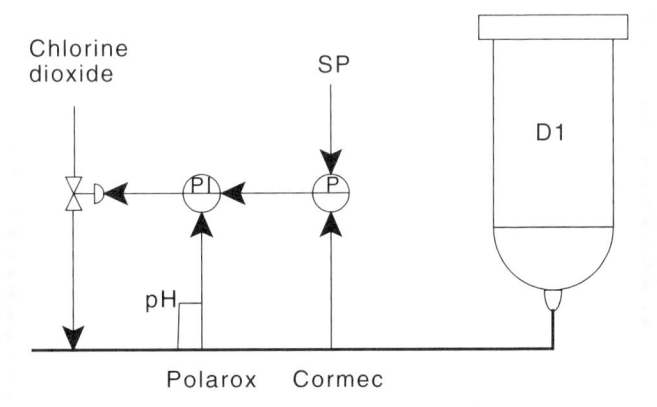

Figure 25. Dual sensor structure for the control of D1 stage[44].

4.5 Control of mechanical pulp bleaching

The dual sensor principle also has use in control of mechanical pulp bleaching[45]. Application has covered almost all types of prevailing mechanical pulping processes. In some cases, additional brightness or residual sensors have supplemented it after the stage.

In hydrosulfite bleaching, the brightness sensor location is before the chemical dosing or immediately after it depending on piping and chemical dosage level. In the first case, only feedforward control from the brightness signal is possible. In the other case, brightness development after a short reaction time is considered as a feedback signal. The residual measurement is placed with a short delay after the chemical dosage. Lower dosage means shorter delays.

The control for hydrosulfite bleaching uses the compensated brightness signal coming from the dual sensor system that considers the brightness development and the concentration of active bleaching chemical. The brightness measurement therefore compensates for the deviations and variations in the incoming pulp brightness. The residual measurement assures reaching the target brightness. The brightness measurement after the stage can also be a feedback. The control diagram resembles the previous ones.

The dual sensor control decreased the brightness deviation by 60% compared with the manual control and 40% when compared with the brightness control. Bleaching chemicals consumption decreased by 40% compared with manual control.

In peroxide bleaching, the brightness sensor should have 5–10 min delay after the chemical dosage. The residual sensor already has a 1 min delay[45]. Sensors are usually very close to each other to avoid time delay compensation between them. Brightness and residual sensors indicate the state of the bleaching process. The main process variables (production, temperature, and tower retention) require integration into the control strategy. The use of both signals is due to the fact that brightness and residual develop in different ways in the bleaching process. The brightness saturates when the chemical dosage increases. Residual increases almost exponentially as Fig. 26 shows[45].

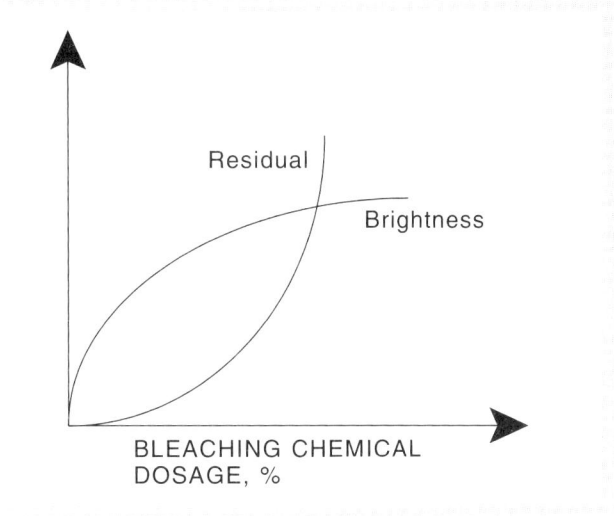

Figure 26. Brightness and residual as a function of bleaching chemical dosage[45].

Peroxide stage control also uses the dual sensor principle. In peroxide processes, the final brightness deviation decreases by 40% and chemical consumption by 25% compared with manual control. Neural networks have also been applied to bleach plant modeling and control[46].

CHAPTER 5

References

1. Sutinen, R., Joensuu, I., Furst, P., et al., 1990 EUCEPA Conference Proceedings, SPCI, Stockholm, p. 253.
2. Sastry, V. A., Pulp Paper Can. 79(5):41(1978).
3. Allison, B. J., Dumont, G. A., Novak, L. H., 1990 Control Systems Preprints, Finnish Society of Automatic Control, Helsinki, p.157.
4. Brattberg, Ö., 1994 Control Systems Preprints, SPCI, Stockholm, p. 298.
5. Jutila, E., Acta Polytechnica Scandinavica, The Finnish Academy of Technical Sciences, Helsinki, 1979, pp. 15–18.
6. Vroom, K. E., Pulp Paper Mag. Can. 58(3):228(1957).
7. Freedman, B. G., Chem. Eng. Progress 72(4):82(1976).
8. Lodzinski, F. P. and Karlsson, T., Tappi 59(9):88(1976).
9. Wells, C. H., Johns, E. C., and Chapman F. L., Tappi 58(8):177(1975).
10. Chari, N. C. S., Tappi 56(7):65(1973).
11. Hatton, J. V., Tappi 56(7):78(1973).
12. Haataja, K., Leiviskä, K., and Sutinen R., 1997 IMEKO World Congress Proceedings, Finnish Society of Automation, vol. XA, Tampere, p. 1.
13. Dyal, B. S., MacGregor, J. F., Taylor, P. A., et al., Pulp Paper Can. 95(1):26(1994).
14. Brewster, D. B.and Robinson, W. I., in Practical Computer Applications for the Pulp and Paper Industry, Miller Freeman Publications, San Francisco, 1975, p. 81.
15. Kallonen, H. and Franzreb, J. P., Pulp Paper Can. 80(12):139(1979).
16. Lemay, R., Pulp Paper Can. 80(6):109(1979).
17. Powell, R. P., Tappi 62(12):21(1979).
18. Dorris, G. M. and Allen, L. H., J. Pulp Paper Sci. 15(4):J122(1989).
19. Wells, C. H., Tappi J. 73(3):181(1990).
20. Saarela, P., Pulp & Paper 57(10):102(1983).
21. Swanson, H. and Thrower, M., Pulp Paper Can. 84(8):46(1983).
22. Brucker, D. S., Kostelic, K., and Voigtman, E. W., Tappi 59(9):93(1976).

23. Tikka, P. O., Kuusela, M. J., and Saarenpää, M. S., 1988 CPPA Control Systems Preprints, CPPA, Montreal.

24. Tikka, P. O. and Piiroinen, P., Pulp Paper Can. 88(2):10(1987).

25. Tikka, P. O. and Virkola, N. -E., Tappi J. 69(6):66(1986).

26. Paulonis, M. A.and Krishnagopalan, A., Tappi J. 71(11):185(1988). .

27. Ryynänen, H. and Sainiemi, J., 1987 Automation Days, The Finnish Society of Automatic Control, part IV, Helsinki, p. 259.

28. Scheldorf, J. J., Edwards, L. L., Lidskog, P., et al., Tappi J. 74(3):109(1991).

29. Uronen, P., Leiviskä, K., and Kesti, E., "Benefits and Results of Computer Control in Pulp and Paper Industry," Ministry of Trade and Industry, Energy Department. Series D:76, Helsinki, 1985, pp. 35–54.

30. Uronen, P., Leiviskä, K., and Sutinen, R., Pulp Paper Can. 86(8):T239(1985).

31. Lundqvist, S. O., Pulp Paper Can. 86(11):72(1985).

32. Turner, P. A., Roche, A. A., McDonald, J. D., and van Heinigen, A. R. P., Pulp Paper Can. 94(9):37(1993).

33. Bender, G., Rochard, L., and Dorn, W., Tappi J. 71(12):115(1988).

34. Sande, W. E., Oestreich, M. A., Poplasky, M. S., et al., Tappi J. 71(3):93(1988).

35. Freedman, B. G., Paper Trade J. (1):22(1975).

36. Rudd, J. B., Tappi 74(10):153(1991).

37. Rankin, P. A. and Bialkowski, W. L., Tappi J. 67(7):66(1984).

38. Kubulnieks, E., Lundqvist, S.-O., and Sandström, P., TAPPI 1988 International Pulp Bleaching Conference, TAPPI PRESS, Atlanta, p. 47.

39. Pettersson, T., Edlund, S. G., and Kubulnieks E., Svensk Papperstidning, (16):14(1985).

40. Corbi, J.-C., Nay, M. J., and Belt, P.B., Tappi J. 69(2):60(1986).

41. Bialkowski, W. L., IEEE J. 28(2):400(1983).

42. Lampela, K., Kuusisto, L., and Leiviskä K., Tappi 79(4):93(1996).

43. Diaz, A. C., Orchard, R. A., Amyot R., et al., Tappi J. 75(11):149(1992).

44. Pääkkönen, T., 1984 Automation Days, Finnish Society of Automatic Control, Helsinki, p.139.

45. Sopenlehto, T. and Moilanen, J., TAPPI 1988 Pulping Conference Proceedings, TAPPI PRESS, Atlanta, book 2, p. 359.

46. Xia, Q., Farzadeh, H., Rao, M., et al., 1993 IEEE Conference on Control Applications, IEEE, Vancouver, p. 593.

CHAPTER 6

Process control in chemical recovery

1	**Evaporation plant control**	**101**
1.1	Control requirements	101
1.2	Control functions	101
2	**Recovery boiler control**	**103**
2.1	Control requirements	103
2.2	Basic control functions	104
2.3	Soot blowing optimization	107
2.4	Fire-room cameras, image analysis, and expert systems	109
2.5	Benefits	114
3	**Causticizing control**	**114**
3.1	Control requirements	114
3.2	Control functions	115
3.3	Benefits	117
4	**Lime mud filter control**	**117**
4.1	Control requirements	117
4.2	Control functions	118
4.3	Benefits	120
5	**Lime kiln control**	**120**
5.1	Control requirements	120
5.2	Control functions	121
5.3	Advanced control systems for the lime kiln	124
5.4	Benefits	125
	References	126

CHAPTER 6

Kauko Leiviskä

Process control in chemical recovery

1 Evaporation plant control

1.1 Control requirements

The evaporation plant is the largest single steam consumer in the pulp mill. Better control could therefore lead to considerable annual savings in energy[1]. The solids control of heavy liquor influences the energy generation at the recovery boiler and the fouling tendency of the heavy liquor evaporation units. Capacity increase is also possible with improved controls by influencing the main disturbances of the operation of the evaporation plant; the control of the solids content in the weak liquor feed, and the decrease of unit washings and other disturbances. In older mills especially, control improvement is a necessity to decrease the environmental load.

1.2 Control functions

Figure 1 shows the control hierarchy of the evaporation control system[1]. The flow of primary steam that is today considered a better alternative compared with pressure control or boiling point rise primarily controls the solids content of the heavy black liquor. Calculation of the energy requirements uses the production rate, dry solids targets of the weak liquor feed and heavy liquor, and the characteristics of evaporator units. The solids determinations use solids or density measurements.

    The solids content of the weak liquor feed from the washing plant must sometimes be increased by mixing heavier liquor to avoid foaming. This is especially true in Nordic countries. In the control system, this has two phases: rough control before the weak liquor tank and more precise control after it. Following the foaming tendency uses conductivity measurements of condensates with correction of the target of the weak liquor solids content when necessary.

    In the evaporation control system, fouling monitoring occurs by calculating heat transfer coefficients and temperature differences and by following the steam pressure. The calculation of heat transfer coefficients uses material, energy, and solids balances and the dependence between boiling point rise and the solids content. Fouling monitoring occurs with the help of fouling reports, trends, balance reports, and alarms. This happens separately for each unit or for the whole plant and at short-term and long-term levels.

CHAPTER 6

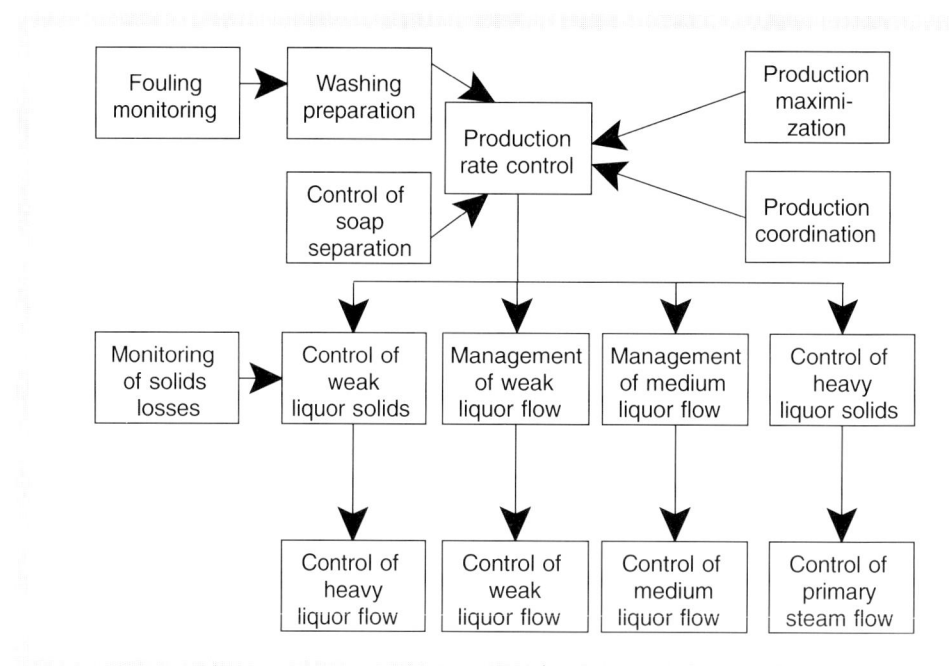

Figure 1. Control functions of the evaporation control system[1].

The washing control of evaporation units also considers the tank situations in the timing of washing operations. The coordination program uses the evaporation efficiency and tries to maximize the production rate of the plant.

Another publication discusses the use of PCGEMS and the Gudmundson model in the calculation of the heat transfer coefficients[2]. The model uses a multi variable correlation of heat flux, feed rate, feed subcooling or superheating, liquor boiling point, and liquor solids content. Heat flux is the most important component because of the nonlinear relationship between heat flux and temperature. Foaming and nonfoaming heat transfer is an additional aspect in the Gudmundson model.

The model needs equipment parameters and process variables for input. The outputs of the simulation are heat transfer coefficients, boiling point elevations, product dissolved solids, vapor dome temperature, and apparent temperature difference in each evaporator unit. Testing of the model showed average error in prediction of 0.8°C.

The same article also reports an application of the real time expert system for evaporation plant control[2]. It uses the simulation and calculation of heat transfer coefficients and the advisory system. Figure 2 shows the decision tree used by the system for troubleshooting of the evaporation plant operation.

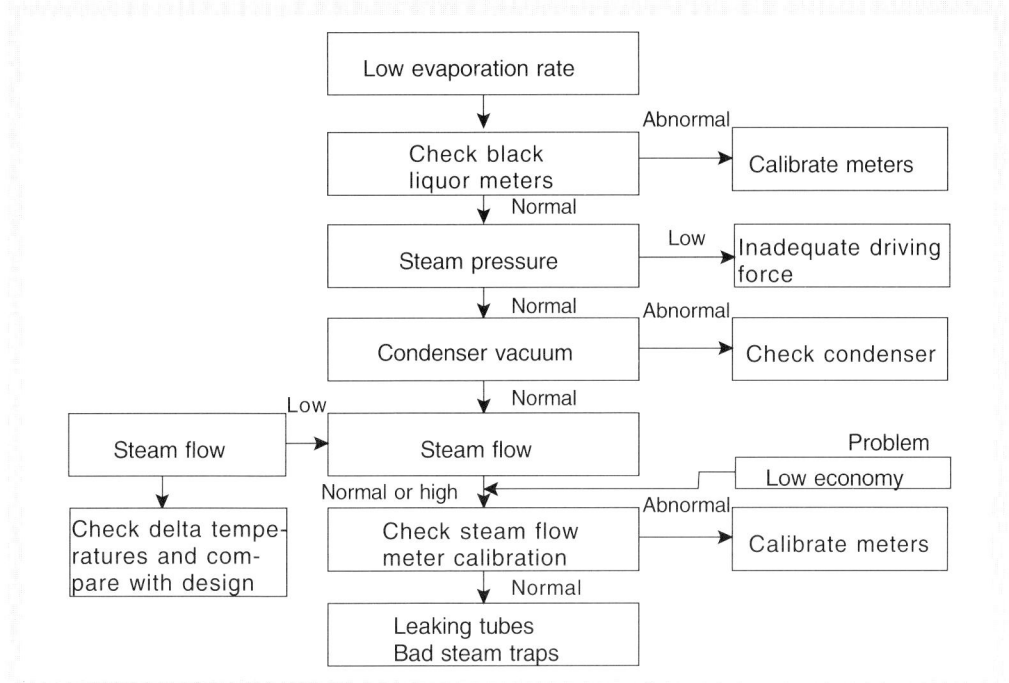

Figure 2. Expert system decision tree for evaporation plant troubleshooting[2].

2 Recovery boiler control

2.1 Control requirements

The recovery boiler fulfills a dual function in pulp mills. It handles chemical recovery of cooking chemicals and heat recovery by burning the organic substance in the black liquor. Achieving and maintaining the high rate of both functions have prime importance. The operation of the recovery boiler also has a considerable influence on the environment. Environmental requirements are becoming increasingly more stringent in every country. They also influence boiler control.

The recovery boiler is an expensive and critical unit process in the pulp mill. In many cases, it is a production bottleneck of the mill. Low reduction degree in the recovery boiler means more dead load to other processes. This can turn them into production bottlenecks. The burning process is also sensitive to external and internal disturbances. Production rate changes and unstable bed conditions cause disturbances in the entire boiler operation. Stable and continuous operation of the boiler by control means therefore has prime importance.

CHAPTER 6

Safety is also an important factor influencing boiler operation and control. High corrosion rates also influence the life of the boiler and cause a safety risk. Improved control can decrease the corrosion.

The first computerized recovery boiler control system started in Finland in 1975–1976. Considerable literature about early systems is available[3-8].

2.2 Basic control functions

The extent and complexity of control systems for recovery boilers are very heterogeneous. An earlier paper[9] divided the control functions into four levels. The first level performs only information tasks such as reporting, alarming, and some routine calculations. The second level is responsible for basic control loops such as the control of black liquor pressure and temperature, control of different air flows based on outside setpoints, etc. The third level includes single-input, single-output control of flue gas compo-

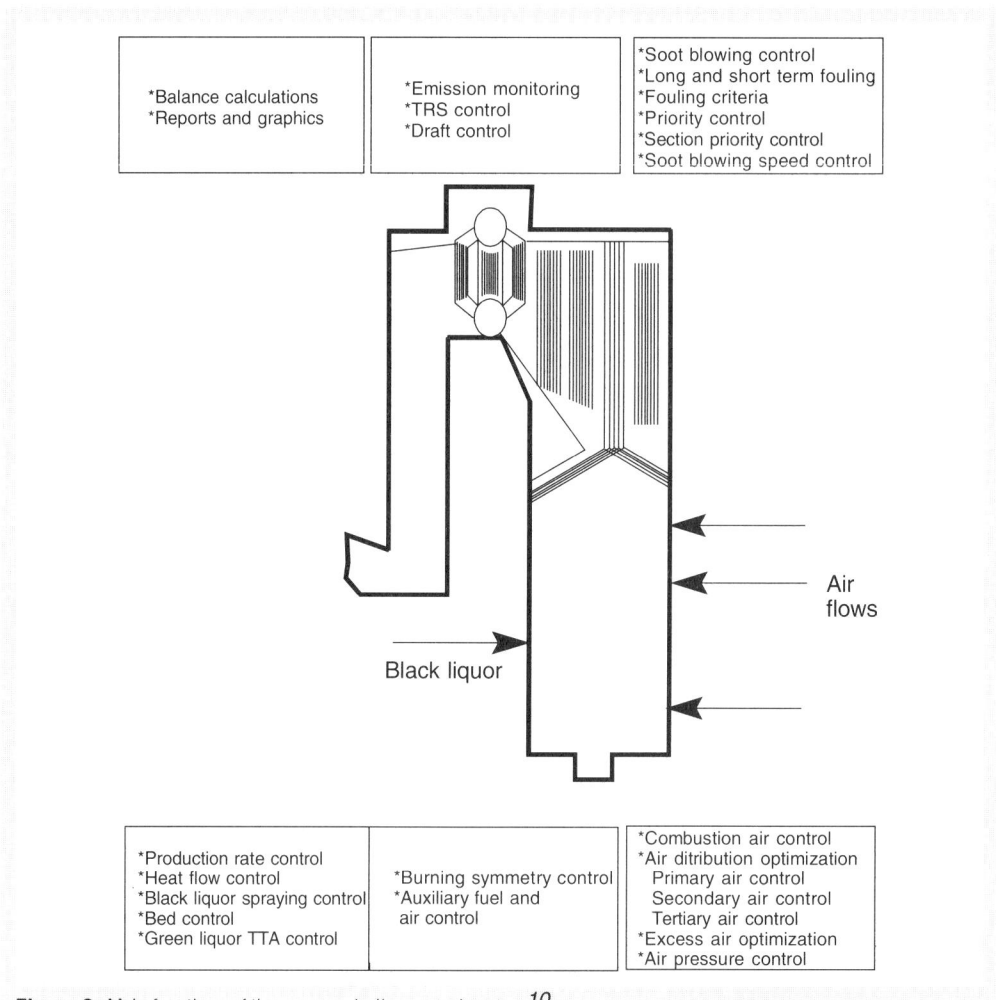

Figure 3. Main functions of the recovery boiler control system[10].

Top boxes:
- *Balance calculations / *Reports and graphics
- *Emission monitoring / *TRS control / *Draft control
- *Soot blowing control / *Long and short term fouling / *Fouling criteria / *Priority control / *Section priority control / *Soot blowing speed control

Side label: Air flows

Middle label: Black liquor

Bottom boxes:
- *Production rate control / *Heat flow control / *Black liquor spraying control / *Bed control / *Green liquor TTA control
- *Burning symmetry control / *Auxiliary fuel and air control
- *Combustion air control / *Air ditribution optimization / Primary air control / Secondary air control / Tertiary air control / *Excess air optimization / *Air pressure control

Process control in chemical recovery

nents, the boiler production, etc. On the fourth level, the multi variable character of the recovery boiler process is totally recognized, and the black liquor and air flows, pressures and temperatures are controlled considering their interactions. A fifth level under consideration at the time of this writing might be the application of image analysis, fireroom cameras, expert systems, and intelligent methods to recovery boiler control.

Figure 3 shows the main control functions.

Liquor feed controls

- Black liquor load control
- Black liquor temperature control.

Liquor spraying requires control so minimum entrainment will occur. This means the average drop size should be sufficiently large without cooling the bed. Burning should also remain in the lower part of the furnace with the temperature high at the bed level. The manipulated variable is the temperature of black liquor or its pressure. Other black liquor properties and the construction of liquor guns and nozzles also influence the drop size. Liquor spraying control relates to the optimum air distribution control.

Kettunen[11] proposed to control the black liquor load according to the amount of organic material in the solids instead of the total solids amount. This stabilizes the heat efficiency of the boiler and the flue gas amount while simultaneously facilitating other control functions. In his approach, the mass flow of organic material is calculated from the black liquor flow and its organic content as follows:

$$m_{org} = \frac{p_{org}}{100} \rho_{BL} f_{BL} \tag{1}$$

where p_{org} is the content of organic compounds in liquor, %
ρ_{BL} the density of black liquor
f_{BL} its volumetric flow rate.

Calculation of the content of organic compounds uses refractive index measurement:

$$p_{org} = f(RI, \rho_{org}, \Delta H_{org}) \tag{2}$$

where RI is the refractive index
ΔH_{ref} the heat value reference.

Refractive index is a common method for measuring organic compounds and solids in liquor. Other methods, such as NIR and coriolis meters are currently used or experimented with.

Boiler load cannot change rapidly without disturbing the entire burning process. Coordinated rate changes occur stepwise with simultaneous control of black liquor flow and total air flow[11].

CHAPTER 6

Air controls

- Total air demand calculation
- Excess oxygen and combustibles control
- Air flow and air distribution control to different air ports.

Calculation of the theoretically necessary total air flow compensates for variations in black liquor solids, black liquor flow rate and its analysis, and for additional fuel and for added salt cake[9]. Feedback from flue gas analyzers guarantees complete combustion. This also means that the thermal efficiency of the boiler remains at maximum with the proper excess air.

When a three level air distribution is in use, the following strategy applies. Combustion in the primary air zone must be maintained so that the largest possible amount of combustible gases rises from the char bed to the secondary air zone. If the primary air is too low, the temperature will decrease. If it is too high, it will burn the char bed. Unnecessary changes in the primary air should be avoided.

The secondary air will burn the gases rising from the bed. This secondary air should have higher pressure than the primary air to ensure good penetration into the center of the furnace.

The tertiary air will assure complete combustion and optimum excess oxygen content in flue gases. The combustion reactions should terminate as close to the tertiary air level as possible. The pressure of the tertiary air should be sufficiently high to assure complete mixing in the upper part of the boiler.

Note that some boilers can even have quaternary air systems.

The air distribution is adaptive according to load. It must be defined empirically for each boiler. Optimization uses empirical air distribution curves, excess oxygen content, uncombustibles, and liquor quality[12]. Bed control changes the air distribution, if necessary, using combined feedback and feedforward control. Feedforward compensates for liquor amount and quality variations, and feedback compensates for changes in liquor burning.

Bed and temperature profile controls

- Bed condition and symmetry control
- Temperature profile control.

Monitoring bed material and energy balances, bed temperatures, and emissions controls the bed conditions and symmetry in the recovery boiler. The corrective actions taken will influence air flows and liquor spraying[9]. Symmetry control can also use fireroom cameras and image processing. Further discussion of this item occurs later. Good results using fuzzy logic control in symmetry control are available[13, 14].

The porosity of the char bed is important for its proper functioning. Excessively large liquor drops will not dry before arrival at the bed, will cause bed cooling, and will increase SO_2 and H_2S emissions. Liquor drops that are too small will burn in the upper

Process control in chemical recovery

parts of the boiler and bring burning particles into the flue gas stream. This causes fouling and losses in thermal efficiency.

The temperature profile in the recovery boiler requires control so the bed temperature remains sufficiently high to obtain high reduction as Fig. 4 shows. This means that burning occurs at the lower part of the furnace. In addition to reduction, this is also good for bed reactions, thermal efficiency, emissions, and fouling of heat surfaces. The higher temperature in the lower bed area makes use of bigger drop size possible. This decreases the amount of particles in the higher parts of the furnace. The composition of fly ash also changes.

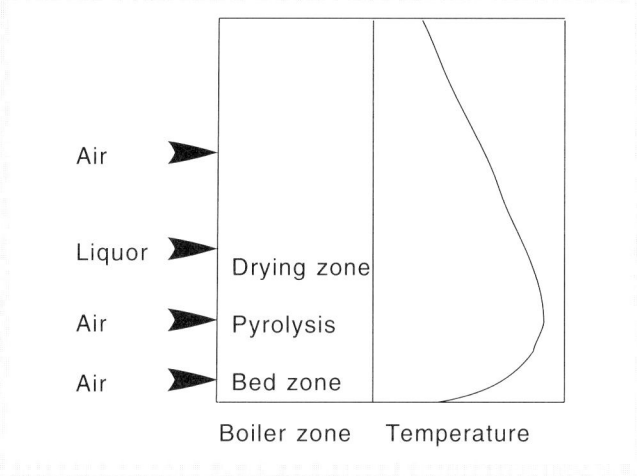

Figure 4. Ideal temperature profile in the recovery furnace.

Another item in control of temperature profile is preventing the temperatures of fly ash at the superheater level from becoming too high. This would cause blocking of the boiler due to sticky ash layers on the heat exchanger surfaces.

Besides conventional methods in temperature profile control, the use of multi variable optimum seeking methods has also been reported[15].

2.3 Soot blowing optimization

Soot blowing maintains boiler efficiency and steam capacity by periodically removing ash and slag from heat transfer surfaces[16]. The convection sectors of the boiler have soot blowers typically activated on a periodic basis. The cleaning process usually occurs after a predetermined operating time that is independent of load or performance of the boiler. The nature of boiler fouling is long-term or short-term as Fig. 5 shows[16].

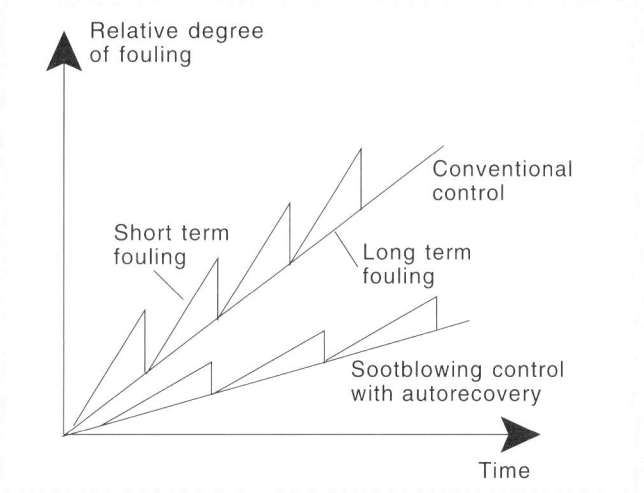

Figure 5. Long-term and short-term components of fouling in recovery boilers[16].

CHAPTER 6

Long-term fouling consists of compounds that cannot be removed by normal soot blowing techniques. Advanced soot blowing systems will reduce both short-term fouling and the rate of long-term fouling.

Deposits in the recovery boiler originate from two main sources: carryover consisting of smelt and black liquor particles and condensation products from volatile compounds. They form in different amounts in different sections of the boiler. In the superheater area, carryover is dominant and forms hard deposits. In the boiler bank and the economizers, deposits consist primarily of condensates. The rate of deposit formation is a function of boiler load, excess air, temperature profile in the boiler, bed conditions, air distribution, etc.

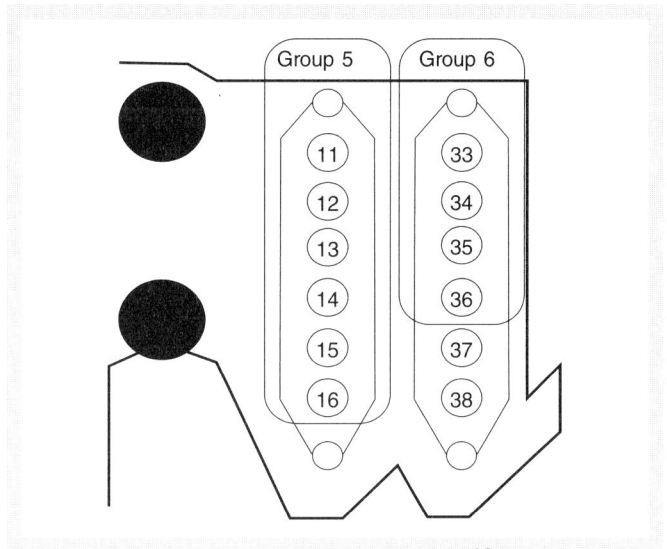

Figure 6. Principle of different soot blowing priorities[16]. Here the soot blowing sequence may be 11, 12, 13, 33, 34, 35, 36, 14, 15, 16, i.e., group 6 has a higher priority than group 5.

Soot blowing controls try to maximize the heat transfer from flue gases to water and steam with optimum consumption of soot blowing steam. Soot blowing optimization uses measured and calculated criteria in different parts of the boiler (super heater zone, boiler zone, and economizer) and soot blowing only on a demand basis. This allows savings of considerable steam. Continuous soot blowing is not necessary in sections where deposits do not accumulate. More frequent soot blowing must occur where massive deposit accumulation happens.

The fouling criteria in one commercially available system are the following[16]:

- Draft losses (pressure loss)
- Heat transfer coefficients
- Elapsed time
- Superheated steam temperature
- Load of induced draft fans
- Special criteria.

Fouling criteria calculations are normalized according to the production rate and long-term fouling characteristics. The importance of each criterion varies with each section.

Process control in chemical recovery

The boiler is usually divided into soot blowing groups according to differential pressure and temperature measurements. Groups have priorities so that the most important groups are at the higher level, and their soot blowing starts as soon as possible after fulfilling the criteria. Single soot blowers are handled individually, and they can be included freely in different groups at the same time[12]. A single soot blower can be started separately. A pair soot blower mode is beneficial in cases with a high demand of soot blowing. This also leads to a shorter soot blowing time and more constant steam consumption[16]. Figure 6 shows an example of use of different priorities in soot blowing control[16].

Soot blowing optimization leads to reduction in the soot blowing steam with less tendency to plug the boiler. The average decrease in steam consumption has been 30%.

2.4 Fire-room cameras, image analysis, and expert systems

Considerable difficulties occur in monitoring smelt bed behavior in the recovery boiler using conventional instrumentation. The bed height depends upon operating variables such as liquor temperature, primary to secondary air ratio, and air pressure. A remarkable improvement potential exists in using digital image processing on fire-room camera information. This allows expanding and improving burning supervision and control[17]. The system efficiency increases by a knowledge-based system[18] and using fuzzy logic in digital image enhancing[19].

Visual image processing typically consists of three main tasks: digitizing, enhancing, and analyzing the image. Image improvement can include noise filtering, contrast to brightness improvement, edge enhancement, area averaging, etc. Figure 7 shows the main stages of digital image processing[10].

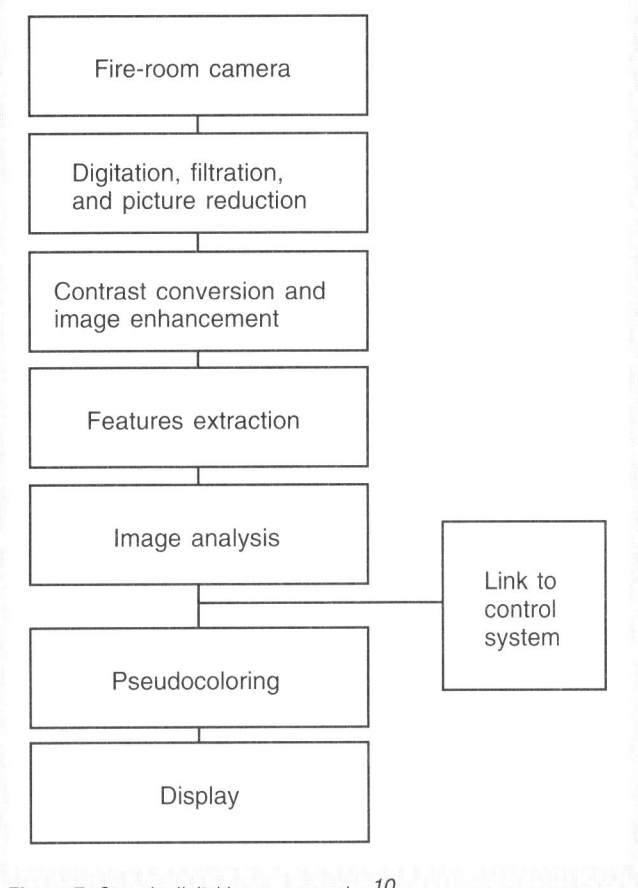

Figure 7. Steps in digital image processing[10].

CHAPTER 6

One commercially available system was originally developed for wood waste burning in grate type boilers. Its good results led to further development for the recovery boiler. It uses the video signal from two smelt bed cameras as the raw information. Their location is such that the bed is visible from both sides.

The application software for this unit depends on the type of process. In recovery boiler operation, the important features are the following[10]:

- Height of the bed
- Horizontal position of the top
- Size of the bed
- Average intensity of the bed
- Shape of the sides of the bed.

The presentation of the analyzed images happens by using synthesized color display images. They give the operator additional information about the process behavior to allow making proper corrections when necessary. Figure 8 shows a sample display of the recovery boiler application[10].

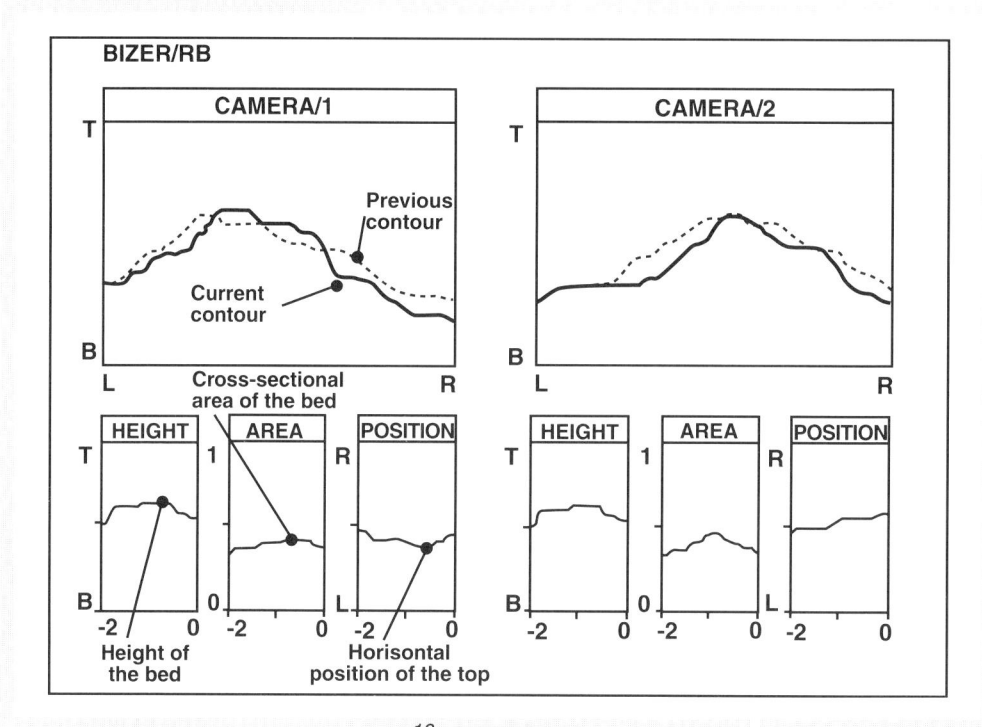

Figure 8. Display for recovery boiler application[10].

The following observations have been reported[10]:

- Smelt bed size changes based on load changes
- Smelt bed symmetry disturbances occur due to load changes
- Reliable height analyses occur during the burning transients
- Average temperature of the bed changes with load changes.

The burning expert system is a knowledge-based advisory system that combines the information given by the above system with other process measurements and examines the status of the boiler according to its rule base[18]. This concentrates on monitoring the factors that today require manual operations. These include air openings and liquor gun plugging management. These functions also have drastic effects on emissions. Some parts of emissions control are included. The system also monitors air penetration, flue gas channeling, and symmetry control.

The target of the rule-based system for recovery boiler control is to connect the image information coming from the above system with other measurements and complement the conventional recovery boiler system with some rule-based functions. It was clear from the start that the rule-based system should concentrate on the functions that are diagnostic in nature or performed manually today, The above definition means that the rule-based system concentrates on the following areas of control:

- Diagnostics of fire-room camera operation
- Follow-up of the function of liquor nozzles and air openings
- Air and flue gas flows inside the boiler
- Emissions control.

Control of liquor nozzles and air openings helps the operator maintain safe and optimum operation of the recovery boiler. This part of the system advises operators how to behave in disturbance situations and does not make any control actions directly. Plugging of air openings results in an increase in emissions, changes in bed symmetry, and increase of air pressure. Symptoms of plugging in liquor nozzles are increase in flue gas combustibles and emissions. The bed height also increases.

CHAPTER 6

Rules are of the common if-else format as follows:

Plugging of liquor nozzles

if   Flue gas combustibles, SO_2, excess O_2, and smelt bed height are increasing and the main steam flow is decreasing.

then   "Black liquor nozzles seem to be plugged."

Figure 9 gives an example of the system's performance displays showing the most crucial burning parameters for current and three previous shifts as "traffic lights."

PERFORMANCE DISPLAY

		Shift		
Quantity	Current	1	2	8
BED HEIGHT (%)	○ 57	◐ 57	◐ 57	● 57
BED AREA (%)	○ 66	◐ 66	● 66	○ 66
BED POSITION (%)	◐ 50	○ 50	○ 50	◐ 50
EXCESS O_2, left (%)	○ 3.4	○ 3.4	○ 3.4	◐ 3.4
EXCESS O_2, right (5)	○ 2.9	○ 2.9	○ 2.9	● 2.9
FG TEMP. BEFORE GEN, left	○ 567	○ 567	○ 567	● 567
FG TEMP. BEFORE GEN, right	○ 593	○ 593	○ 593	○ 593
FG COMBUSTIBLES (ppm)	○ 255	● 255	○ 255	○ 255
FLUE GAS SO_2 (ppm)	○ 67	○ 67	○ 67	○ 67
SOOTBLOWING STEAM (kg/s)	○ 8.6	○ 8.6	○ 8.6	○ 8.6
FLUE GAS ASH pH	○ 10.1	◐ 10.1	○ 10.1	◐ 10.1

● Poor (red)   ◐ Acceptable (yellow)   ○ Excellent (green)

Figure 9. Example of the performance display [18].

Ozaki et al.[20] have described a system where the bed shape from the image analysis system is first classified by three-layer back propagation neural network into three classes as Fig. 10 shows. The fuzzy controller of Fig. 11 changes the air flows into the recovery boiler by using the classifier information. ZO = zero, PS = positive small, PB = positive big.

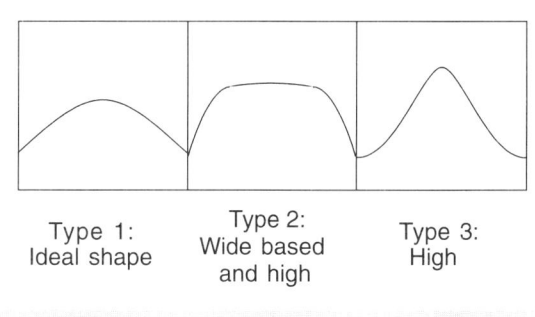

Type 1: Ideal shape
Type 2: Wide based and high
Type 3: High

Figure 10. Classification of char bed [20].

Process control in chemical recovery

One group proposed the combination of simulation and real time expert systems to recovery boiler control[21]. The recovery boiler model uses a steady-state modular simulator. An adiabatic free energy calculation predicts the conditions in the combustion zone. Radiation and convection heat transfer models describe the firebox and individual tube sections. Furnace temperature profile, heat fluxes, and dust accumulation are estimated with the complete energy and material balance.

Type 1	Type 2	Type 3
1L:ZO	1L:PB	1L:PS
1H:ZO	1H:PB	1H:PS
2L:ZO	2L:PB	2L:PS

ZO = zero, PS = positive small, PB = positive big.
1L, 1H, and 2L are air control actuators.

Figure 11. Fuzzy rule base for the visual control of the char bed shape[20].

A real time advisory control integrates process knowledge in the form of rules and simulated values with process data as Fig. 12 shows. Once a disturbance is detected, the advising messages are shown to the operator. Rules use statistical process control (SPC) methods to reveal when a variable is in abnormal condition. This makes detection of sensor problems and process upsets possible in real time easily understood. SPC criteria include one, two, and tree sigma confidence intervals for sample average and range. A rule to detect the shift in the sample average is also included. SPC confidence intervals depend on the process state.

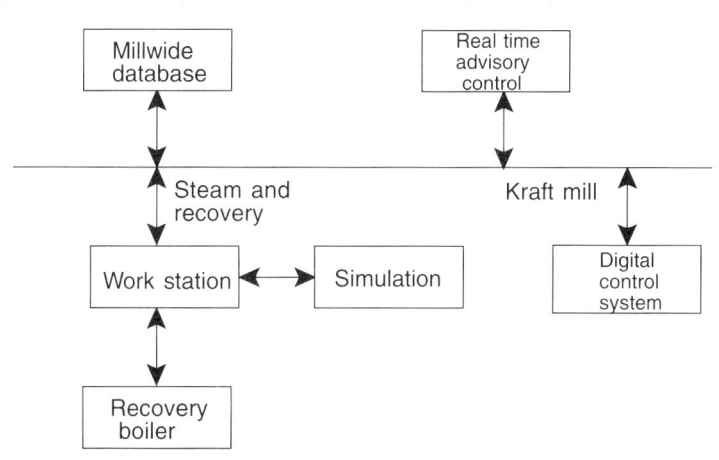

Figure 12. Structure of the recovery boiler expert system[21].

CHAPTER 6

2.5 Benefits

Table 1 gives some results from the benefits and savings reached by the recovery boiler control systems[9].

Table 1. Benefits from recovery boiler control[9].

Increased capacity	2–6	%
Decrease in soot blowing steam	18–50	%
Increase in reduction	1–7	percentage units
Increase in thermal efficiency	1–3	percentage units
Decrease in flue gas temperature	5–35	°C
Decrease in excess O_2 target	0.4–1.9	percentage units
Decrease in SO_2 emissions	100–300	ppm
Decrease in H_2S emissions	30–90	ppm
Decrease in fan power	7–35	%

Remember that these results show considerable variance, and all the benefits are not possible simultaneously for the same boiler. The results depend heavily on the base line and the actual operating conditions. Besides the above figures, a remarkable decrease in disturbances and shutdowns can occur.

A considerable decrease in standard deviation of the reduction degree by using fuzzy logic control is possible[13]. It decreased from 1.3% to 0.5% while simultaneously increasing its average value by one percentage unit to 97.5%. The deviation in excess oxygen also decreased to 0.2% (at the level of 3% excess oxygen).

3 Causticizing control

3.1 Control requirements

Causticizing and lime mud washing strongly influence lime kiln control. Disturbances in the lime kiln can also influence operation of the causticizing plant. Coordinating the control of these two pulp mill departments is therefore advantageous. The first system to do this came into use in the late 1970s[22].

In causticizing control, the following primary requirements are necessary[1]:

– The reaction equilibrium in the causticizing reaction must clearly be on the product side; i.e., the mill must be operating at the highest possible causticity level as follows:

$$CaO + H_2O \rightarrow Ca(OH)_2 \quad (3)$$

$$Ca(OH)_2 + Na_2CO_3 \leftrightarrow 2NaOH + CaCO_3 \quad (4)$$

- The highest possible white liquor concentration with minimum variations must occur to decrease the energy costs of the entire mill.
- Lime mud must be easily separated from white liquor.

The requirements partly conflict with each other. Their solution is a typical optimizing problem that must consider the entire mill.

3.2 Control functions

Figure 13 shows the control hierarchy of the causticizing controls in the computerized control system.

The green liquor flow from the dissolving tank defines the production rate of the causticizing plant. It must balance with the production requirements of the entire mill. The total alkali content of green liquor is usually controlled in the dissolving tank. Sometimes the fine control occurs after the green liquor storage tank. The control system supervises the effective alkali content of white liquor by controlling green liquor density according to the alkali model. This control works in two ways. Besides controlling the white liquor concentration, more uniform white liquor quality also results because deviations in white liquor quality are often the result of green liquor density variations.

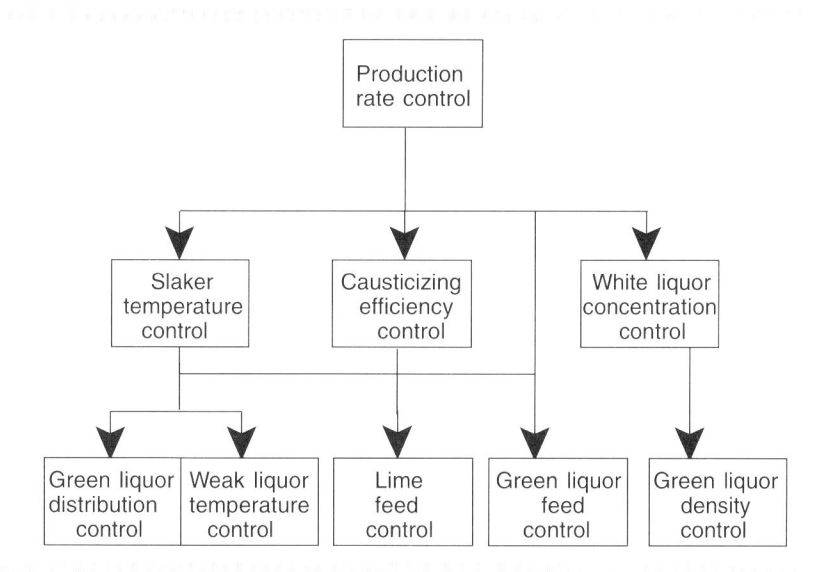

Figure 13. Control hierarchy of causticizing controls[22].

CHAPTER 6

A major variable controlled in the causticizing plant is the causticizing efficiency (CE). Figure 14 shows the conductivity based causticizing control.

In manual operation, the operator titrations made every one to two hours control the CE. The control variable is the lime feed to the slaker. Besides conductivity measurements, the causticizing control also uses a predictive model that considers sulfidity, green liquor total alkali content, green liquor feed distribution in the slaker, and slaker temperature. The model is calibrated automatically using laboratory analyses. Again, use of NIR analyzers is rising for direct measurement and control of liquor components.

The slaker temperature control forms the basis for successful operation of CE control. It allows immediate reactions to nonmeasurable disturbances in lime quality. In addition to lime feed, the slaker temperature control occurs with green liquor temperature and the distribution of green liquor to the slaker. These factors also influence the settling rate of the lime mud.

Other publications have introduced conductivity based CE control[23–25].

The CE control can improve if white liquor analy-

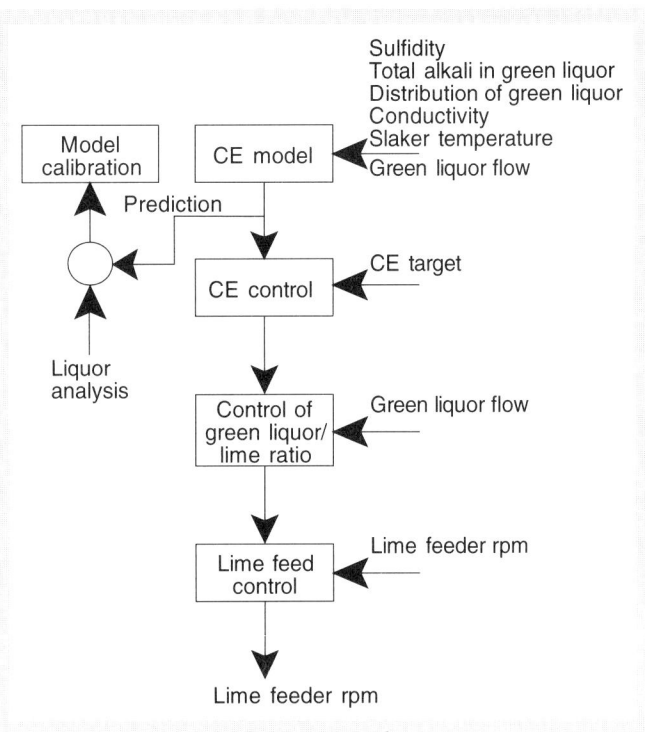

Figure 14. Control of causticizing efficiency[1].

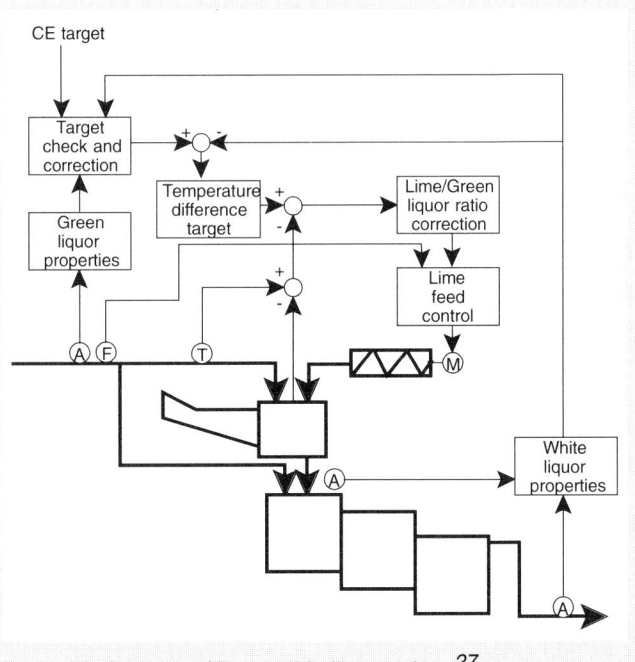

Figure 15. Controlling CE using white liquor analyses[27].

ses are available on-line[26-28]. The CE model is unnecessary if the white liquor analyses come in 5–15 min intervals. The actual CE control then proceeds directly by changing the lime to green liquor ratio or using temperature difference between green liquor and slaker as an intermediate variable as Fig. 15 shows[27].

In lime mud separation and washing, different strategies are necessary considering the process equipment in use. Usually, the operator sets a target for lime mud density. Liquor or wash water flow rates control this. If a vacuum filter is used, the vacuum, wash water flow, and filter rpm are controlled to achieve low residual alkali in lime mud while obtaining sufficiently high white liquor concentration.

In lime mud filters, filter rpm and wash water flow rates control the solids content of lime mud to the lime kiln. This control must also consider the residual alkali content. Residual alkali and free lime in lime mud influence the operation of the kiln. Optimum values for alkali content vary from 0.3% to 1.5%[22] as Na_2O. Under- or overshooting this has caused dusting or ball and ring formation. The control system alarm sounds when the precoat filter is clogged, and the precoat filter is blown clean when receiving a command from the operator. Today, continuous precoat renewal systems are available to reduce the number of blowoffs and also help to control TRS emissions.

3.3 Benefits

With more accurate control of the slaker, approaching the theoretical lime requirements is possible. In manual control, overliming occurs especially when the lime and green liquor qualities fluctuate.

In one case[22], the lime feed to the slaker decreased by about 0.6 lb/ft^3 of white liquor. This results in kiln fuel savings or an increase in extra capacity of the kiln per ton of pulp. The reduction in lime circulation can decrease the chemical losses at a corresponding proportion. Makeup lime consumption actually decreased by 24% during nine months under computer control when calculated as a 12-month floating average. When the lime availability increased due to temperature control and lime mud quality, circulation of inert material decreased.

With improved green liquor quality and slaker control, white liquor quality is more uniform. CE standard deviation of about 1.9% compared with 4% before the installation of the control system is possible. As mentioned, CE can increase with computer control without excess lime dosing or lowering of white liquor quality. An increase in CE also results in an increase in the capacity of the evaporation and recovery boiler. This is approximately 1.5% for this example.

4 Lime mud filter control

4.1 Control requirements

The goals of lime mud filter control are higher mud solids content, steady process operation, low alkali content, and low maintenance costs. Specifications are as follows[29]:

- Control the rotational speed of the filter to reach the optimum solids content
- Control the wash water flow to keep alkali in mud sufficiently low

CHAPTER 6

- Control the density of the feed flow
- Monitor the filter efficiency and start the blowing sequence when required
- Adapt to changes in feed quality
- Adapt to control according to the changing filter performance
- Provide fault detection and reorganize the control during instrument failures.

The most significant process disturbances in the mud filter are the following:

- Production rate changes
- Density changes
- Alkali content changes
- Impurities in mud
- Particle size changes.

Dirt formation on the filter cloth is also a disturbance. The operational parameters such as the vacuum and wash water pressure sometimes change due to disturbances in other parts of the mill.

After blowing of the filter, the new cake may form irregularly leading to poor mud quality. The same problem occurs if the scraper is not working properly. Sensor faults and malfunctions are also common sources of control problems.

4.2 Control functions

From the control point of view, the lime mud filter is an interconnected multi variable process according to Fig. 16[29]. The possibilities of multi variable control, self-tuning control, and internal model control (IMC) are available in another publication[29]. Figure 17 shows a typical control strategy for lime mud filters.

The density target usually remains constant with control by dilution water. Here the feed valve controls the filter vat level. Control of the wash water flow stabilizes the alkali content of lime mud. The base level of the drum speed is calculated from the production rate, but it is also used to control the mud moisture content. The wash water also naturally influences the mud moisture.

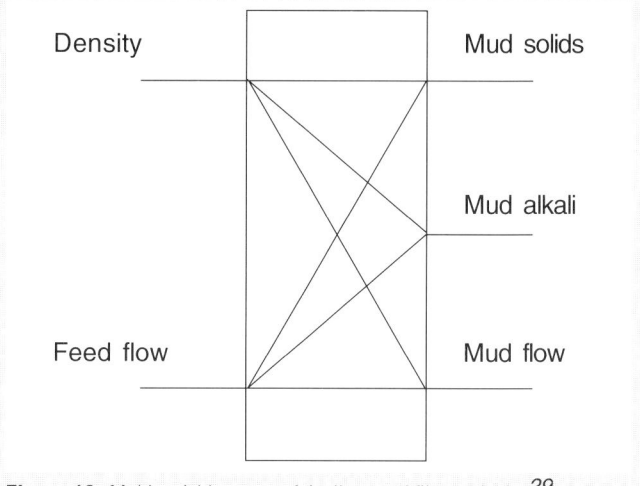

Figure 16. Multi variable nature of the lime mud filter variables[29].

Process control in chemical recovery

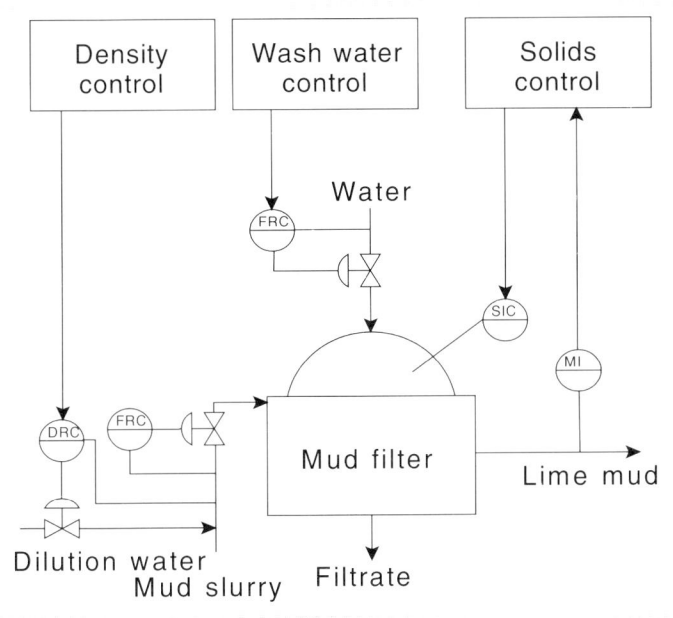

Figure 17. Typical control strategy for lime mud filter[29].

The monitoring of filter performance can use filtration and washing efficiency factors defined as follows:

$$f_d = 1 - \frac{Q_{ck}}{Q_{in}} \tag{5}$$

where Q_{ck} is the water flow in mud cake
Q_{in} the water flow in mud feed.

$$f_w = 1 - \frac{x_{ck}}{x_{in}} \tag{6}$$

where x_{ck} and x_{in} are alkali contents in mud cake and mud feed, respectively.

The alkali contents can be calculated indirectly from conductivity measurements.
Davey et al.[30] have given another modeling approach for the lime mud filter. It starts from the question how to maintain the moisture content of the lime mud and its alkali content close to their targets. The control parameters are the following:

- Drum speed
- Wash flow rate
- Pressure differential
- Volume rate of production
- Blade position.

They suggest using blade position in control because both output variables show the same kind of behavior when the blade position changes as Fig. 18 shows[30].

4.3 Benefits

With better control of the lime mud filter, the standard deviation of the moisture and alkali contents of lime mud can decrease at least one-fifth of the original[29]. This would mean considerable energy savings if used as increased target moisture. The chemical savings are more difficult to estimate, but they consist of decreased dust losses and better lime quality. This means less makeup lime and sodium hydroxide. A remarkable decrease in TRS emissions from the kiln is also possible simultaneously.

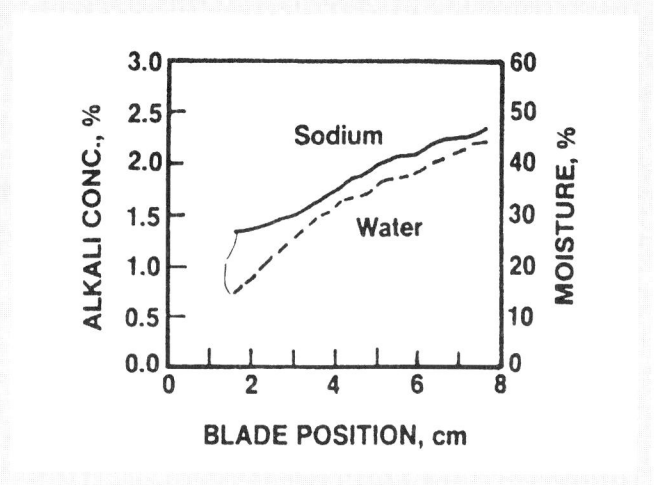

Figure 18. Alkali and moisture variations vs. blade position[30].

5 Lime kiln control

5.1 Control requirements

Lime kilns in the pulp industry convert lime mud to lime in the recovery cycle. Because of the large quantities of lime mud handled, using an effective control scheme for the economical production of good quality lime is desirable. The control of the lime kiln is multi variable by nature. The reaction time in a rotary kiln system is long and variable. A large quantity of lime mud and lime are in process within the system at any given time. Classical model based supervisory control of the process is nearly impossible, and the experience of veteran operators is often the best solution. Control changes should use small increments to allow sufficient time to observe the effect of each change. Overreaction leads to overcompensation that generally causes oscillations and upsets. These may take hours to return to normal operation.

The target of lime mud burning is to produce constant lime quality for causticizing with minimum energy consumption. As already mentioned, the causticizing control and the quality of the lime mud feed influence the control of the kiln. The following requirements are necessary for lime kiln control:

- Maintain heat losses via flue gases as low as possible
- Maintain the temperature profiles in the drying and heating zone of the kiln (Fig. 19) to control pelletizing, dusting, and ball and ring formation
- Maintain correct decomposition reaction temperature and sufficient time for this reaction to occur to control residual carbonate and lime availability to their target values with minimum variations as follows:

$$CaCO_3 \rightarrow CaO + CO_2 \qquad (7)$$

- Minimize emissions
- Maximize energy use.

The main difficulties in lime kiln control are due to the variations in lime mud feed and its moisture and alkali content. Difficulties in measurements such as the burning zone temperature disturb the control. The variables are lime mud moisture, feed-end temperature, middle-zone temperature, burning-zone temperature, pressure in the flue gas chamber, pressure after the flue gas fan, power demand of kiln drive, and amount of excess oxygen in the flue gas.

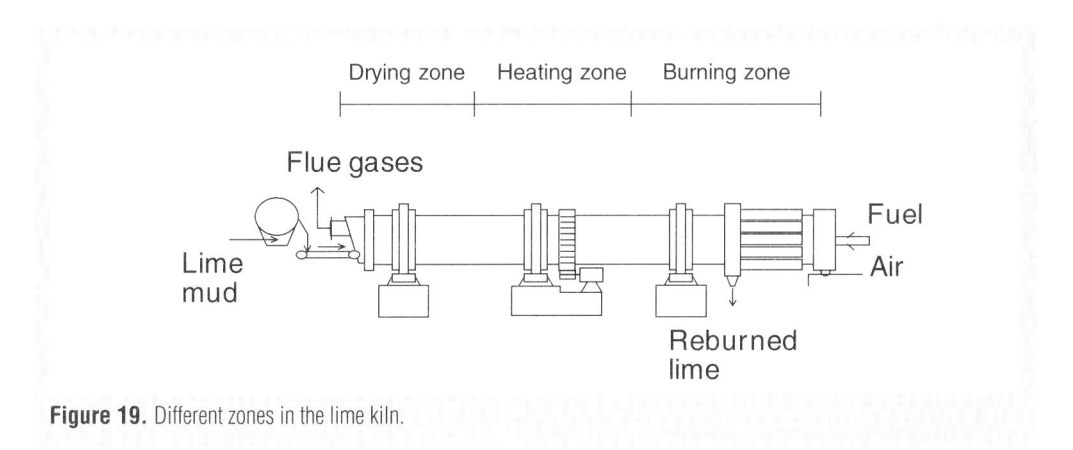

Figure 19. Different zones in the lime kiln.

5.2 Control functions

Figure 20 shows the control hierarchy of the lime kiln control system[22].

The controlled variables are lime mud feed, fuel flow, draft, primary air flow, and kiln rpm. The system guides the temperature profile to its correct shape. The production rate expressed as reburned lime is controlled to the given target by the lime mud flow to the filter. Monitoring of the power demand of the drive prevents overloading.

Residual calcium carbonate analysis in lime monitors the quality of reburned lime. Changing the temperature profile corrects this. Temperature profile measurement uses at least three points in the kiln. The flue gas flow (the draft fan rpm or the position of the flue gas valve) controls the feed end temperature. The fuel flow rate controls the burning temperature. The interactive nature of the kiln requires consideration as does the actual control strategy in a mixed feedback-feedforward type unit with a multi variable nature.

Many items can cause incomplete burning and increase total reduced sulfur (TRS) emissions from the kiln. The excess oxygen control that ensures maintaining excess oxygen at predetermined limits controls this. Corrections change the temperature profile. If analyzers for TRS and CO exist, they complement the control strategy and also serve as feedback to the lime mud filters.

CHAPTER 6

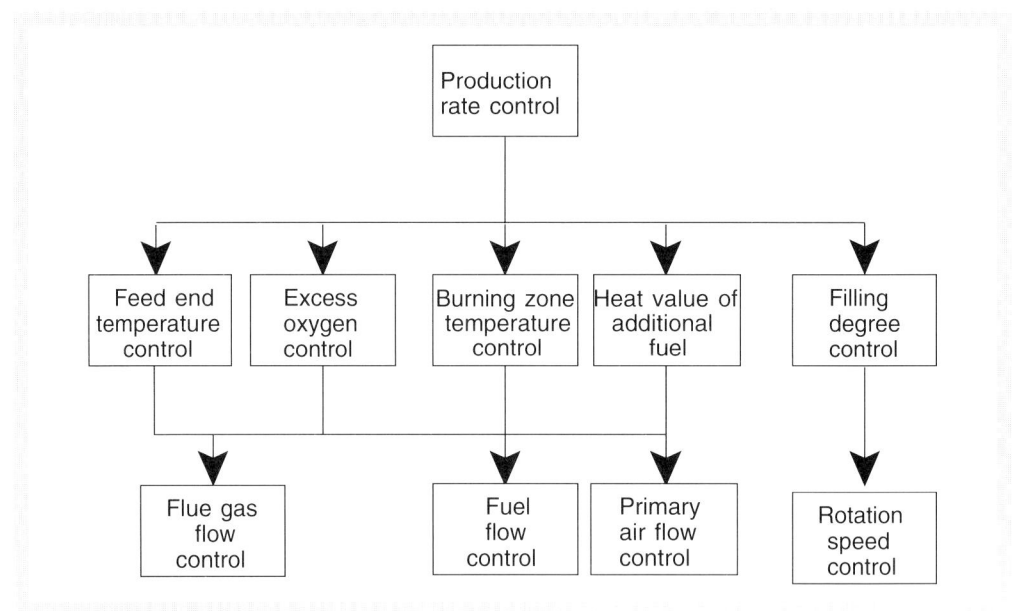

Figure 20. Control hierarchy of the lime kiln control system[22].

Sensor based applications[31] are also possible. Control of excess oxygen content has remarkable effects on the energy economy of the lime kiln. To guarantee complete combustion, a small amount of excess air is necessary. Increasing the excess air level increases the heat losses from the lime kiln. Decreasing excess air level increases incombustibles. The optimum amount of excess air is probably the point where CO starts to exist in the flue gases. According to mill tests, the excess O_2 limit where CO started to exist in the flue gas varied from 1.2% to 2.0%. The measurement of O_2 only therefore cannot guarantee the optimum excess air level.

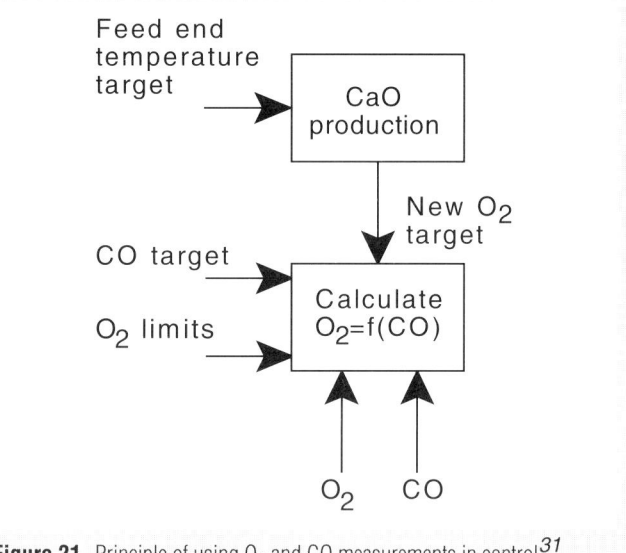

Figure 21. Principle of using O_2 and CO measurements in control[31].

The best control strategy is to determine the optimal excess O_2 target using the selected low CO target and keep it by controlling the kiln draft. Another possibility is to control the kiln draft based on CO measurement directly. This could lead to difficulties because a slightly higher excess oxygen content will sometimes be necessary to keep the feed end temperature from decreasing excessively especially during disturbances. Figure 21 shows the basic control principle.

The possibility of using a dust analyzer to minimize the dust circulation seemed limited[31]. The analyzer is useful in process monitoring. In disturbances, it could also have use to determine alarm limits so that the dust circulation remains in moderate proportions. According to mill tests, a clear correlation exists between carbonate remaining in the lime and the CO_2 content of flue gas. Unfortunately, simple material and energy balance calculations reveal that 2% change in remaining calcium carbonate means only 0.45% change in CO_2 content. This is only 2% of the measuring scale. This measurement can therefore only have use to find larger disturbances in calcining.

Mill tests have also revealed that the CO content starts to increase before a radical increase in H_2S content occurs[31]. Under certain conditions at least, this indicates that predicting the H_2S formation with the help of CO and O_2 measurements is possible. This is true when the soluble alkali content in the lime mud feed is low. In the opposite case, one should use the H_2S analyzer.

An increase in the dry solids content of lime mud will improve energy economy of the lime kiln. At higher dry solids content, dust losses from the kiln will increase. According to mill tests[31], the optimum is probably 80%–85% solids content. The limiting factor in the studies was the constant length of the kiln chain section. The optimum will be different if the length of the chain section changes. From a control point of view, the target for the lime mud solids content should occur at the supervisory control level because it determines the basic level of the dust circulation. In all cases, the lime mud solids content should be stabilized so the kiln temperature profile is stable. Control of the feed end temperature is especially difficult if large variations occur in the lime mud solids content. The control block in Fig. 21 is a part of the feed end temperature and excess oxygen optimization block. If the feed end temperature target is set too high, the kiln draft increases with simultaneously increasing dust circulation.

Uniform and correct burning temperature control makes refractory life longer in the burning zone. This can be very significant if the mill has operated at excessively high burning temperature to control residual carbonate.

The first commercial control system for controlling the lime kiln came into operation in 1976—1977[22].

Several kilns in the United States installed commercially available lime kiln control systems during the 1980s[23]. Fuel savings ranging from 7%–20% have occurred with remarkable capacity increase in some cases. These systems aim for good overall control of the kiln. They reduce disturbances caused by manual control actions and running the kiln with an excessively high burning temperature. The operator only gives targets for production rate and lime quality. Systems provide other targets such as temperatures, etc.

According to a reference[23], the control uses multi variable algorithms consisting of feedback and feedforward parts. Feedforward is used in compensating disturbances. No exact form of algorithm is available in the reference, but one can assume that the multivariable behavior is considered using software rather than with an actual multi variable control algorithm.

Blevins and Rice[32] reported an application for lime kiln control. In this system, the control strategy differs from the usual approach. Fuel flow controls the flue gas temperature, and the kiln draft controls the burning temperature. The system also tries to stabilize excess air inside the kiln.

Another control system for lime kiln control uses the usual basic control strategy[15]. It also uses a model that calculates the temperature and concentration profiles inside the kiln and predicts the peak wall temperature and the residual carbonate in reburned lime. These values are used to change the target of the burning temperature. The model also estimates the burning temperature and the resulting carbon dioxide content of flue gases used in following the model correctness. The system also includes production optimization and coordination features.

5.3 Advanced control systems for the lime kiln

The first lime kiln control system based on fuzzy logic was installed in a Swedish paper mill at the end of 1979[33]. Other installations followed later using a system that aims to stabilize and optimize lime quality. It also minimizes fuel consumption and optimizes production. The control itself has burning zone control and combustion control.

The burning zone control includes the following objectives:

- Control very cold kiln
- Stabilize temperatures
- Correct quality
- Optimize production.

Combustion control aims for the following:

- Control very high CO
- Control low O_2
- Control exit temperatures
- Control high O_2.

The operation of the kiln has been very stable[34]. Variations in residual carbonate have decreased by two-thirds. The supply of makeup lime has decreased from 10% to 5%.

A fuzzy expert system application for lime kiln control is available[35]. In the beginning of the 1990s, good results occurred using fuzzy control in an actual rotary lime kiln[36]. At the time of this writing, testing of fuzzy lime kiln control using linguistic equations is occuring[37].

The internal model control strategy for the lime kiln control has been suggested[38]. It uses multilayer, feedforward back propagation network with 8 inputs, 2 outputs, and 2 hidden layers with 20 and 10 nodes. Figure 22 shows the principal control strategy.

Process control in chemical recovery

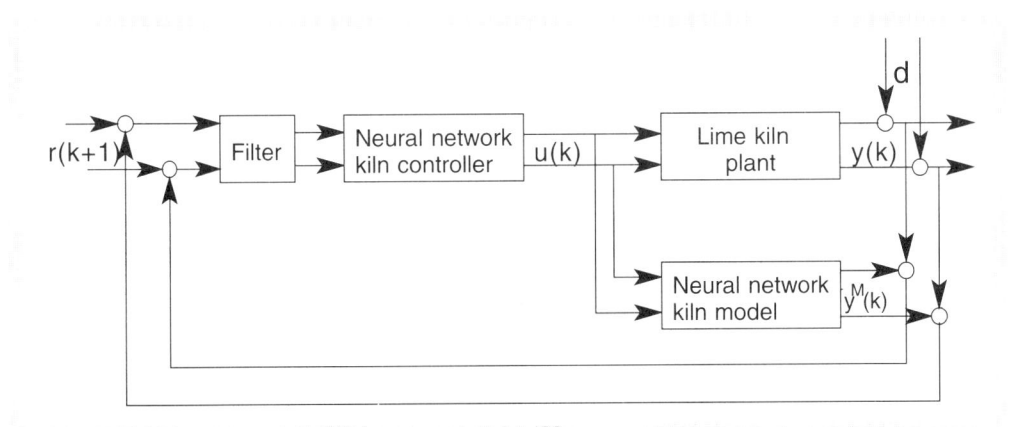

Figure 22. Neural network based control strategy for lime kiln[38].

Model predictive control has also been applied to control the lime kiln[39]. This approach keeps the feed end and burning end temperatures at some desired setpoints by manipulating the fuel flow rate and the air damper. The controller requires dynamic models that describe the relationships between the manipulated variables and the controlled outputs. The models have been identified with the aid of step tests. The major advantage of the method is that it can function satisfactorily in an uncertain process environment dominated by disturbances and lack of good models. It can also be tuned to produce dynamic behavior acceptable to the process operator.

5.4 Benefits

Lowering the residual carbonate level does not have a substantial influence on energy consumption, but benefits can result as more even white liquor quality and improved slaker control. Decreasing the residual carbonate level by about 1% is possible. As noted earlier, a reduced variation is usually more essential.

An increase in lime kiln capacity was 16.5% in one case[22]. Raising the capacity of white liquor production by 6.8% was possible. Six systems for lime kiln control were analyzed[40] with an average 8.7% capacity increase and 10% energy savings.

CHAPTER 6

References

1. Sutinen, R. and Koskela, O., 1981 Automation Days, Finnish Society of Automatic Control, Helsinki, p. 33.
2. Brooks, T. R. and Edwards, L. L., Tappi J. 75(11):131(1992).
3. Honka, I., Kuukkanen, K., Aksela, R., et al., Chemical Engineering Progress 79(9):31(1983).
4. Jutila, E., Pantsar, O., and Uronen, P., Pulp Paper Can. 79(4):21(1978).
5. Jutila, E., Uronen, P., Huovinen, N., et al., Pulp Paper 55(7):51(1981).
6. Sutinen, R., Aksela, R., and Huovinen, N., 1983 PRP-5 Preprints, IFAC, Antwerp, p. 107.
7. Sutinen, R., Telimaa, E., and Koski, S., 1982 Black Liquor Recovery Symposium Proceedings, Helsinki, Finland, p. 132.
8. Uronen, P., Jutila, E., and Pantsar, O., 1978 IFAC 7th Triennial World Congress Preprints, Pergamon Press, Oxford, p. 249.
9. Uronen, P.and Leiviskä, K., TAPPI 1985 International Chemical Recovery Conference Proceedings, TAPPI PRESS, Atlanta, book 1, p. 127.
10. Tuhkanen, M., 1992 Latin American Recovery Boiler Conference Preprints, ATCP, Conception, p. 18.
11. Kettunen, H., 1995 Automaatio Proceedings, The Finnish Society of Automation, Helsinki, p. 334.
12. Mikkonen, J., 1993 Automaatio Proceedings, Finnish Society of Automation, Helsinki, p. 225.
13. Lampela, K., Mikkonen, J., and Leiviskä, K., 1996 TOOLMET - Tool Environments and Development Methods for Intelligent Systems Proceedings, University of Oulu, Oulu, p. 120.
14. Mikkonen, J., Koskinen, J., and Hanhinen, J., Paperi ja Puu 76(5):5(1994).
15. Rastogi, L. K., Spannbauer, J. P., Goto, K., et al., TAPPI 1982 Pulping Conference Proceedings, TAPPI PRESS, Atlanta, p. 492.
16. Sutinen, R. and Koskela, O., TAPPI 1992 International Chemical Recovery Boiler Conference Proceedings, TAPPI PRESS, Atlanta.
17. Sutinen, R., Huttunen, R., Ollus, M., et al., Pulp Paper Can. 92(4):T83(1991).
18. Hosti, T., Leiviskä, K., and Sutinen, R., 1994 Control Systems Preprints, SPCI, Stockholm, p. 313.

19. Murtovaara, S., Juuso, E. K., and Sutinen, R., 1996 IWISP Proceedings, Elsevier Science, Manchester, p. 423.
20. Ozaki, N., Yamazu, S., Tateoka, K., et al., 1994 Control Systems Proceedings, SPCI, Stockholm, p. 307.
21. Smith, D. B., Edwards, L. L., and Damon, R. A., Tappi J. 74(11):93(1991).
22. Elsilä, M., Leiviskä, K., Nettamo, K., et al., Pulp Paper 53(12):152(1979).
23. Bailey, R. B. and Willison, T. R., TAPPI 1985 Pulping Conference Proceedings, TAPPI PRESS, Atlanta, p. 619.
24. Bertelsen, P., 1986 Causticizing–Lime Kiln Symposium, SPCI, Stockholm, p. 54.
25. Lounasvuori, R., Sutinen, R., Juuma, M., et al., TAPPI 1985 International Chemical Recovery Conference Proceedings, TAPPI PRESS, Atlanta, p. 419.
26. Hultman, B., and Berglund, A., 1981 CPPA/TAPPI International Conference on Recovery of Pulping Chemicals Preprints, TAPPI PRESS, Atlanta, p. 95.
27. Leiviskä, K., Savukoski, M., and Uronen, P., "Causticizing Reaction and Methods to Analyze its Proceeding," Report 105, Department of Process Engineering, University of Oulu, Oulu, pp. 22–34.
28. Rasmusson, L., 1986 Causticizing-Lime Kiln Symposium, SPCI, Stockholm, p. 67.
29. Haataja, R., Leiviskä, K., and Uronen, P., 1985 Conference on Digital Computer Applications to Process Control Proceedings, International Federation of Automatic Control, Vienna, p. 125.
30. Davey, K. R., Vachtsevanos, G., Cheng, J. C., et al., Tappi J. 72(8):150(1989).
31. Uronen, P., and Leiviskä, K., Pulp Paper Can. 90(9):113(1989).
32. Blevins, T. and Rice, R., Tappi J. 66(3):103(1983).
33. Ostergaard, J. -J., 1993 EUFIT Congress Proceedings, ELITE Foundation, Aachen, p. 552.
34. Nilsson, L. and Langkjaer, T. F., 1996 EUFIT Congress Proceedings, ELITE Foundation, Aachen, p. 1901.
35. Leiviskä, K., Huttunen, R., and Uronen, P., 1988 Conference on The Use of Computers in Chemical Engineering (CHEMDATA 88) Proceedings 2, EFCE, Gothenburg, p. 379.
36. Haataja, K. and Ruotsalainen, J., 1994 MEPP Conference Proceedings, Mariehamn, Finland.
37. Juuso, E., Ahola T. and Leiviskä K., Control Systems '98 Proceedings, Finnish Society of Automation, Helsinki, Finland, p. 54.
38. Ribeiro, B., Dourado, A., and Costa, E., 1993 International Conference on Artificial Neural Networks and Genetic Algorithms Proceedings, Springer Verlag, Innsbruck, p. 117.
39. Charos, G. N., Arkun, Y., and Taylor, R. A., Tappi J. 74(2):203(1991).
40. Uronen, P., Leiviskä, K., and Kesti, E., "Benefits and Results of Computer Control in Pulp and Paper Industry," Ministry of Trade and Industry. Energy Department. Series D:76. Helsinki, 1985, pp. 49-51.

CHAPTER 7

Process control in mechanical pulping

1	**Refiner control**	**129**
1.1	Control requirements	129
1.2	Control functions	130
1.3	Quality control	133
1.4	Applications of intelligent control	136
1.5	Results	138
2	**Grinding control**	**138**
2.1	Control requirements	138
2.2	Control functions	139
	References	143

CHAPTER 7

Kauko Leiviskä

Process control in mechanical pulping

1 Refiner control

1.1 Control requirements

Two important disturbances of different types influence the control in thermomechanical pulp (TMP) plants: slowly occurring wearing of refiner plates and more rapid variations in raw material quality. Raw material variations lead to changes in the mass flow rate in the chips fed into the refiner. Even in normal operating conditions, these disturbances can be 10%–15%. Variations in chip moisture and chip size also lead to consistency changes in the refining zone inside the refiner. Steam flows inside the refiner complicate the control. These variations require compensation to produce pulp of high quality. The problem is that variations and their influences on quality variables are very difficult to measure on-line. Laboratory testing usually measures quality. This information is always a few hours old and therefore insufficient for good quality control.

The time scale of disturbances is also very wide. Quality of chips can vary rapidly. Annual variation of moisture, chip size distribution, and temperature can also occur. Wearing of refiner plates changes the achieved quality of pulp under the same process conditions.

The main intent of refiner control is a constant quality at the highest possible level with minimum energy consumption. This usually means leveling the freeness variations. A strong correlation exists between refiner pulp quality and the specific energy consumption in refining. Specific energy usually controls quality. The difficulties in defining production rate and the specific energy consumption based on it limit control. Consistency is also an important variable from the control point of view. Methods that can measure it directly are therefore essential for control.

Good control of basic process variables is necessary for successful quality control and optimization of the entire refiner line. Control should try to keep the refiner operation in an allowable area for avoiding disturbances. When the refiner operates steadily and disturbances are effectively compensated, the final results and the quality of refiner pulp are easier to control. In this case, the changes in setpoints of process variables proposed by the quality control occur rapidly and with the least possible disturbances on the plant operation. The results give optimum production capacity and energy consumption.

CHAPTER 7

The control of refiner lines is multi variable by nature. It includes control of freeness, long fiber content, and shives content. By controlling these variables, the papermaking properties of refiner pulps such as strength and optical properties remain at their optimum values or at least inside allowable ranges. In practice, the selection of quality control strategies depends on what quality variables in each mill are measured.

Four generations of refiner control systems have existed[1]. The first generation used laboratory freeness measurement and the relationship between freeness and specific energy. Production rate used the calibration of feeding screws and estimates from chip moisture and bulk density. The second generation used the development of in-line sensor technology and the quality feedback if provided. Disc clearance control with position measurements was characteristic for the systems of this generation. The third generation systems used the feedback from pulp quality monitor (PQM) measurements. Quality feedback manipulated the specific energy setpoint and the power split between the primary and secondary refiners. Adding to this, the fourth generation fully uses model-based control. Use of new sensor technology and intelligent methods has influenced refiner controls.

1.2 Control functions

Figure 1 shows the total control hierarchy of refiner control[2]. The control of dilution waters, hydraulic pressures, rotational speed of the feeder, and other functions at the basic control level are not in the figure. The adaptive basic controls provide these control loops with their setpoints. The following presentation shows realization of the control functions in Fig. 1. The actual installation obviously varies with mills.

Quality controls Freeness control Control of long fiber content	30 min -1 h
Specific energy and power split controls Stagewise power controls	5 min
Adaptive controls Consistency control Fiber rate control	10-20 s

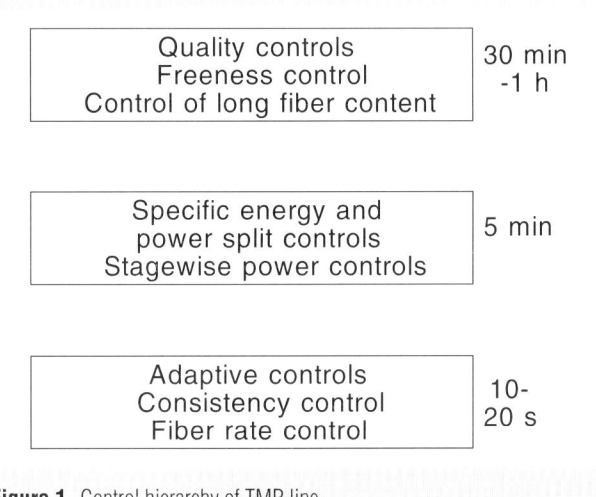

Figure 1. Control hierarchy of TMP line.

The chip feeder speed controls the production rate of the refiner line as Fig. 2 shows. Calculation of the production rate in tons of pulp uses chip feeding screw speed. Short-term variations in chip feed occur first as changes in the refiner power intake. The fiber rate controller that changes the chip feeder speed momentarily within a predefined range compensates the variations. This uses the fact that chip feed changes directly influence the power intake if the basic control loops of the refiner (pressure, hydraulic pressure, and consistency) are operating well. The recommended range of change in the chip feeder speed is a few percent. This strategy eliminates the effects of chip quality variations in the frequency of 0.1–20 minutes that are outside the scope of quality measurements.

Process control in mechanical pulping

Two main variables influence refiner consistency: flow rate of the dry wood and total water entering the chip refiner[3]. The latter includes the chip water and the various water flows including seal water. The blow line consistency depends on all these flows into the refiner.

Consistency variations during normal operation are at least 4%–6% or even higher[4]. A typical consistency range within which to select the consistency target is 10%. Outside this area, major difficulties occur in operating the refiner and assuring pulp of acceptable quality.

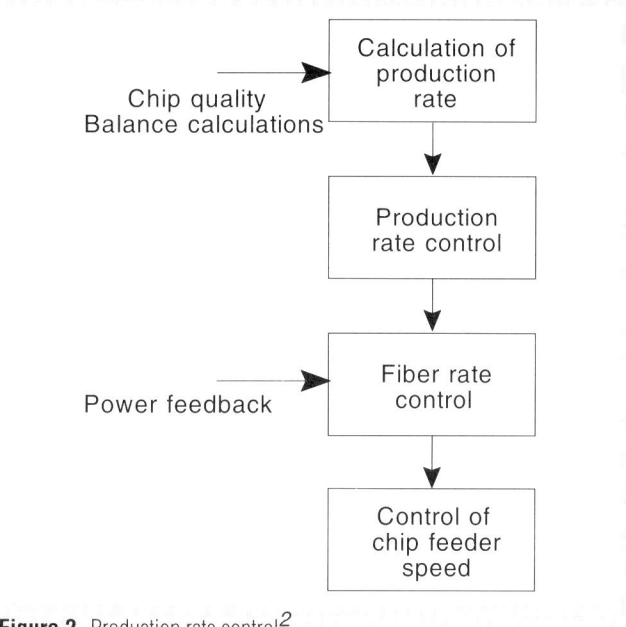

Figure 2. Production rate control[2].

A strong interaction exists between consistency control and fiber rate control. Figure 3 shows the basic control strategy[3,5]. It includes two control loops. One loop is for the consistency using the dilution flow as the control variable, and the other loop is for fiber flow rate adjusting chip screw speed according to variations in the power intake.

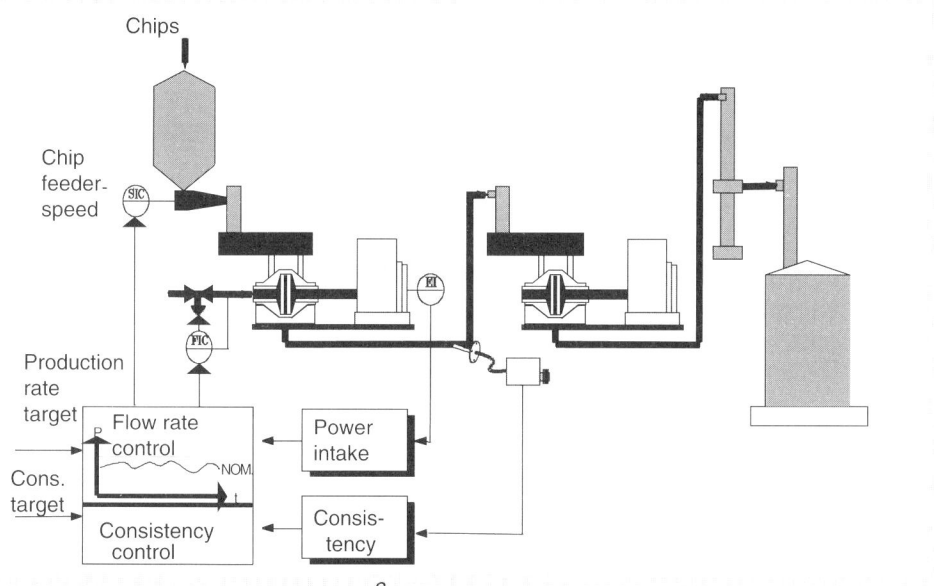

Figure 3. Production rate and consistency control[3].

CHAPTER 7

The consistency control of the first stage refiner uses on-line consistency measurement. Control uses an adaptive multi variable controller as in Fig. 4[5]. It adjusts itself continuously to changes in process dynamics and also includes a built-in dead time compensator and a continuous adaptive feedforward control. Due to the latter, the controller learns the dynamics of the external influences and compensates for them before manipulating the controlled variable. It also has proportional control that acts when the error becomes too large.

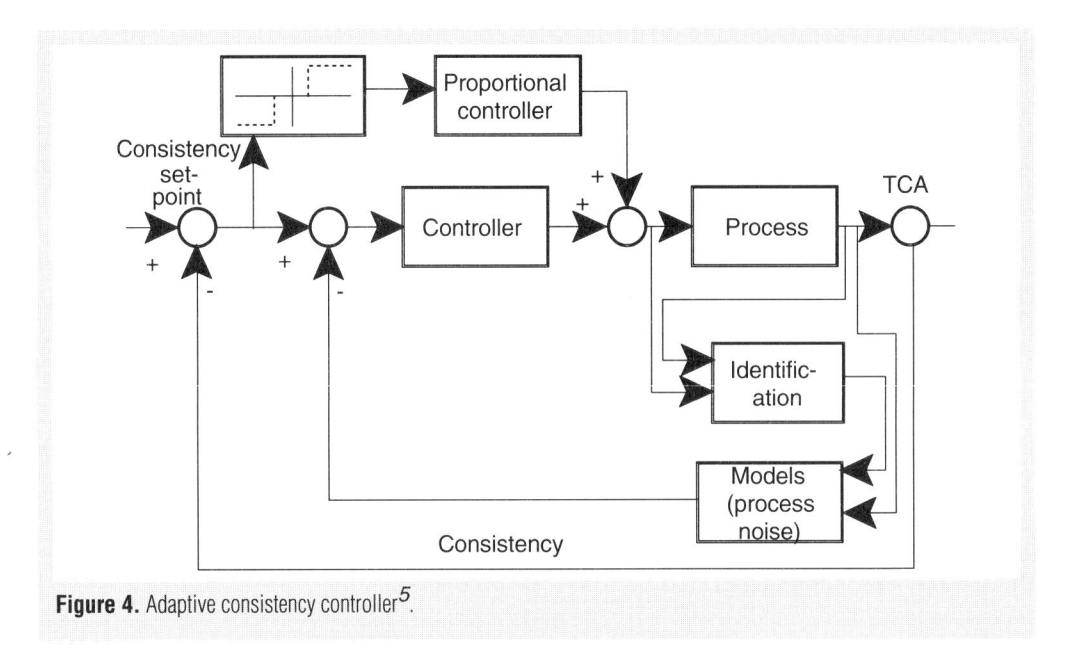

Figure 4. Adaptive consistency controller[5].

The correlation between consistency and refiner load also depends on the refiner gap. Figure 5 shows this correlation[4]. The shaded area shows the recommended gap vs. consistency area.

Proper control of consistency is also important in the secondary refiners, since they are typically subject to the same disturbance patterns as the primary refiner[5]. The load variations at the secondary refiner are even larger than at the primary refiner. This emphasizes

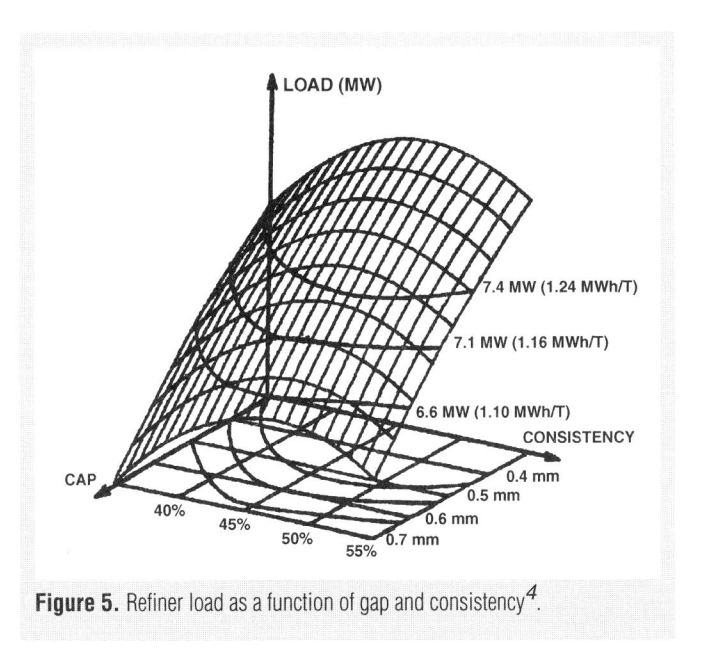

Figure 5. Refiner load as a function of gap and consistency[4].

good control of the primary stage as a means of avoiding amplified disturbances in the secondary stage.

The consistency control of the second refiner stage uses feedforward control from the first stage consistency measurement with the feedback from the material and energy balances of this stage. In the second stage, the final consistency control is also a function of the dilution water flow.

Another publication[6] introduces a model-based consistency control strategy. It uses a factor network model for predicting blow line consistency. The model is a combined mechanistic-statistical one. In the first stage, it calculates a theoretical consistency using material and energy balances. Then it compares theoretical consistencies with additional process data in a factor network model. The second stage is necessary to check the correctness of assumptions made concerning chip moisture, motor idling losses, seal water leakage, and steam splits in the balance models.

The control system uses the model output to remove the long-term variations (over 5 min) in blow line consistency.

Specific energy consumption tells energy consumption when producing one ton of refiner pulp. Specific energy control aims to reduce variations in specific energy consumption to improve pulp quality. Specific energy consumption changes by changing the power intake of the refiner (refiner gap) or by changing the production rate. The first strategy helps to keep the production rate as constant and stable as possible.

Cluett et al.[7] have proposed a control strategy where feed screw speed controls the refiner load. Experience showed that the relationship between the feed screw speed and the load was relatively time invariant. This strategy considers variations in chip quality (moisture, density, and wood species) better than the first strategy because the relationship between the refiner gap and the load varies over time.

Another specific energy control strategy is possible[8]. It strives for minimum energy consumption by operating with the smallest possible gap and highest possible production rate on a constant freeness level according to the following principle:

– Constant gap (as small as possible)

– Constant consistency in the blow line (calculated from material and energy balances)

– Constant freeness controlled by the feed screw speed. (Freeness is calculated from the model and compared with the on-line freeness measurement.)

Quality control determines the setpoint of the specific energy consumption. The power split control divides this setpoint further to each single refiner. The setpoint for the power split is also calculated at the quality control level or set by the operator.

1.3 Quality control

The success of quality control depends on how well the production rate control and the consistency control operate. The results to be gained also depend very much on the conditions at the mill, primarily on the extent of raw material variations. These variations

CHAPTER 7

set a natural limit to the improvements that can result from more efficient control.

Description of the quality of refiner pulps uses three factors: freeness, fiber length distribution that in practice means the long fiber content, and shives content. Changing the refiner operation can continuously control the first two factors. Shives control usually means the changes in the pulp flow to the reject refiner. A definition of "good" pulp at a given freeness level can be as follows[9]:

- Low shive content
- High long fiber content
- High strength
- High scattering coefficient.

Three different control strategies can usually control the quality variables:

- Control using laboratory analyses
- Model-based control
- Analyzer-based control.

Figure 6 shows the time spans of these alternative control strategies[2].

The two stage refiner case leads to a situation with two quality variables (freeness and long fiber content) controlled by four control variables (consistencies in each stage, power intake or specific energy consumption, and power split). This supposes that the production rate is not part of the quality control strategy. In principle, the situation shown in Fig. 7 is the result. This seems to offer several possibilities to combine control variables.

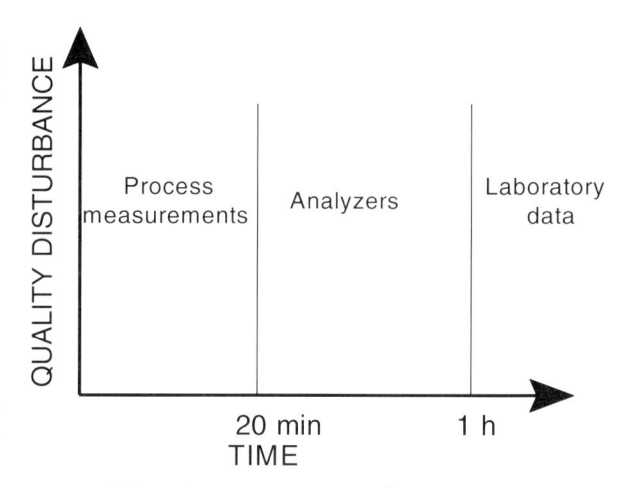

Figure 6. Time spans of quality disturbances[2].

The selection of the final control strategy uses mill evaluations according to experience, mill tests, or both. In practice, all possible combinations in Fig. 7 are not possible. Some lead to excessively complicated control systems.

In principle, freeness control can occur stage wise. The first stage consistency handles fine control, and changing the power intake handles bigger changes. Changes in consistency naturally influence the power intake in the first refiner stage and thus the specific energy consumption. If only freeness control is operative, the power split remains unchanged unless the power intake in one of the refining stages meets its upper or lower limits.

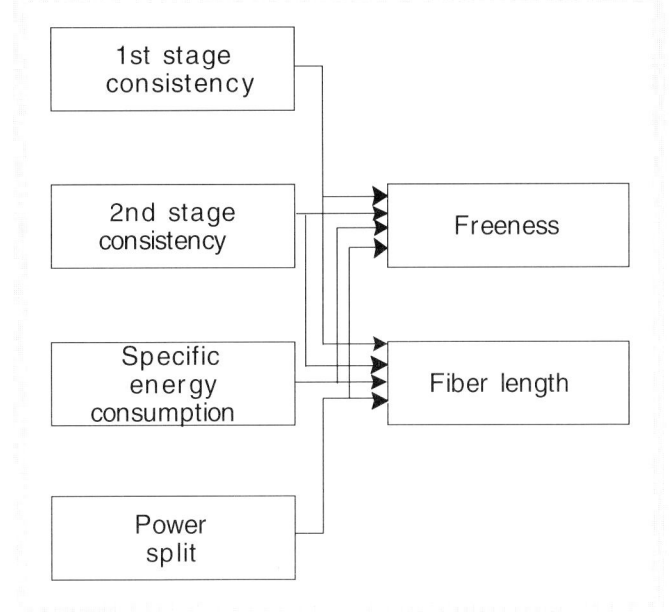

Figure 7. Dependencies between quality and process variables[2].

This procedure guarantees continuous maintenance of the optimum value of the first stage consistency and quick corrections for changes in freeness. Other strategies are also possible[8]. The highest possible consistency is usually advantageous for quality[9]. This

- Reduces pulp freeness slightly
- Reduces shive content
- Raises long fiber content
- Improves tear and tensile strength
- Improves scattering coefficient.

The operation of the first stage refiner primarily determines the fiber length distribution of refiner pulps. Changing the power split between the stages controls the distribution by searching for the combination of consistency and gap that maintains the required freeness with maximum fiber length. Consistency in the second stage refiner also has some effect on the fiber length. Fiber length control is naturally a constrained control case. The system tries to maintain the fiber length as high as possible.

In practice, the situation is not as easy as this. The specific energy consumption, power split, or the second stage consistency control the long fiber content depending on the situation. If the freeness control is off, the specific energy consumption controls the long fiber content. If freeness control is on, small variations in the long fiber content can be compensated by changing the setpoint of the second stage consistency. Changing the power split as Fig. 8 shows compensates larger variations.

Larger changes in the shives content always mean process disturbances that are not manageable by control engineering methods. Raw material quality and general refiner operations require inspection. The process operation requires tuning into its nominal operation area. The only possibility is to correct the quality variations in the reject refining by increasing the amount of reject pulp and increasing the specific energy consumption of the reject refiner if necessary.

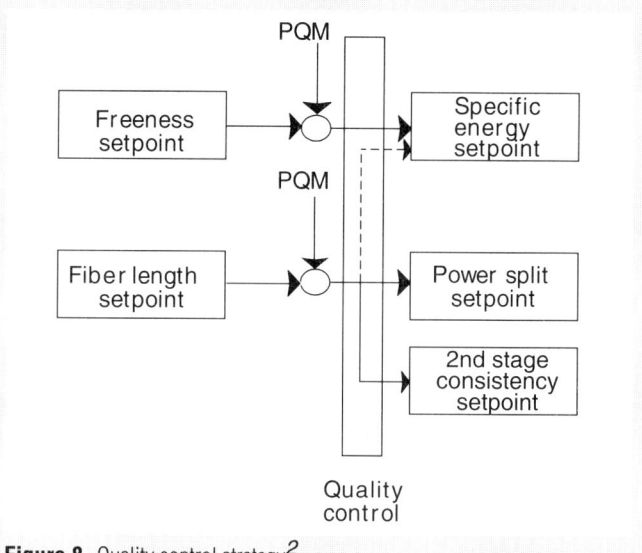

Figure 8. Quality control strategy[2].

The discussion above concerns refiner pulp quality. Controlling this naturally influences paper properties. Average fiber length gives a good estimate of tear strength, density, and light scattering of paper. Freeness is a good tool to estimate tensile and bursting strength[10].

One publication[11] introduces a prototype system that predicts on-line the tensile and tear index based on specific analyzer readings. Accuracy of the estimate is close to laboratory testing accuracy. The models used in prediction undergo continuous updating from the laboratory results.

1.4 Applications of intelligent control

The fuzzy control strategy based on freeness and mean fiber length control with specific energy using on-line freeness and fiber length analyzers is available[12]. Two fuzzy controllers are used in cascaded mode. The first controller observes the fiber length and smoothly begins to raise freeness setpoint if fiber length starts to become too short. The controller checks if the fiber length allows reduction of the freeness setpoint. The second controller keeps freeness on a desired level by changing the setpoint of total specific energy. The controllers use floating averages of on-line freeness and fiber length measurements. Figure 9 shows the control strategy.

Multi variable fuzzy quality controller was tested at the refiner line as an advisory controller. It does not control the process itself. Operators must fulfill the suggestion of the controller. The fuzzy freeness control notices freeness level changes much earlier than operators react to them.

Figure 9. Control strategy of fuzzy quality control[12].

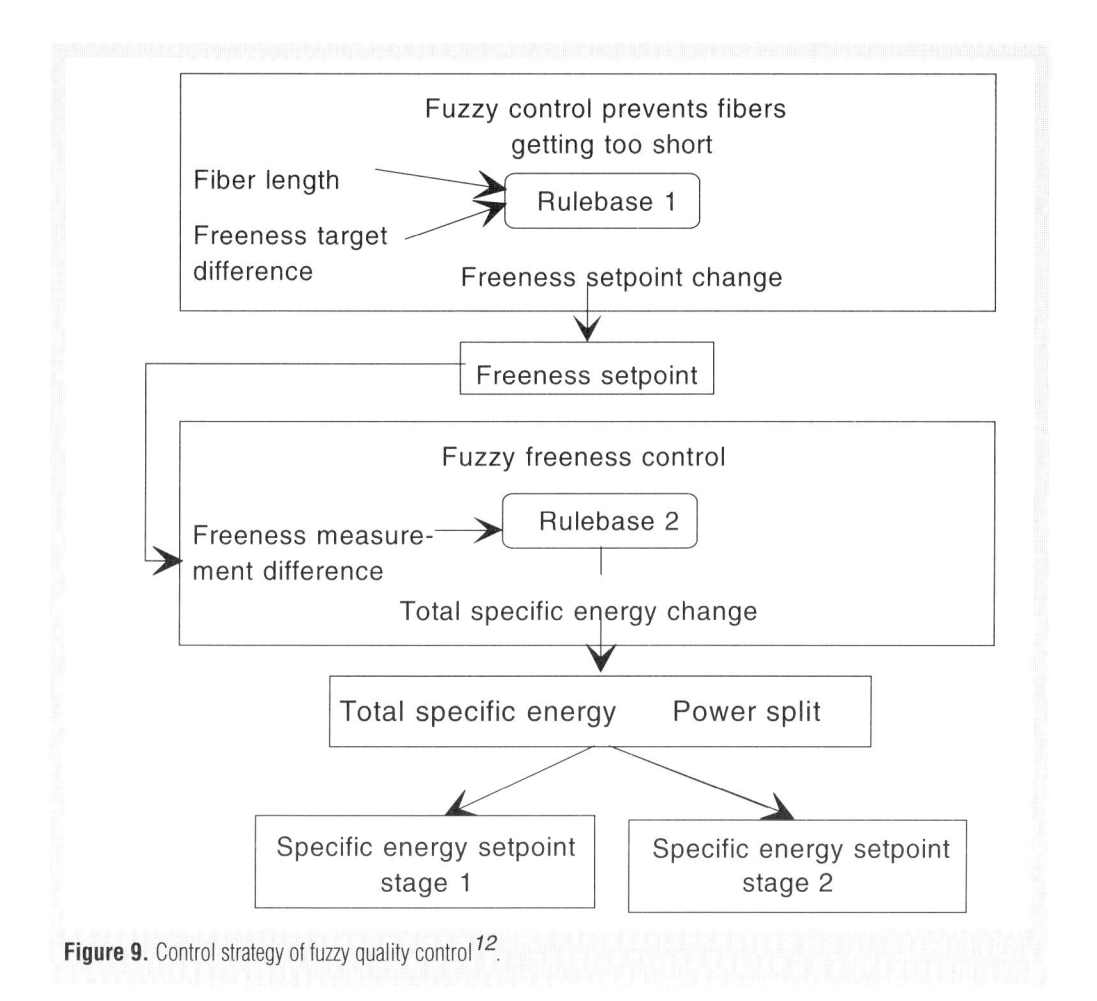

Another publication[1] discusses the application of intelligent methods in refiner control. Their system consists of the following:

- Thermodynamic model that calculates the blow line consistency every 10 s
- Neural network model that predicts pulp quality at 10 s intervals and trains and tunes models in real time
- Fuzzy logic control for freeness and fiber length that moves all refiner gaps simultaneously
- Rule-based system for power saving analysis and maximizing refiner production during off-peak time
- Rule-based monitoring of stability of the refining process
- Neural network model for estimating stock consistency using fiber properties, stock linear velocity, stock temperature, and dilution flows.

CHAPTER 7

Application of back propagation neural network to replace the self-tuning controller in specific energy control is possible[13]. The system is tested by simulations.

1.5 Results

Good results from the refiner control are available in the literature. Consistency variations have decreased by 50%–80% and variations in refiner energy consumption decreased by 30%–70%[3]. The same article also reports tear and tensile strength improvements of 5%–10% and 5%–15%, respectively. Freeness variations decreased up to 70%, and variations in other quality variables decreased as much as 80%. One publication[6] reports 35% decrease in consistency variations at the consistency level of 60%. Other work[1] shows the standard deviation of pulp freeness decreased from 15–30 mL to 3–10 mL. Variation in fiber length also decreased.

Strand et al.[8] reported decrease of freeness variations by 75% and energy savings of 5%.

2 Grinding control

2.1 Control requirements

Kärnä and Liimatainen[14] set the general goal for grinding control as follows: A group of grinders should produce the desired amount of pulp at the specified freeness level with the highest possible quality. Proper control of a pocket grinder requires the following[15]:

- Grinding is done at appropriate target values
- Equal grinding conditions are maintained in both pockets
- Target values are maintained within narrow limits.

The target setting in grinding depends heavily on the selected strategy that depends on desired paper quality. Targets include pulp quality, constant or maximum production, and constant or minimum energy consumption.

A specific feature influencing operation and control of the grinder is the slow dulling of the grinding stone over time and its effects on production rate and quality[16]. Figure 10 shows this behavior[17].

The figure shows two sharpness levels where S1>S2 and two operating points, A and B, with constant specific energy. When the stone is dulling and specific energy must be constant, the operation point is

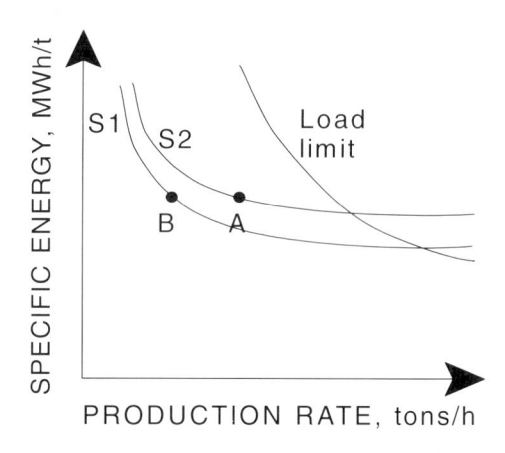

Figure 10. Effect of pressure and sharpness on production rate and specific energy[17].

moving from B toward A. It finally reaches the load limit value. Then the specific energy must increase or the production rate will decrease. Obviously, sharpening the stone will restore the production rate.

Stone sharpening causes another disturbance to the system. It is difficult to sharpen stones to the same degree each time. With constant freeness level, long fiber content and the strength properties vary for a long time after sharpening[18]. According to the article, one should avoid coarse sharpening because it makes the long fiber content remain low for several days.

Besides disturbances from sharpness development, grinding controls are subject to variations in wood properties.

2.2 Control functions

The grinding controls have two hierarchical levels according to Fig. 11: the control of individual grinders and the grinding room controls.

The first integrated grinding control system was developed in Finland in the early 1970s[19, 20]. It tried to control the grinding room as one entity to meet the requirements set for pulp quality, production rate, and energy consumption. Figure 12 shows the basic principles.

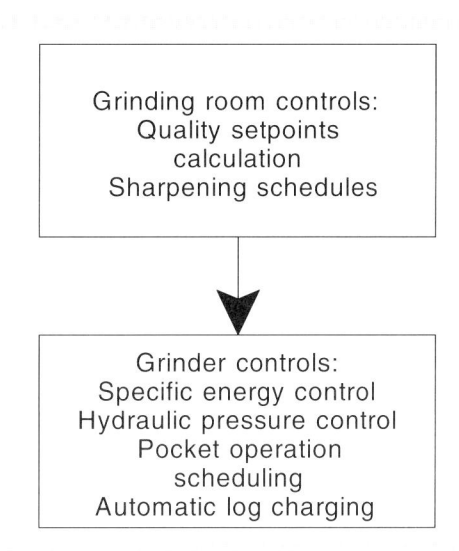

Figure 11. Hierarchy of grinding controls.

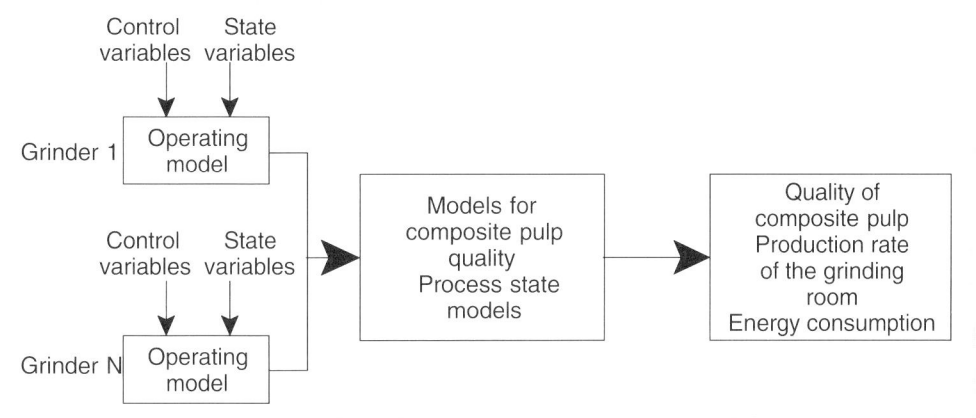

Figure 12. Basic principles in control grinding system[16].

CHAPTER 7

The control of individual grinders uses the dependence between grinding variables, production rate, load, and sharpness according to the following equation:

$$m = SP^{\alpha} \tag{1}$$

where m is the production rate, [tons/h]
          P load, [MW/h]
          S sharpness.

Constant α has a value of 1.4[17]. When graphing the dependence between the load and production rate at constant sharpness, the specific energy consumption results from the slope of the line through the origin at each operating point as Fig. 13 shows[17].

In practice, the grinding temperature and pressure remain constant at certain levels, and the same

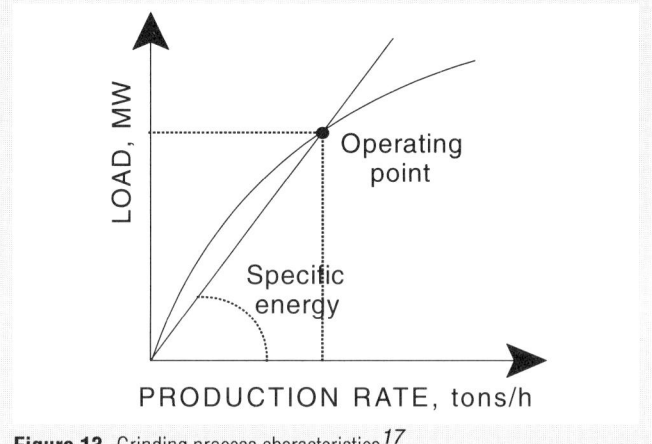

Figure 13. Grinding process characteristics[17].

is true for the flow of shower water, the distribution and practices of showering, and the depth of submersion of the stone. This means that the above mentioned requirements must be fulfilled primarily by one control variable. This is the hydraulic pressure (motor load) in each individual pocket that presses the logs against the grinding stone. In the control sense, speed of shoe can substitute for motor load[14]. This means that two modes of control for a grinder are available:

- Constant motor load

- Constant speed of shoe (rate of production).

Earlier workers[14] tested a third mode that is constant specific energy consumption (CSPC).

Each strategy has a different effect on pulp quality and rate of production[21]. Controlling the motor load yields the largest quantity of pulp but with the largest variations in properties over the sharpening cycle. Specific energy control produces less pulp and also gives reduced variation in quality. Controlling production rate means that pulp is midway between these extremes in quality and quantity.

Modern control systems operate with specific energy control with or without freeness feedback. This requires reliable measurement of the actual production of grinders. The strategies try to maintain a constant specific energy input regardless of the stone

Process control in mechanical pulping

sharpness level[15]. Good controls make it possible to decrease the level of specific energy. This means reduced energy costs or increased production depending on the limiting factors of grinders. Hill has reported 11% savings in energy and a 9% potential production increase[15]. Another approach is also possible[22].

A grinder control mode using constant specific energy consumption was another proposal[14]. Figure 14 shows the basic principles. In practice, a grinder operates at a selected specific energy consumption (SPC in Fig. 14) only when the point of operation remains within the chosen limits of minimum production rate and maximum motor load. Otherwise, the control automatically goes over to the modes of set limits.

Specific energy control considerably reduced the process variations. Figure 15 shows the effect of different grinder control modes on variations in specific energy consumption[14].

Grinding room control is a different but equally complicated problem. Here a balance must occur between the operational targets set for the entire grinding room and the operation of separate grinders. The required production rate and the quality of composite pulp must be fulfilled. The control possibilities are limited to setpoints of quality (specific energy) variables and the sharpening actions.

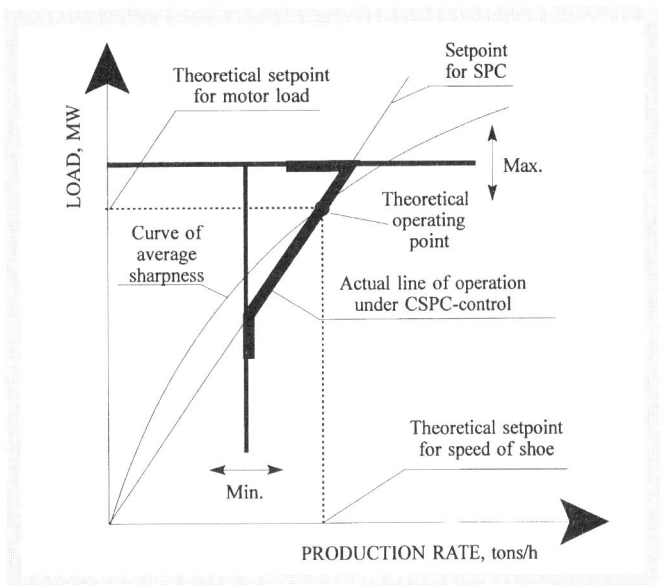

Figure 14. Scheme of operation of control at constant specific energy consumption[14].

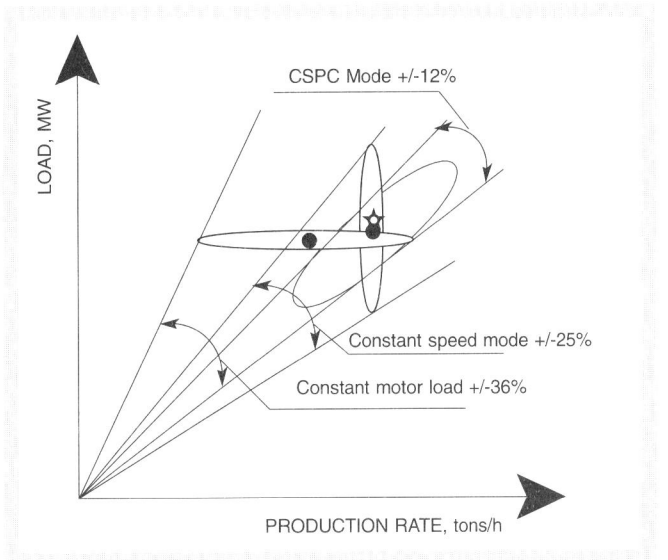

Figure 15. The effect of different control modes on the variation in specific energy consumption[14].

CHAPTER 7

This problem has two parts[17]:

- The effect of these variables on composite freeness, production rate, and load are estimated with predictive models or simulators.
- New targets for the above mentioned control variables are calculated to optimize the operation of the grinding room. The optimization must consider production and its costs (primarily energy costs). This kind of system does not operate in a closed loop manner. The operator actually closes the loop. An example is available in the literature[17]. This has also been the main operational principle of the previously mentioned Finnish control system.

Other researchers used this approach[14]. For a group of grinders, a target area in proportion to the rate of production and quality was formed. The intent was to keep the composite blend in this area. For each stone of the group, corresponding target areas for motor load and rate of production were similarly determined. Figure 16 shows this schematically[14].

The setpoints of the speed of the shoe, motor load, and specific energy consumption are determined inside this rate of production and motor load window. The operation of each stone and the quality they produce requires a separate model. After modeling, the stones receive ranking according to the pulp quality. Certain rules and priorities must be followed. One is that the "best stone is loaded most heavily"[14].

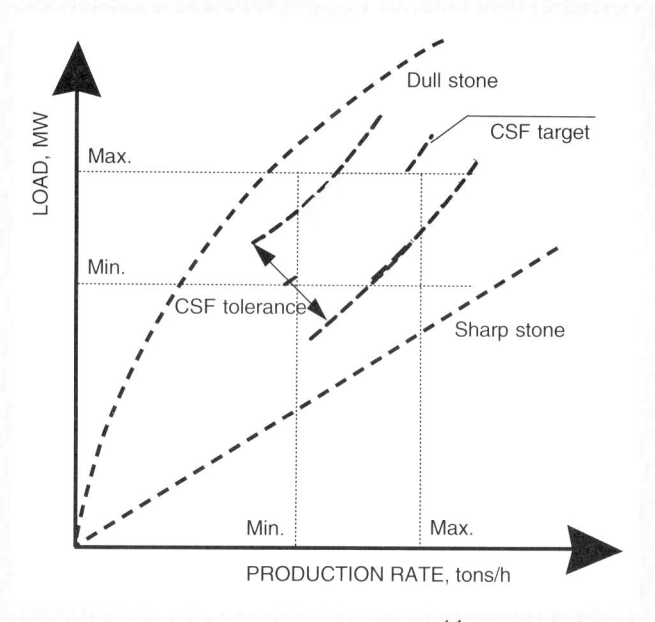

Figure 16. Target window for operation of a grinder[14].

Several factors require consideration in grinding room control. Uniqueness of each stone, variable rates of dulling, and deviations from average area of operation are such factors. Screening, cleaning, and reject refining also require consideration as does the level of automation in these processes.

References

1. Anon., "Oliver's TMP refiner control system." Available [Online} <http://members.tripod.com/~olivers2/review.htm>[Feb. 24, 1998].

2. Leiviskä, K., Sutinen, R., and Saarinen K., 1993 18th International Mechanical Pulping Conference Proceedings, EUCEPA, Oslo, p. 245.

3. Sutinen, R., Saarinen, K., and Leiviskä, K., ABB Review (9):10(1994).

4. Hill, J., Saarinen, K., and Stenros, R., Pulp Paper Canada 94(6):T165(1993).

5. Sutinen, R. and Saarinen, K., 1992 International Conference New Available Techniques and Current Trends Proceedings, SPCI, Bologna, p. 297.

6. Strand, B. C., TAPPI J. 79(10):140(1996).

7. Cluett, W. R., Guan, J., and Duever, T. A., Pulp Paper Canada 96(5):5(1995).

8. Strand, B. C., Mokvist, A., Falk, B., et al., 1993 International Mechanical Pulping Conference Proceedings, EUCEPA, Oslo, p. 143.

9. Ryti, N. and Manner, H., Paperi ja Puu 59(10):640(1977).

10. Kaunonen, A. and Luukkonen, M., TAPPI 1991 International Mechanical Pulping Conference Proceedings, TAPPI PRESS, Atlanta, p. 109.

11. Strand, B. C. and Mokvist, A., TAPPI 1991 International Mechanical Pulping Conference Proceedings, TAPPI PRESS, Atlanta, p. 101.

12. Myllyneva, J., Leiviskä, K., Kortelainen, J., et al., 1997 International Mechanical Pulping Conference Proceedings, STFI, Stockholm, p. 381.

13. Kooi, S. B. L. and Khorasani, K., Tappi J. 75(6):156(1992).

14. Kärnä, A. and Liimatainen, H., Paperi ja Puu 65(2):78(1983).

15. Hill, J., Pulp Paper 55(7):69(1981).

16. Paulapuro, H., Vaarasalo, J., and Mannström B., in Puumassan valmistus II (N. -E. Virkola, Ed), part 1 Teknillisten tieteiden akatemia Turku, 1983, pp. 533–637.

17. Brewster, D. B. and Wells, C. M. in Pulp and Paper Manufacture (D. B. Brewster, Ed.), The Joint Textbook Committee of the Paper Industry, Montreal, vol. 1, 1993, pp. 158–175.

18. Jönsberg, Ch., Hoydahl, H. E., and Solheim, O. J., Pulp Paper Can. 82(5):T180(1981).

CHAPTER 7

19. Ryti, N., Paulapuro, H., and Manner H., *Paperi ja Puu* 55(11):811(1973).

20. Talvio, P. and Korhonen, J., *Pulp Paper Can.* 75(7):106(1974).

21. Anon., Untitled. Available [Online} <http://www.nrcan.gc.ca/es/etb/cetc/facts/cetc02gl.htm> [Feb. 25, 1998].

22. Makkonen, H., Tiikkaja, E., Saloranta, E., et al., 1983 PRP 5 Conference Proceedings, International Federation of Control, Antwerp, p. 175.

CHAPTER 8

Process control in paper mills

1	**Paper machine control**	**147**
1.1	Control requirements	147
	1.1.1 Quality variables	147
	1.1.2 Reasons for variability in the paper machine	148
	1.1.3 Measurement principles	149
	1.1.4 Interactions in paper machine control	150
	1.1.5 Effect of time delays	152
1.2	MD controls	152
1.3	CD controls	153
	1.3.1 Measurement and control configurations	153
	1.3.2 Implementation of CD controls	155
	1.3.3 Elimination of interactions	158
	1.3.4 Basis weight control	159
	1.3.5 Moisture control	165
	1.3.6 Combined CD controls	166
	1.3.7 Web tension control	167
2	**Calender control**	**168**
3	**Coater control**	**173**
3.1	Coating weight control	173
	3.1.1 Scanner synchronization	174
	3.1.2 Dynamic mapping	174
	3.1.3 Control and tuning	175
	3.1.4 Coater startup	175
3.2	CD moisture	176
4	**Stock preparation and wet end control**	**178**
4.1	Stock preparation control	178
4.2	Wet end control	178
5	**Quality inspection systems**	**179**
6	**Grade change automation**	**182**
6.1	Mass production vs. flexible production	182
6.2	Grade change process	183
6.3	Time required for grade change	186
	6.3.1 Mill operations policy	186
	6.3.2 Mill structure	187
	6.3.3 Mill infrastructure	187

CHAPTER 8

6.4 Grade change controls 188
    6.4.1 Alternative approaches 188
    6.4.2 HMPC (Horizon multi variable predictive control) 190
    6.4.3 ADAPLANC (Adaptive planning controller) 191
References 193

CHAPTER 8

Kauko Leiviskä and Timo Nyberg

Process control in paper mills

1 Paper machine control

1.1 Control requirements

1.1.1 Quality variables

Paper machine controls try to keep quality variables at their target levels with minimum variability. Each paper grade has its specific targets and limits for many quality variables such as the following:

- Basis weight
- Moisture
- Caliper
- Ash content
- Smoothness
- Gloss
- Formation
- Strength properties
- Fault distribution.

Most variables are measured on-line and therefore can theoretically have automatic control. Automatic quality control of paper machines has two divisions: machine direction (MD) control and cross direction (CD) control.

Figure 1 shows that several stages in the paper machine process influence the variables[1]. The figure shows that

- One process stage influences the CD profiles of several paper properties
- The CD profile of one particular paper property is influenced in several locations in the process
- The degree of influence depends on the process stage and the particular paper property.

CHAPTER 8

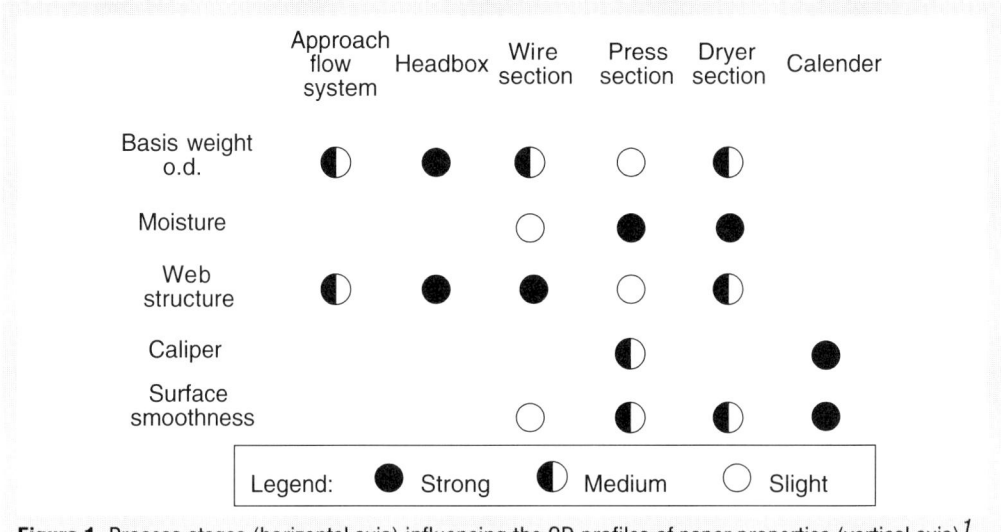

	Approach flow system	Headbox	Wire section	Press section	Dryer section	Calender
Basis weight o.d.	◐	●	◐	○	◐	
Moisture			○	●	●	
Web structure	◐	●	●	○	◐	
Caliper				◐		●
Surface smoothness			○	◐	◐	●

Legend: ● Strong ◐ Medium ○ Slight

Figure 1. Process stages (horizontal axis) influencing the CD profiles of paper properties (vertical axis)[1].

1.1.2 Reasons for variability in the paper machine

Two sources of variability exist that disturb paper machine control. Raw materials entering the process are one source, and the process itself is the other source. The first is due to raw material variability and poor mixing. The second is the result of ineffective control systems, nonuniform fluid flows, mechanical forces, and other technical process reasons in the paper machine.

Compensating for the causes of paper product variability is difficult because the variability spreads over a large frequency area. Table 1 shows some examples[2].

The variability connected with process control starts at seconds and goes to hours per cycle. The most significant sources of variability are level controls and pulper operation. The first is due to the way the level controls are tuned, and the cause of the second is that pulper operations influence consistency control at frequencies not easily controlled[2].

Table 1 gives a slight idea about the frequency range of variations in paper machine operations. Each machine and grade have their own characteristics that influence the process control in their own way. If compensation for variations in the downstream processes in the paper machine is not possible, the variability caused by the raw material variations and other factors will be visible in the final product. Obviously, no single sensor and process control system can cover the entire spectrum of variations that influence paper quality. Several combined approaches are necessary.

Table 1. Paper machine variability from very high frequency to very low frequency range (values calculated for machine speed of 1000 m/min.)[2].

Item	Period	Frequency
Furnish properties		
Width - softwood fiber	1.8–3 µs	0.33–0.56 MHz
Length - softwood fiber	0.18–0.3 ms	3.3–5.6 kHz
Wire mark	0.08–0.15 ms	6.7–13 kHz
PM design/Maintenance		
Mechanical vibration	5–100 ms	10–200 Hz
Hydraulic pulsations	0.02–0.2 s	5–50 Hz
Clothing rotation	2–6 s	0.16–0.5 Hz
Process control problems		
Beta gauge bandwidth	100—500 ms	2–10 Hz
Flow controller problems	3–100 s	0.01–0.3 Hz
Consistency controller	30 s–5 min	0.003–0.03 Hz
Basis weight feedback	12–40 min	0.4–1 mHz
Level controller problems	48 s–1 hr	0.3–21 mHz
Laboratory testing		
Per reels - c/2 reels	90 min	0.19 mHz
Per day - c/2 days	48 h	6 µHz

1.1.3 Measurement principles

Good measurements are necessary for successful control. Several quality variables require measurement and control in the paper machine. Various quality sensors usually exist along the paper machine to control the paper properties in the machine and cross machine directions. Section 4.3 presents the basic operational principles of these sensors.

Figure 2 shows that three possibilities exist for paper sheet measurements. The conventional technique is to measure MD and CD variations by scanning the sheet with a single sensor using a certain speed. The sensor follows a diagonal path with an angle that depends on the relationship between the speed of the paper machine and the scanner. This relationship also defines measurement resolution. This is important in MD control systems. It is even more important in cross machine controls because the statistical significance of the measurement depends on the number of scans included in the profile calculations.

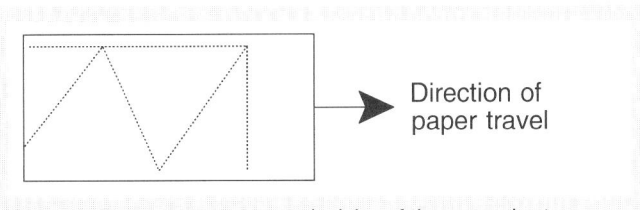

Figure 2. Three measurement principles of the paper sheet.

CHAPTER 8

The second possibility is to use the fixed array type of measurement. This can measure a quality variable across the entire sheet at the same time or scan the sheet so rapidly that the effect of MD movement on measurement is negligible. This kind of sensor improves measurement resolution and also solves some interaction problems existing in CD control systems.

The third possibility is a point-wise, nonscanning measurement in a certain position of the paper web.

1.1.4 Interactions in paper machine control

Two problems make paper machine control difficult from the control engineering point of view: severe interactions between the controlled variables and long time delays for controlling some variables.

In MD control, the most common interaction is between basis weight control and moisture control as Fig. 3 shows. Consider the example when the basis weight controller increases the stock flow. The amount of water to be evaporated increases. Moisture content will also increase. If steam flow increases to correct the moisture, the basis weight will decrease. Control engineering techniques must decouple such an interaction.

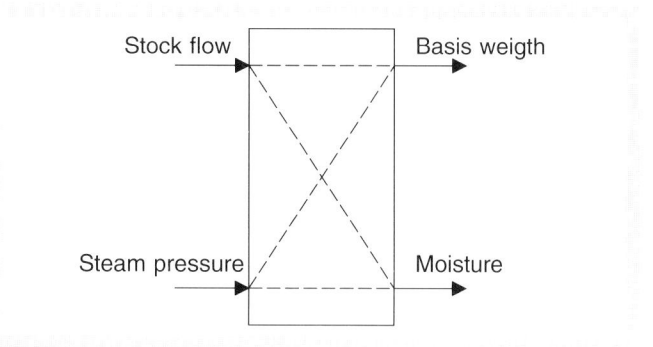

Figure 3. The multi variable interconnected system in basis weight and moisture control.

CD control introduces even more severe interactions. Figure 4 shows a matrix describing the complexity of CD profile control using printing paper as an example[1]. It shows how control of seven controlled CD profile variables on the vertical axis influence CD profiles of six paper properties on the horizontal axis. Obviously, basis weight CD profile control influences all other CD profiles to some degree. CD control of moisture, web structure, caliper, and surface smoothness also has a strong influence on the CD profiles of other variables. The CD profile of printability is the total effect of the CD profile control of all these properties.

The figure shows that an improvement in the uniformity of CD profiles of the paper machine requires the papermaker to minimize or at least avoid the negative side effects of a certain CD profile control to other CD profiles. This is possible as follows[1]:

- Sound knowledge of the interactions leading to these negative side effects

- Knowing the admissible range of CD deviations of the various properties in the final product

- Application of adequate machine components, CD actuators, and control equipment.

	Moisture	Strength	Caliper	Smoothness	Curl	Printability
Basis weight o.d.	◐	◐	◐	◐	○	◐
Moisture		◐	◐	◐	◐	○
Formation		○	○	○	○	◐
Web structure		◐	◐	◐	●	●
Strength					○	○
Caliper		○		●	○	◐
Smoothness		○	●			●

Legend: ● Strong   ◐ Medium   ○ Slight

Figure 4. Interactions between CD profile control of one paper property and the CD profiles of other properties. The influencing profile is on the vertical axis (rows), and the influenced ones are on the horizontal axis (columns)[1].

Another interaction in CD control arises from the principle of scanning measurements. Each scan across the sheet proceeds diagonally. Figure 5 shows that the measurements collected this way include MD and CD variations[3]. Because of this interaction, determining an accurate CD profile with the information of only one scan across the moving sheet is impossible. To calculate an accurate or "true" CD profile, averaging profiles of many scans is necessary to separate the CD component from the composite effect. The most common method is exponential trending. This combines the newest profile measurement with the old "true" profile. In changing situations, this adds an additional delay in the control loop. The delay obviously depends on the scanning speed and the machine width.

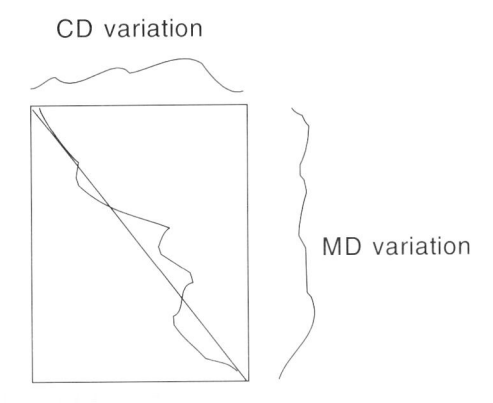

Figure 5. The contribution of MD and CD profiles to the composite variability in paper[3].

Tuning of the exponential filter is an optimization problem[3]. If the filtering is too light, the new profiles are reflected rapidly in the "true" profile. If MD variations increase, they will erroneously occur in the filtered profile. This will lead to the situation that the CD profile control will introduce variability in the final product. If the filtering is too heavy, the time for profile changes to be reflected in the "true" profile will be long, and the profile control will become sluggish. One way to solve this problem is to adapt the filtering factor for each data box separately.

Wallace[3] reported that with faster scanning speed and adaptive tuning of the filtering factor the time to determine "true" profiles decreased by about 25%. A subsequent section of this chapter covers the actuator interaction typical in CD control.

1.1.5 Effect of time delays

Time delays in the control loops depend on the location of actuators relative to the sensor arrangements. Basis weight control is a good example. A thick stock valve before the fan pump usually controls it. The effect of vessels, piping, and process parts before the response is visible on the basis weight sensor. In the tuning of the control loop, this very complex system is usually described with a simple system using dead time and one time constant. The slow dynamics are usually handled by dead time compensating algorithms such as Dahlin's algorithm or by model reference controllers.

The scanner and scan averaging also contribute to the slow dynamic behavior by adding an excessive time delay. Delays can exceed several minutes in slow machines. An additional feature is that the delays are time dependent and change when the machine speed is changing.

1.2 MD controls

Figure 6 shows the general principle of MD controls that has been used since the very beginning of quality control systems for paper machine control[4]. The controller (The type and structure depends on the variable controlled.) calculates the output signal to the actuator using the deviation between the target value and the scan average that describes the state of the paper machine relative to the variable in question. The actuator causes the process to respond, and the scanner measures this response. At the end of each scan, the signal processing averages the scanned profile, and the controller uses the results once again.

The performance of this kind of controller depends primarily on two items:

- Dynamic properties of the feedback loop (As noted earlier, these are limited by long process delays.)

- Scanning speed and the time consumed for the signal processing. (As mentioned above, speeding these two operations improves performance.)

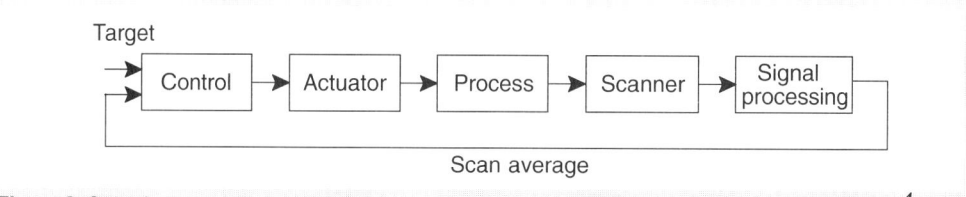

Figure 6. General representation of MD control loop. The detailed presentation depends on the quality variable[4].

Process control in paper mills

This control obviously cannot compensate for the disturbance over the entire frequency range, but its effects are visible for many low frequencies.

As noted, interactions in MD control occur when two or more variables such as basis weight and moisture in Fig. 3 are controlled simultaneously, and the control loops influence each other. Figure 7 shows the general technique to cope with this situation for a two variable case. Crossing arrows from one controller to another denote the multi variable nature of the control. Compensation usually uses the decoupling techniques discussed in Chapter 3 or a good pairing of control variables.

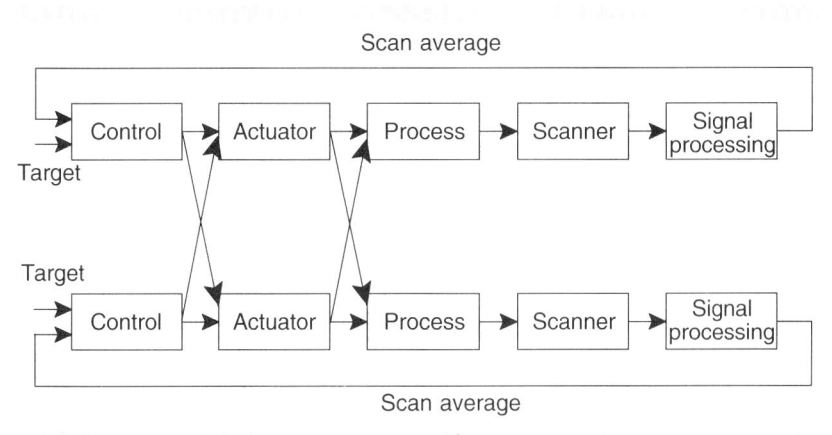

Figure 7. Principle of decoupling two interactive control loops[4].

1.3 CD controls

1.3.1 Measurement and control configurations

Figure 8 shows a typical measurement and control configuration of a CD controller[5]. Data from a scanner is filtered and assembled into a measurement profile. The output signal of the measuring device is usually divided into sections in the cross direction varying from 1 cm to several cm called measurement boxes. The averaging occurs inside the data box, and the box average forms CD profile information used in monitoring and control. The box size is an important factor for system resolution. For reporting and control, the CD profile requires several repeated measurements. The most important factors for accuracy of measurement are the following[5]:

- Number of scans
- Scan direction
- Size and frequency of MD variation
- Signal processing methods (the filter factor discussed above).

There is a conflict situation between the system's ability to detect fast changes and filter out unwanted variations.

CHAPTER 8

For the actuator, important factors are the actuator size, control span, and control resolution. Note that the actuator size (L1) does not have to be the same as the size of the measurement box (L2). Figure 8 shows an example for a one-row actuator. With multi-row actuators, a control block including the load distribution between actuator rows should be added to the above diagram.

A transformation (mapping) algorithm is necessary in cases where the size of the measurement box is not the same as the actuator size. Some profile information is lost in this spatial transformation (see later discussion).

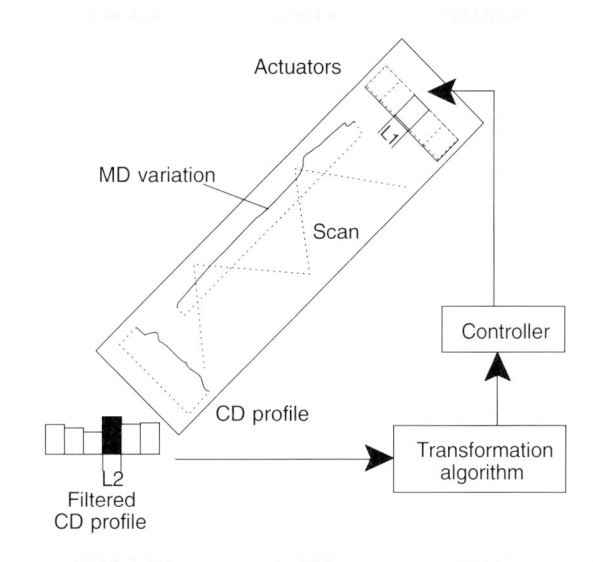

Figure 8. Measurement and controller configuration in MD control system[5].

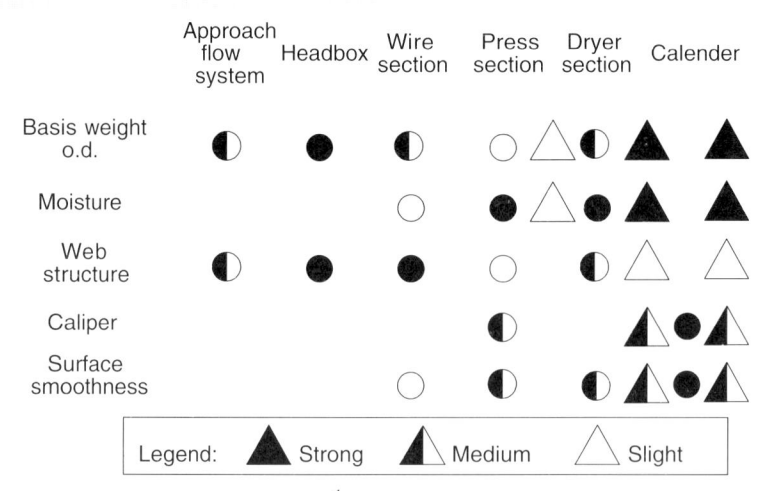

Figure 9. On-line CD profile measuring possibilities[1]. Triangles show good, medium, and poor measurement possibilities, respectively.

Figure 9 shows the measurement places for the profiles of the various paper properties[1]. Many excellent possibilities exist to measure different quality variables on-line. This is obviously a condition for good control.

Process control in paper mills

A major negative aspect is that CD profiles are usually measured only at the end of the papermaking process when the paper is already dry. This means long delays and also few possibilities to contribute to the attenuation of variations. One control vendor is experimenting with the use of a CD weight measurement at the wire of Fourdrinier machines. This measurement would solve the time delay problem. The wire sensor will have to be continuously calibrated using the reel scanner measurement, but preliminary results indicate that this technology will work.

The profile measurements show only the total effects of all influencing variables in the various process stages. No indication of possible negative influences of control interactions is visible. As noted, these interactions always exist.

Figure 10 shows where and by what means the CD profiles of different paper properties are controlled[1].

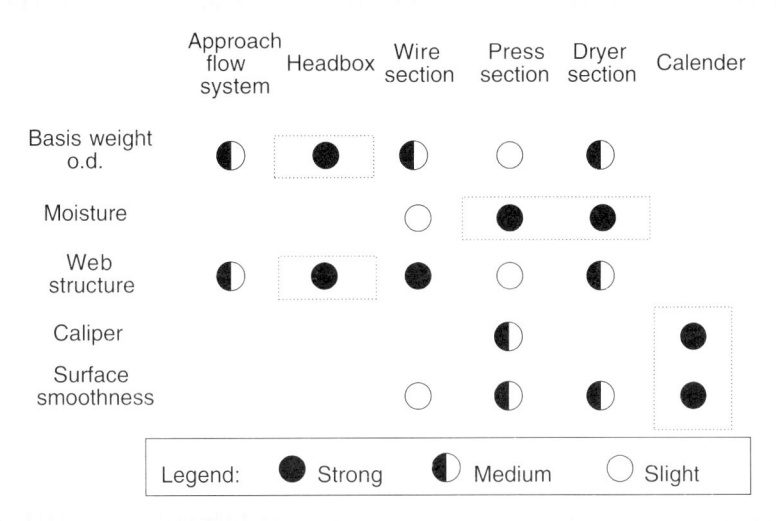

Figure 10. Location of control for CD profile control[1].

1.3.2 Implementation of CD controls

Multiple actuators controlling a row of measurements usually use a corresponding number of controllers (control loops) for regulating the setpoints of the actuators. Each controller and actuator is treated as a single input single output (SISO) application. The most common implementation of CD profile control is essentially a parallel array of single loop MD controllers resembling the one in Fig. 6. Figure 11 shows two adjacent loops in such a CD array[4]. This structure may be compared with the multi variable MD control in Fig. 7. The functions of the controller, actuator, and process are identical. The operation of the scanner, signal processing, and profile processing blocks are different.

155

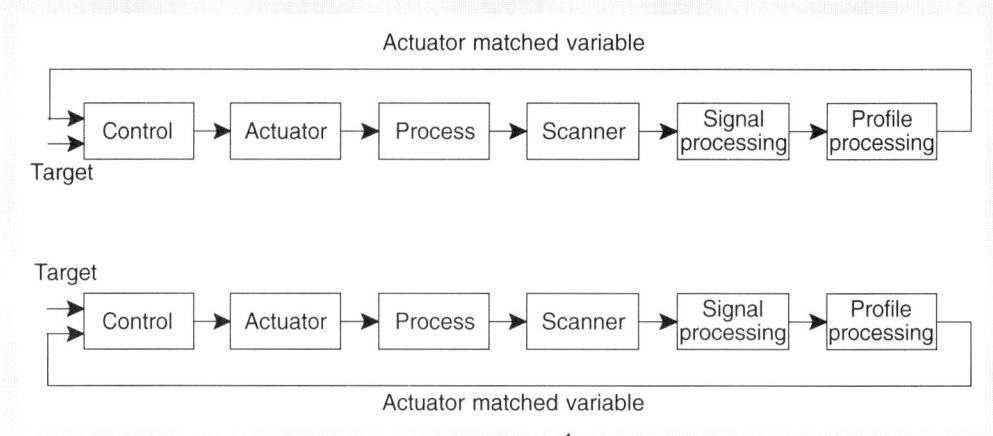

CHAPTER 8

Figure 11. Diagram of the operation of the single loop CD control[4].

The scanner samples the variable by collecting the data over the width of a predetermined measurement box. The signal processing block averages these values over the indicated zone and saves the respective value until the end of the scan. Subtraction of the scan average then forms a deviation. These deviation values from successive scans are filtered to compensate for short-term variations.

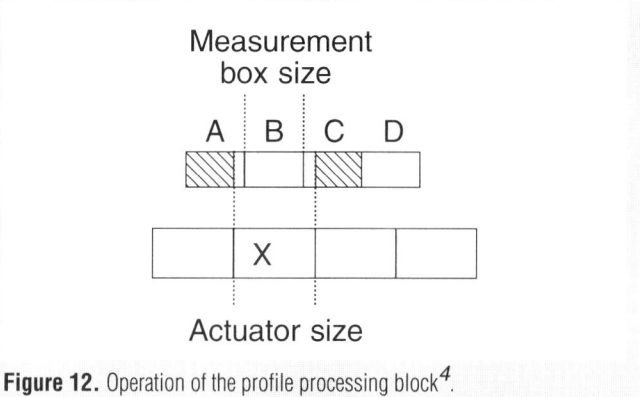

The filtered value goes to the profile processing block that calculates the actuator matched variable. The calculations are complicated by the fact that the size of the measurement box and the actuator size are usually different. Then the profile processing block must calculate the average profile for the actuator size using one or more adjacent measurement boxes as Fig. 12 shows[4]. In Fig. 12, three measurement boxes are averaged.

Figure 12. Operation of the profile processing block[4].

This simple and direct control approach is adequate for taking some performance benefits expected from using multiple actuators. This simple strategy has some shortcomings[6]:

- Different actuator and measurement resolution and response width
- Contradicting and interacting control efforts
- Actuator saturation and different actuator dynamics.

If these issues do not receive proper handling, the results may appear as undesirable transient response, controllable profile errors left uncontrolled, or wasted actuation energy[6].

The situation in Fig. 12 is an example of actuator and measurement interaction. It arises because the width of the measurement box does not exactly equal actuator size. The error caused by this is greatest when the measurement width is greater than the actuator width. The error decreases as the ratio between actuator size and measurement box increases. This means that the size of the measurement box should be as small as possible. Some factors limit the reduction of the box size[4]:

- Sensor "beam" size
- Sensor noise statistics
- Signal processing electronics
- Scanner speed
- Scanner alignment stability
- Computer speed and memory
- Display resolution.

Another interaction occurs between the adjacent actuators as a "spill-over" from one actuator to another. This is spatial interaction or actuator/actuator interaction[4]. This interaction is negligible when the sheet is fully formed, and the actuator operates directly on the sheet. Correspondingly, Fig. 13 shows that this effect is very high in the headbox area controls. Major interactions require more extensive modeling and tuning work. In practice, the existence of this interaction also smooths the response. Spatial actuator interactions are discussed again under CD basis weight control.

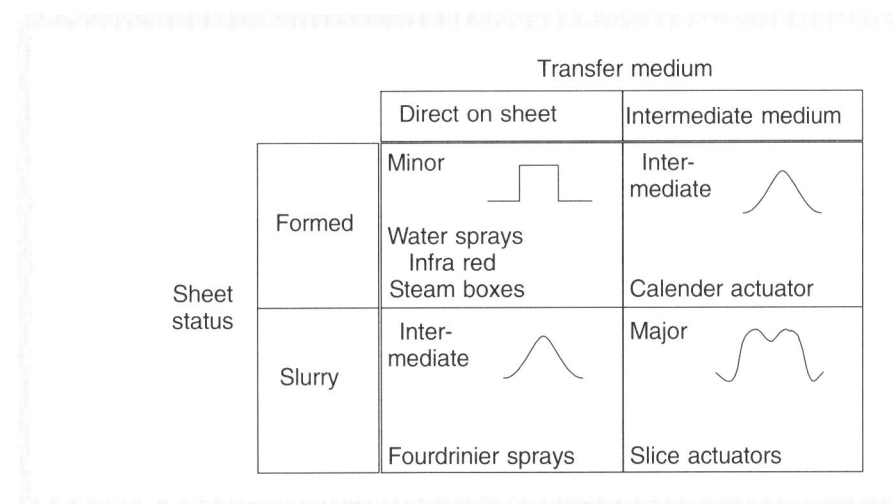

		Transfer medium	
		Direct on sheet	Intermediate medium
Sheet status	Formed	Minor Water sprays Infra red Steam boxes	Intermediate Calender actuator
	Slurry	Intermediate Fourdrinier sprays	Major Slice actuators

Figure 13. Actuator spatial interactions[4].

CHAPTER 8

1.3.3 Elimination of interactions

Interactions between multiple CD actuators are addressed by introducing coordination techniques between the multiple feedback control loops. The coordination of multiple CD control loops is a higher level function in the hierarchy of CD control. It is used in the form of profile prediction, actuator midranging, and control synchronization. The coordination techniques are extended from the block diagram illustrated in Fig. 11. The control coordination does not require redesign of the feedback controllers. The CD controller in Fig. 11 remains the same and continues to provide feedback control[6].

Estimators predict process changes resulting from each individual closed-loop control action. The estimator output decouples control actions of the multiple actuators; the prediction of a controller output is used as a feedforward signal to other control loops. This minimizes or eliminates duplication of control efforts and gives way to more aggressive tuning of individual control loops[6].

Figure 14 illustrates the use of prediction estimators with standard application of feedback control.

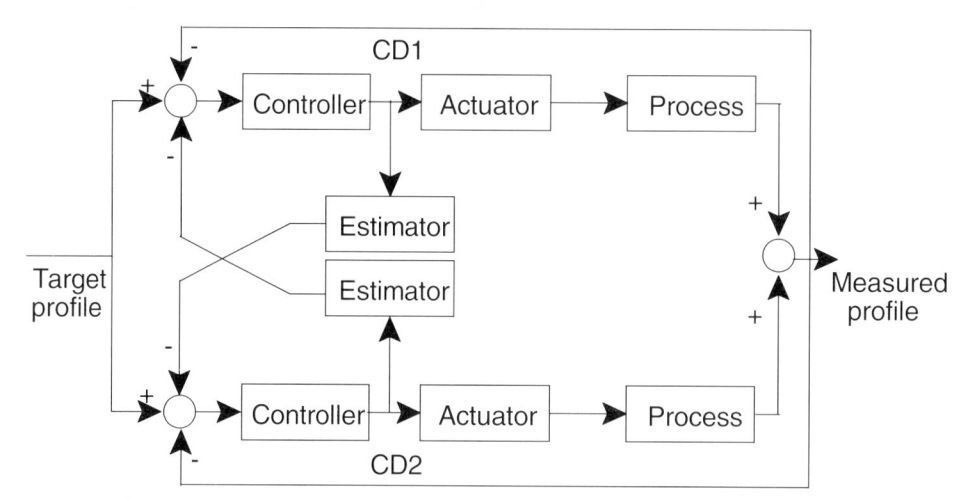

Figure 14. Control coordination using prediction estimators[6].

MD and CD effects of the process response must be separated in the estimator design. The CD effect is modeled by mapping all the profile data points to the actuator set points as Fig. 12 shows. The MD effect is modeled by relating the individual profile data points to the actuator setpoints. A diagonal transfer function matrix usually represents this relationship. The simplest representation is first order dynamics with dead time transfer function[6] (See Chapter 10.2.8.).

Controller synchronization means a sequential execution of the control loops. Within any given computational cycle, a control loop estimates its predicted profile and feedforwards it to all the recipient controllers. When the next control loop executes, it can therefore decouple the control actions of all the loops that have already executed

Process control in paper mills

within the computational cycle. For example, an actuator with a wide response may be allowed to execute before an actuator with a narrow response[6].

In all practical applications, an actuator has a finite operating range. Saturation can occur if the actuator has an insufficient size for the magnitude of the process error. The appropriate solution is to midrange the actuator when it approaches saturation, i.e., move the saturated setpoints away from the limit while maintaining the setpoint shape of the actuator[6].

1.3.4 Basis weight control

Slice lip control

The CD profile of basis weight is usually controlled at the headbox by local adjustment of the slice opening and sometimes by variation of the distribution overflow. Side effects such as change in local jet velocity and change in local jet direction can have negative effects on other paper properties.

A traditional way of performing cross direction basis weight control is to change the position of the slice lip of the headbox in the cross machine direction. Several methods exist: motors, thermal and hydraulic actuators, and motorized robots. Thermal and hydraulic actuators use thermal expansion. The first type uses a metal rod, and the second one uses a fluid.

Although the slice lip is the most suitable method, it has certain characteristics that require consideration in control. The most serious is the interaction between neighboring slice lip positions as shown in the weight profile of Fig. 15[7]. This coupling between adjacent control positions can change according to the location across the machine and time. It can also influence other variables such as the orientation angle.

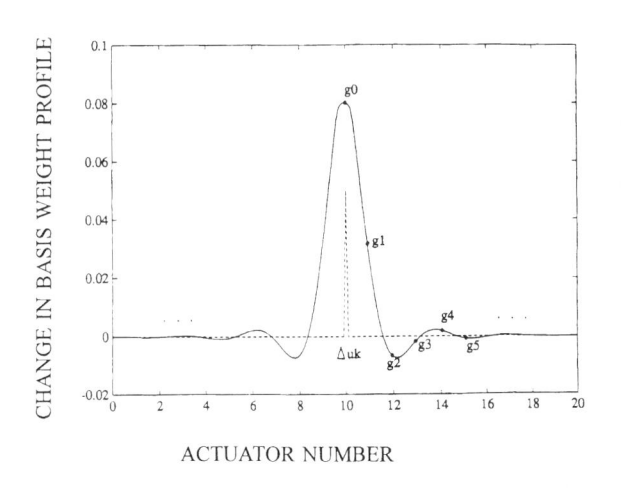

Figure 15. Effect of slice lip change on basis weight[7].

For good basis weight control, precise information on the position of the slice lip in the cross machine direction is also necessary. Using the slice lip with care is important to avoid causing damage by forcing the slice into a position where it becomes permanently deformed and to prevent continuous cycling of the slice position[8].

CHAPTER 8

Decoupling control is necessary to eliminate spatial interaction between adjacent slice lips. The following simple model is suitable[8]:

$$\Delta y = A \Delta u \quad (1)$$

where Δu are changes in slice position at each actuator
        Δy changes in the weight profile at corresponding positions.

If the responses to all actuators are similar and symmetrical around the actuator positions, the matrix, A, will be band diagonal and symmetric as follows:

$$A = \begin{bmatrix} a & b & c & 0 & 0 & 0 & . & . & . & 0 \\ b & a & b & c & 0 & 0 & . & . & . & 0 \\ c & b & a & b & c & 0 & . & . & . & 0 \\ 0 & c & b & a & b & c & . & . & . & 0 \\ . & & & & & & & & & \\ . & & & & & & & & & \\ . & & & & & & & & & \\ 0 & . & . & . & b & & & c & b & a \end{bmatrix} \quad (2)$$

Here, a, b, and c denote the profile response to slice changes at five adjacent positions. Calculating the control uses the following equation:

$$\Delta u = k A^{-1} \Delta y^* \quad (3)$$

where Δy^* is   the desired change in profile
        k a scaling gain factor usually less than 1.

The CD basis weight control is complicated by the fact that A matrix depends on grade, machine speed, wire conditions, stock drainage, and headbox conditions. Model identification is necessary, and self-tuning methods are essential in actual control.

The fact that basis weight influences most sheet characteristics makes CD weight control a prime candidate for coordinated control. Figure 16 shows a basis weight control algorithm that allows for two inputs[9]. These two inputs can be weighted to emphasize one effect of the weight control or another. The use of integral control in the weight feedback prevents any long-term error from developing in the weight profile and simultaneously permits the second input to adjust the weight profile within the dead band to achieve a second objective. This scheme for controlling moisture and weight has had use for many years with good results.

Figure 16. Multiple input CD weight control[9].

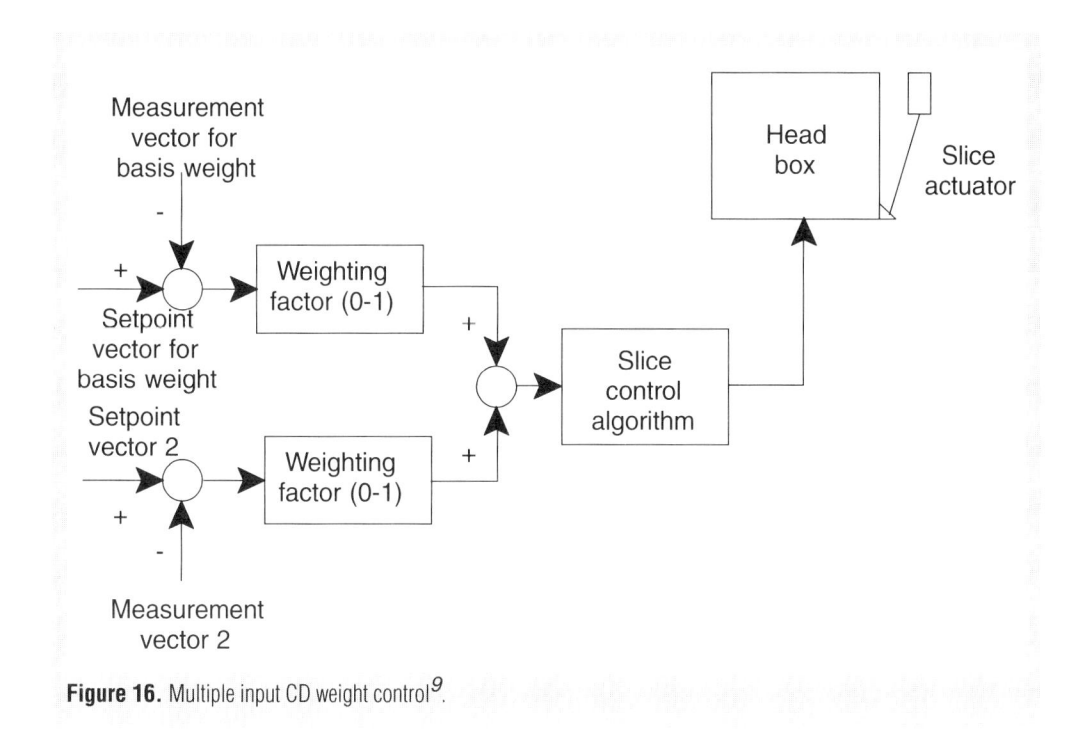

Dilution headbox control

Dilution headboxes (or consistency profiling headboxes) present several advantages over conventional headboxes for control of dry weight profile. In particular, the dry weight profile response to a local dilution change is narrower than a response to a slice lip movement. It is also uniform in sign. Dilution zones are commonly 40–60 mm wide, and slice actuators are usually at 75–100 mm intervals. The dilution system therefore has much better bandwidth for control of dry weight profiles, but it requires more accurate mapping information[10].

In contrast to the slice lip control that changes the amount of material locally discharged in the jet, dilution control changes the composition of the material. When using white water dilution, the dilution profile may therefore influence the ash profile differently from the dry weight profile especially at high dilution ratios. No such effect exists with clear water dilution[10]. Large differences between adjacent dilution flows can also have an impact on the quality of sheet formation.

Mapping of actuators to profiles is more demanding in dilution systems than in conventional slice control due to the closeness of dilution zones and the narrowness of responses. Accurate measurement of the sheet edges is imperative at the scanners and any trim positions. The edges require measurement simultaneously at both sides of the sheet especially on wider machines where some web drift may occur during a single scan. The accuracy must be better than one-quarter of a profile cell. Better than one-eighth of a profile cell is preferable[10].

CHAPTER 8

By controlling the weight profile with dilution, introducing fiber orientation to avoid dry weight problems is no longer necessary. This provides considerable improvement in fiber orientation profile compared to the traditional headbox where the dry weight profile control used the slice lip shape. Use of dilution to control the dry weight profile does not guarantee good fiber orientation profile. Thermal deformations and minor defects in the headbox and former can cause local jet misalignment. This misalignment can vary with operating conditions. Jet misalignment leads to fiber orientation problems especially at low rush to drag values.

Consider the results of dilution headbox control for a 7.9 meter supercalender (SC) paper machine[10]. The SC machine was started with manual control of dilution and then immediately switched to automatic when no further manual improvements were possible. Figure 17 shows the best dry weight profile achieved with manual control of dilution. The figure also contains a typical profile from the first day of automatic control[10].

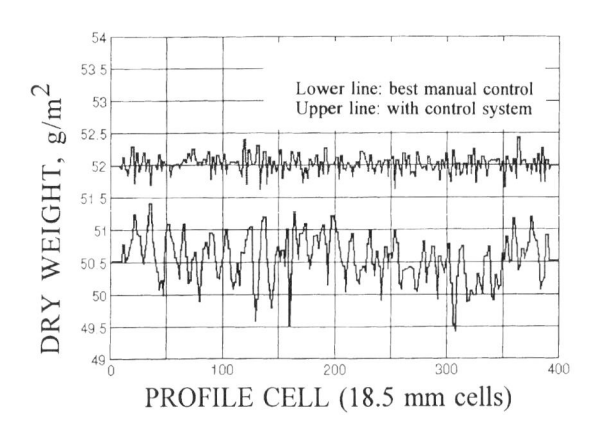

Figure 17. High resolution dry weight profiles in manual and automatic control[10].

Most dilution headboxes have a slice lip that has automatic adjustment in a coordinated manner with the dilution actuators. The slice lip can finely control fiber orientation by adjusting the local jet angle. The dilution system can easily compensate for consequent changes in dry weight.

Figure 18 shows an example of such a system[11]. This controller allows automatic decoupled control of a laboratory or on-line measured fiber orientation profile and the measured weight profile. For laboratory measurements, the fiber orientation profile is entered by the laboratory and converted by the control system into the appropriate resolution objective function. The decoupling vector between fiber orientation and weight changes is a function of CD position and is determined during the control commissioning and tuning phase using a set of on-line control tuning tools. This vector can be compensated as required for such operating factors as flow, speed, slice opening, etc.

Another possibility is to simplify the algorithm[12] shown in Fig. 19 that allows control of the fiber orientation profile alone by manipulating the shape of the slice lip of the dilution headbox with a prediction of consequent changes to the dry weight profile. The dilution system can easily compensate for moderate slice-induced changes to dry weight.

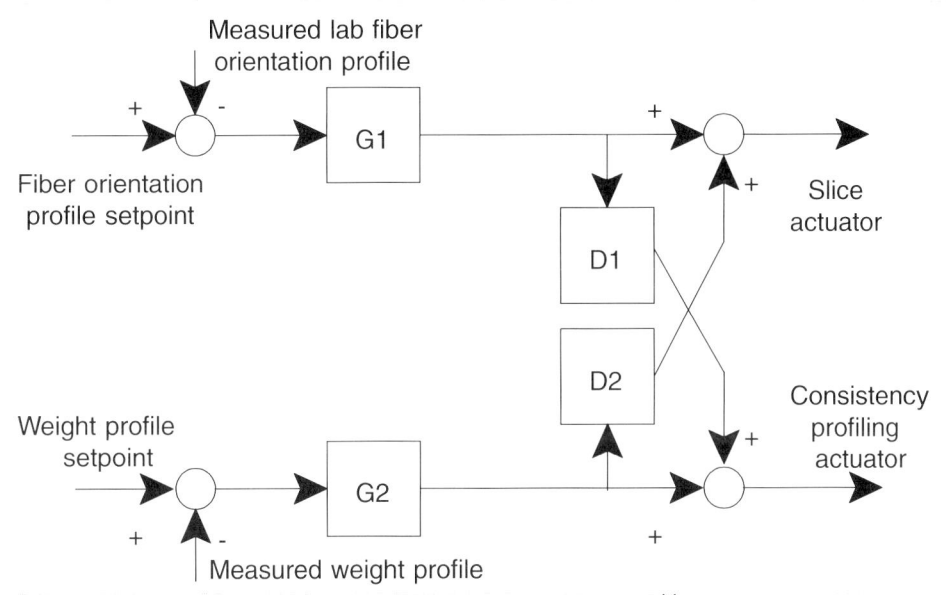

Figure 18. Coordinated, decoupled control of fiber orientation and weight profile[11].

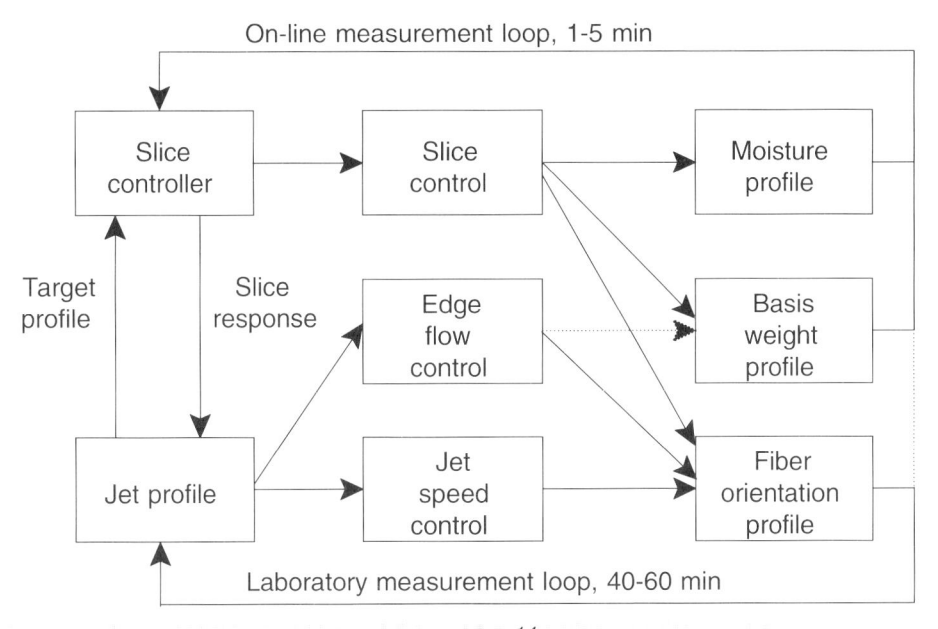

Figure 19. Control system for basis weight and fiber orientation[14].

CHAPTER 8

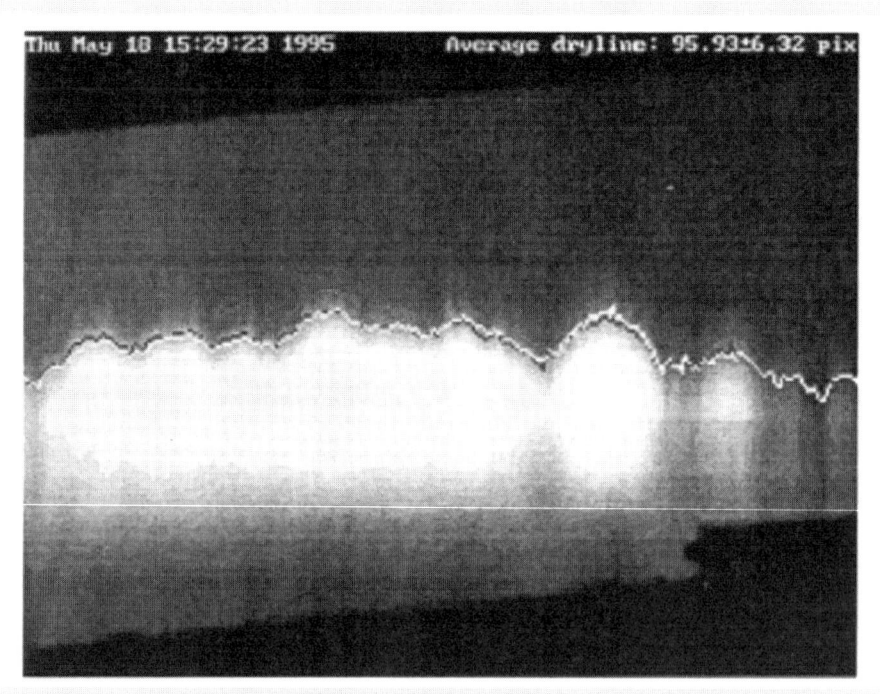

Figure 20. Rectified, smoothed wire image showing the dry line. Wire runs upward in the picture, and the dry line is indicated by the white trace. The average position of the dry line is given in picture element coordinates[13].

Consistency profiling headboxes can achieve improved profiles over conventional slice lip control provided the entire system has proper design, implementation, and control. Proper prediction of performance requires high-resolution profile analysis and consideration of the nature and distribution of streaks and process variability[11].

Dry line in basis weight control

Viewing pulp on a fourdrinier wire with the naked eye to catch early information gives reflections of light that reveal parts of the dry line. This line relates to the disappearance of liquid water from the pulp surface. Although such observations are never complete and cannot be properly quantified, all process operators at all levels of machine automation try to trace it and then adjust the control inputs available to them[13].

Two methods have been reported to detect the dry line with camera systems and image processing methods[13].

In the first method, the light sources and camera are located so that virtually no specular reflections from the wire hit the camera. Instead, the camera receives light from the pulp downstream of the dry line. This has a matte surface of diffuse reflectivity. The dry line is observable as the transition from dark to light surface[13].

In the second method, the wire is not illuminated directly by the primary sources of light but indirectly by a large surface of low, homogeneous luminosity. The specular reflection of this surface by the wet pulp is effectively used for formation of the image in the camera. Drier pulp after the dry line diffuses the light in all directions above the wire transmitting less light into the camera. In this case, the dry line is detected as the transition from light to dark surface[13].

Figure 20 shows a typical, geometrically realigned terminal image displayed to the operator in the control room with the digitally extracted dry line data on the average dry line location. It also includes the time of the last image[13].

The system reported in[13] uses a standard 0.5 in. Charge-Coupled Device (CCD) camera. The detection algorithm of the system has a modular structure to consider the different paper machine environments and the instrumental implementation. After the geometrical realignment, the resolution of the total image is 128 x 128 picture elements, and the momentary dry line data therefore consists of 128 points. Each picture element corresponds to 4 x 5 cm^2 rectangle on the wire.

Tests have shown that dry line measurement immediately after a disturbance has entered from the headbox can predict changes of the dry basis weight profile. The two dry line measurements can apply to the feedback control of the dry line location in a straightforward way. In CD control, the adjustments necessary at the slice lip are solved by deconvolution in terms of the CD coordinate. This is true if the responses of the dry line to the settings of the individual actuators are known whenever a change of the dry line is detected by measurements. The dry line profile control can be extended to cascade control of the dry basis weight profile. In both cases, the actuators would be the same as those used today for control of basis weight and its profile, i.e., the stock valve and the slice screws. Dilution water and valves controlling its distribution also seem to apply to dry line control equally as for control of dry end quantities[13].

Dry line information will essentially speed the correction of errors in the final quality variables because its speed depends only on the time taken by the image analysis and the control algorithm. The time of transport on the wire influencing the existing controls is eliminated. The performance of control systems today can obviously be improved by incorporation of the dry line measurement and control facility[13].

1.3.5 Moisture control

CD moisture profile control in the press section uses rolls with a crowning that can be adjusted to obtain the required line force profile. Steam boxes that heat the paper in sections to improve the local mechanical dewatering can also control CD moisture. Sectionally adjustable blow boxes in the drying section control the amount and location of air blown into the web. Use of segmented infra red dryers is also possible. Water sprays can also rewet the web.

CD moisture profile control influences local CD dry content of the paper sheet. This results in local changes in stress and strain and can lead to curl problems in the final product. Density changes also influence any density dependent quality variables such as stiffness, porosity, and strength properties[1].

The interaction of the various moisture CD control actuators is well understood. A machine with steam box, water spray, and segmented IR dryers can achieve profile correction with any of these devices. Cost and resolution balance applies when using these actuators. For example, the steam box is a cost effective method of achieving better drainage in a section of the machine wire or in a press. The zone width is typically several times wider than that of a water spray[9].

Coordination of the CD direction moisture control can use a steam box on the wire and a water spray at the end of the dryer[9]. The reported system gave an improvement in the profile from 1.03 two sigma to 0.83 two sigma while reducing the steam use by 4000 lb/h. In this case, the water spray achieved the high-resolution control, and the steam box gave reduction of general areas of high moisture.

1.3.6 Combined CD controls

Changing the slice lip profile on a conventional headbox changes the basis weight profile and the fiber angle profile. Other control devices on the headbox such as edge flows also influence these properties. A further coupling results from the shrinkage profile on the machine.

To produce a flat basis weight profile at the reel, the weight profile on the wire must be lighter toward the edges to compensate for uneven shrinkage. Figure 21 illustrates this[14]. The edges are closed, and the middle of the slice is open more. This puts less stock near the edges and more in the middle. It also causes jet misalignment that causes the fibers to become misaligned especially near the edges of the machine. This further influences stiffness, strength properties, curl, and dimensional stability of the paper. Jet velocity variations also cause CD and MD variations in paper tensile strength.

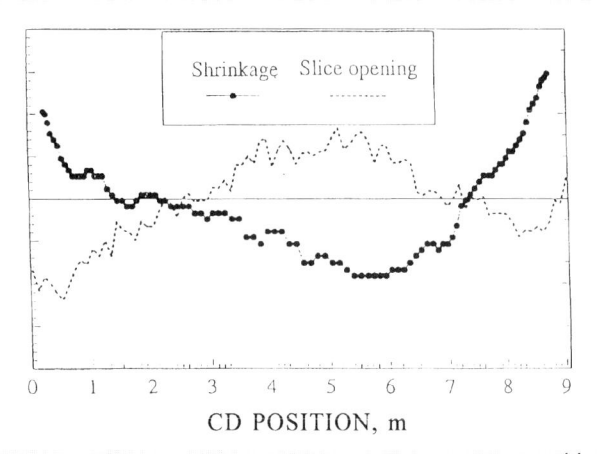

Figure 21. Relationship between slice opening and shrinkage profile[14].

Figure 22 shows the mechanisms leading to interactions in basis weight and fiber orientation profiles.

Figure 19 showed the principal block diagram of the CD control system that controls fiber orientation and basis weight. The control algorithm uses the simplified physical model presented in the previous diagram. The slice profile influences jet thickness profile and therefore changes basis weight. Simultaneously, it also changes the jet misalignment and velocity profiles influencing fiber orientation. This effect is usually omitted, but this system considers it since slice response influences control of the jet profile. On-line optimization produces the best compromise between fiber orientation and basis weight.

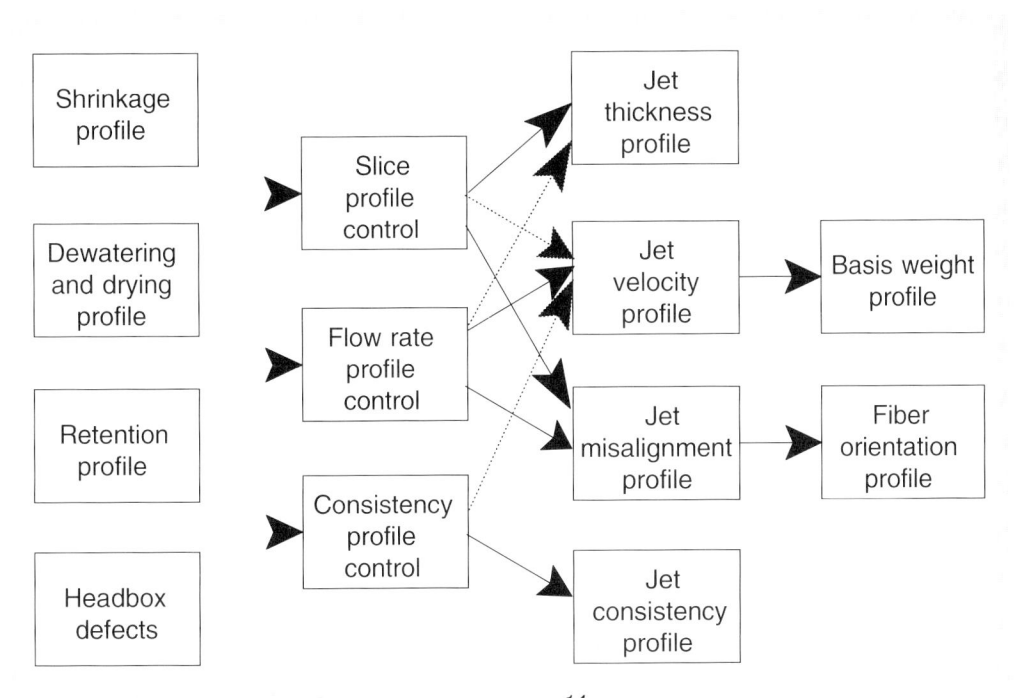

Figure 22. Interconnections of basis weight and fiber orientation [14].

1.3.7 Web tension control

Several web tension measurements are available [15]. One system measures the counter-pressure of compressed air blown to the web. This correlates with web tension. Moving the sensor measures the web tension profile. A hand-held tension meter measures the reaction force of the web when the meter presses against the web. One system [16] uses a laser beam to measure the passing time of a propagating membrane wave in the web. The system can be used independently of other measurements. The only information it needs is basis weight. The system optimizes the frequency and amplitude of the sound burst for every paper quality and converts the propagation delays to corresponding absolute tension values. The sensor mounts on a specially constructed beam that allows scanning the sensor across the web. This provides easy web tension profile measurement. Another unit uses a machine-wide roller divided into several rotors. Each rotor has an independent air bearing. Tension profile results from the load indicated by each rotor.

Another measuring method has a beam used to deflect the web from its normal direction [17]. The measurement is not in direct contact with the sheet because of a thin layer of air between the running web and the rounded surface of the beam. The pressure of this air layer is measured through several small perforations of the beam connected to the pressure sensors. The pressure correlates with web tension. Selection of the distance between two measuring points in the cross direction can be any desired value.

Testing of this system on a pilot winder used different paper grades: newsprint, LWC-base, and coated fine paper. Trial running speeds were 600–2000 m/min[17]. The linearity was especially evident. Paper properties such as basis weight, porosity, and air permeability do not noticeably influence the method. The system can measure rapid dynamic changes. Two on-line installations tested the measurement system. The first prototype used a newsprint paper machine, and the second system used a printing machine[17].

Machine direction web tension is a critical runability parameter for a paper machine. An uneven tension profile can lead to runability problems on the paper production line and at the printing press. The most common problems are sheet breaks and wrinkles caused by slack edges of the paper web[17].

The primary factors influencing web tension profile are the moisture profile and its evolution in the drying section. An effective moisture control requires measuring stations immediately after the press section and a measuring technique that can cover the entire web at the same time[15].

The supercalender also has a strong influence on the tension profile and whether the plastic deformation of the paper web and the changes in the structure of the paper will change the profile. Differences in plastic elongation may also influence tension. After supercalendering, differences in stress relaxation on the reel influence the tension profile[15].

CD control of web tension requires a control strategy that includes coordination of several CD controls and simultaneous optimization of several paper properties. Similar strategies to that shown in Fig. 19 are possible.

2 Calender control

Many paper properties change substantially in calendering or supercalendering operations. The thickness (caliper), roughness, gloss, porosity, and printing properties all change, but they cannot change independently in normal calendering operations. These properties correlate strongly. A change in one influences all the others. Viewing the calendering operations using one variable and determining the others through their relationships with this primary property is therefore important[18]. A convenient choice is the bulk (thickness divided by basis weight). Most modern paper machine operations measure thickness, basis weight, and moisture.

Nip load, paper temperature, and paper moisture content influence the reduction of thickness or bulk in a calender or supercalender. An increase in any of these variables will lead to reduced bulk and roughness, increased gloss, and usually to improved printing properties. Nip load and temperature can be changed selectively across the width of the calender. They are used to control the CD uniformity of bulk reduction. Moisture content is not usually manipulated in the calender, but moisture variations in

Process control in paper mills

the web entering the calender with basis weight and temperature variations can be a major source of disturbances that a CD control system must eliminate[18].

The calender itself can contribute to CD thickness variations. Calender rolls may sag under their own weight and the weight of the rolls above. They can deform with an externally applied load. The weight of journals and bearings also causes the roll to deflect. CD variations in roll diameter can be due to faulty grinding or a poorly designed roll heating system.

Cooling or heating of calender rolls and locally changing their diameter controls the CD thickness profile of paper in a machine calender. Cold air blows onto the calender where thicker paper is desired, and hot air applications are at locations where thinner paper is required. Air is heated selectively by electric heaters at the desired CD positions. Air temperatures as high as 300°C find use. Another method is the use of induction heaters where high frequency current passing through an actuator coil creates a varying magnetic field that generates alternating current in the roll. The dissipation of these currents heats the roll[18].

Air showers have not found use in CD thickness control of supercalenders. The reason is that nip load change generated by heating or cooling is too small to be effective if one roll is soft. Variable crown rolls with CD zone control also have use. The force applied on the journals of the rolls adjusts the average thickness. Figure 23 shows an example of using variable crown rolls in control. New methods for zone controlled rolls are available[19].

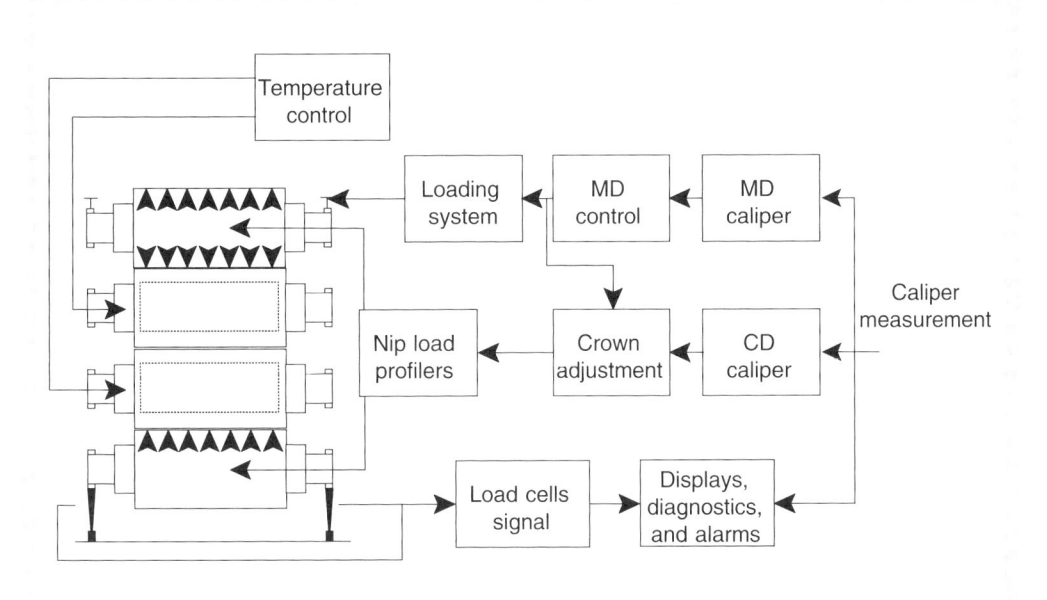

Figure 23. Variable crown rolls with CD zone control used in caliper profile adjustment[18].

CHAPTER 8

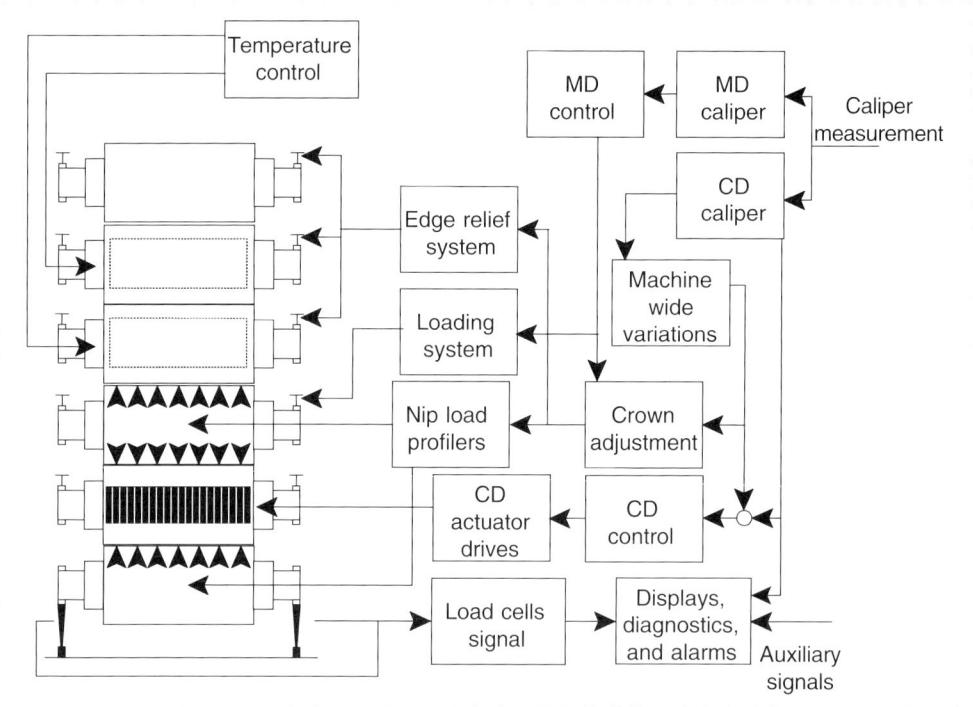

Figure 24. Machine calender equipped with CD and MD caliper control systems[18].

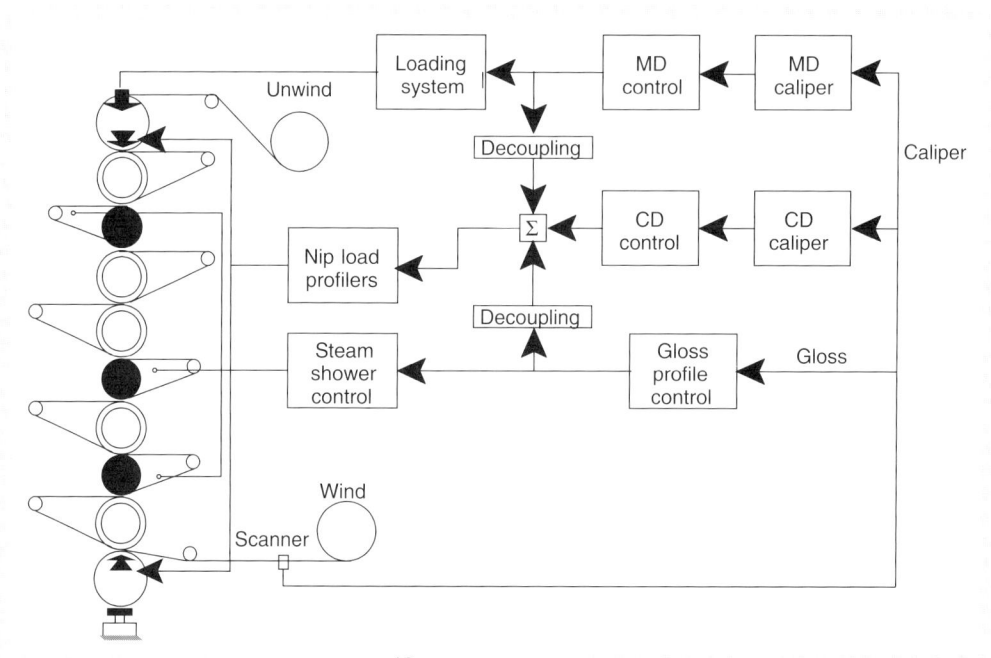

Figure 25. The supercalander control system[18].

Process control in paper mills

Figure 24 shows an example of a machine calender with MD and CD controls. Several possibilities exist to change the average thickness of the paper leaving the calender:

- Change the temperature of the calender rolls
- Change the load in the two bottom nips by loading or relieving the intermediate variable crown roll.

Figure 25 shows a supercalender equipped with caliper and gloss control systems. Squeezing the calender between the top and bottom variable crown rolls controls the MD and CD caliper. Applying steam showers selectively on each side of the web adjusts gloss. Figure 26 shows a gloss control system. It also uses steam showers as manipulated variables. When the steam pressure increases, the profile control range increases. To maintain profile control stability, the CD controller gain automatically retunes each time the pressure changes[20].

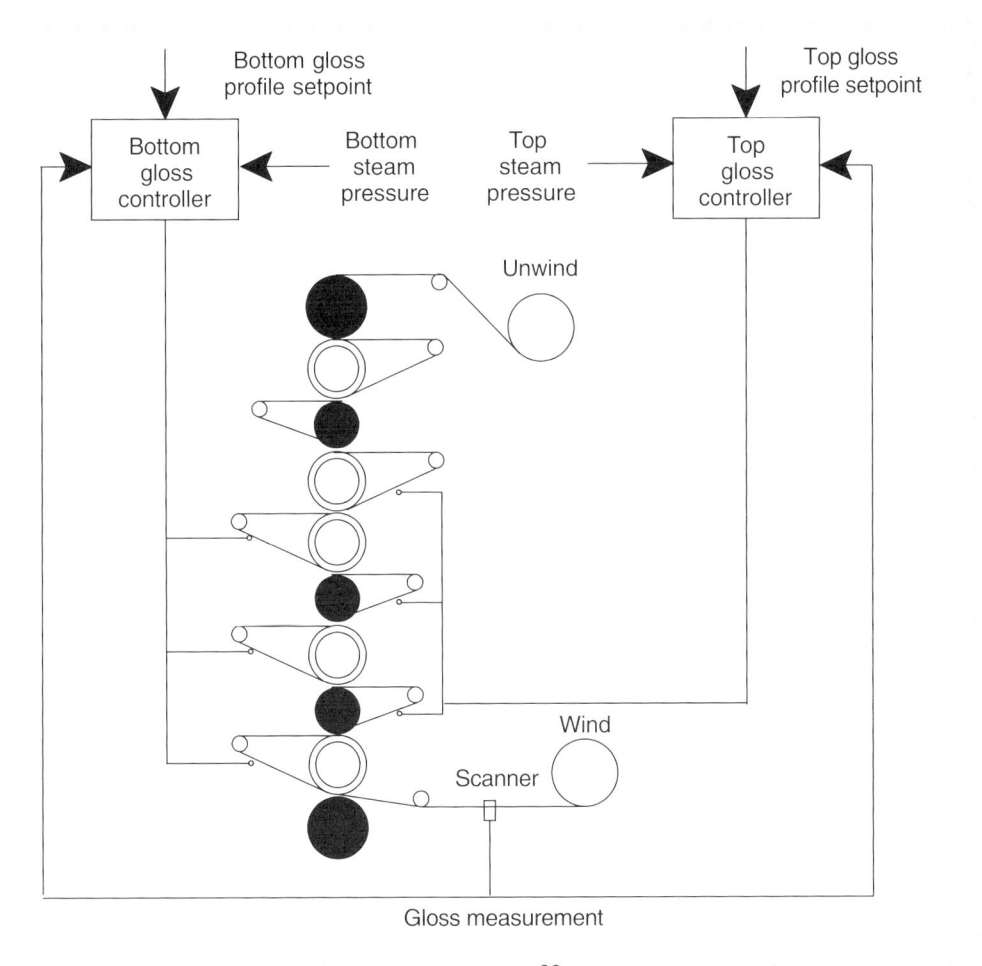

Figure 26. CD gloss control with automatic control gain retuning[20].

CHAPTER 8

Figure 27 shows the principle for the startup and reel change sequence. The system ramps the speed and steam pressure to minimize the effect of speed on gloss. The control ramps the speed to the break-point value and then uses an optional lower ramp rate until the transition to the final operating speed[20].

The interactions between caliper and surface smoothness profile control are available in another publication[1]. This indicates that final CD profiles depend on the following parameters as Fig. 28 shows:

- CD profiles of various paper properties entering the nip such as basis weight, caliper, bulk, moisture, and temperature
- CD profile of calender nip as defined by crowning and roll bending due to the line force, grinding accuracy, wear, CD profiles of the roll temperatures, elasticity, and surface characteristics.

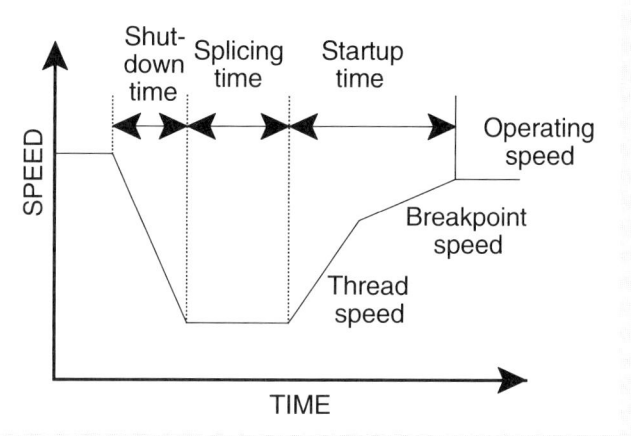

Figure 27. Start-up/reel change control speed ramping sequence[20].

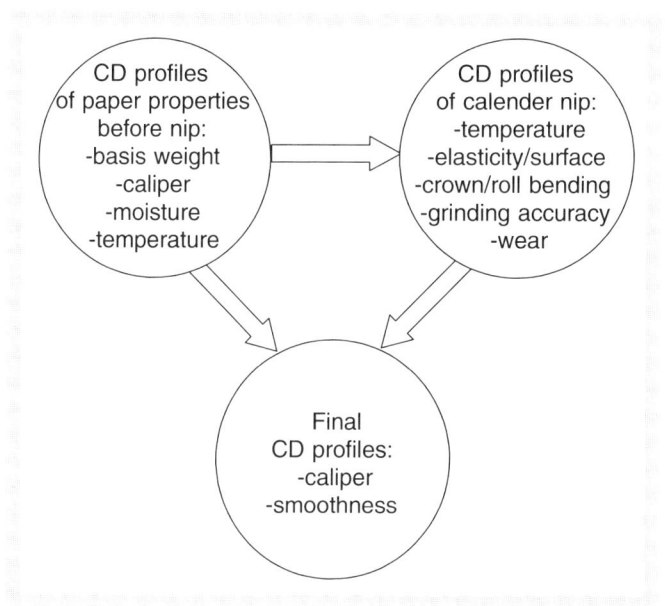

Figure 28. Interactions between CD profiles of paper properties entering the nip, calender nip profile, and the final CD profiles of caliper and smoothness[1].

Process control in paper mills

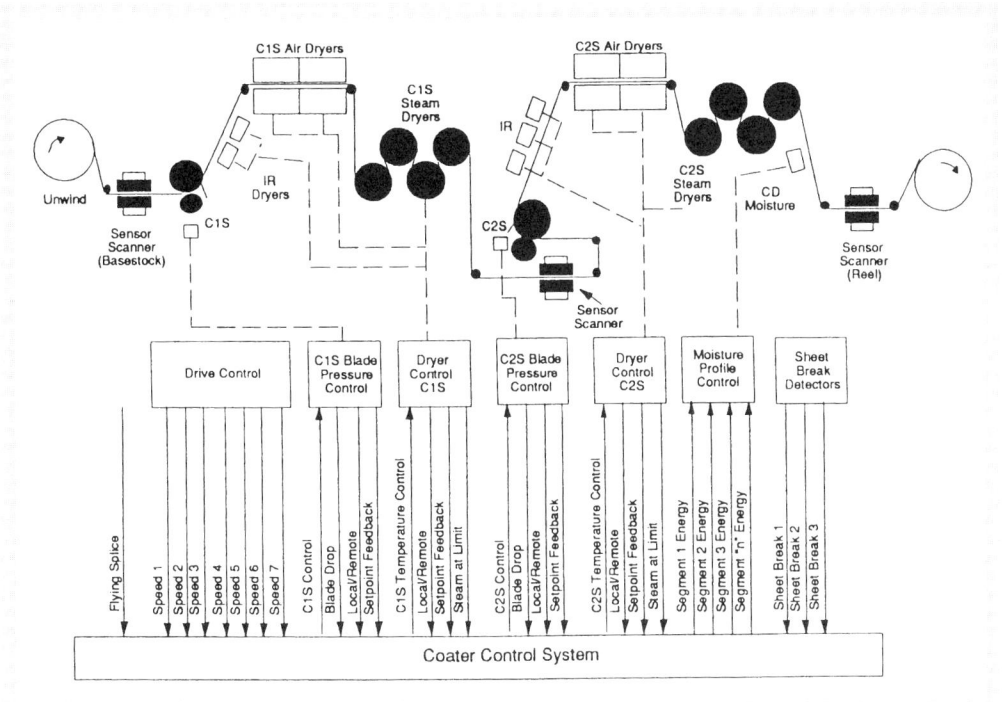

Figure 29. Coater instrumentation and control systems[21].

3 Coater control

Figure 29 shows the basic instrumentation and control functions for an off-machine coater[21]. With the exception of the flying splice instrumentation, the on-machine coater looks very similar.

3.1 Coating weight control

Many factors influence the quality of a coated sheet. Coating and drying methods, surface properties of the base sheet, and the composition of coating materials are the most important. Good coating control requires expertise in the coating process, actuators, measurements, and computer control design techniques. The CD control strategy on coaters tries to reduce variation after blade changes, grade changes, process upsets, and sheet breaks. It also tries to maintain the CD profile variation within the desired limits.

The primary actuator in coating weight control is blade pressure. Secondary actuators depend on the machine and blade angle, coating solids, coater speed, air knife pressure and angle, and roll speed.

Adjusting rods similar to slice screws in paper machine headboxes have been installed to modify blade pressure at intervals across the web. CD coating weight control is also possible in the calender using a row of narrow CD actuators such as hot or cold air showers or electric or induction heaters.

CHAPTER 8

Accurate determination of the coating weight usually involves measuring the properties of the sheet before and after the coating station. Figure 30 shows the block diagram of a coating weight control package[22]. It has four major functional blocks: coating process, profile measurement system, profile processing unit, and control algorithm. The unique features of this package include scanner synchronization for accurate determination of coating weight profile, dynamic mapping for on-line compensation of linear shrinkage and sheet wandering without bump testing, and sophisticated control algorithm handling the changing process conditions[22].

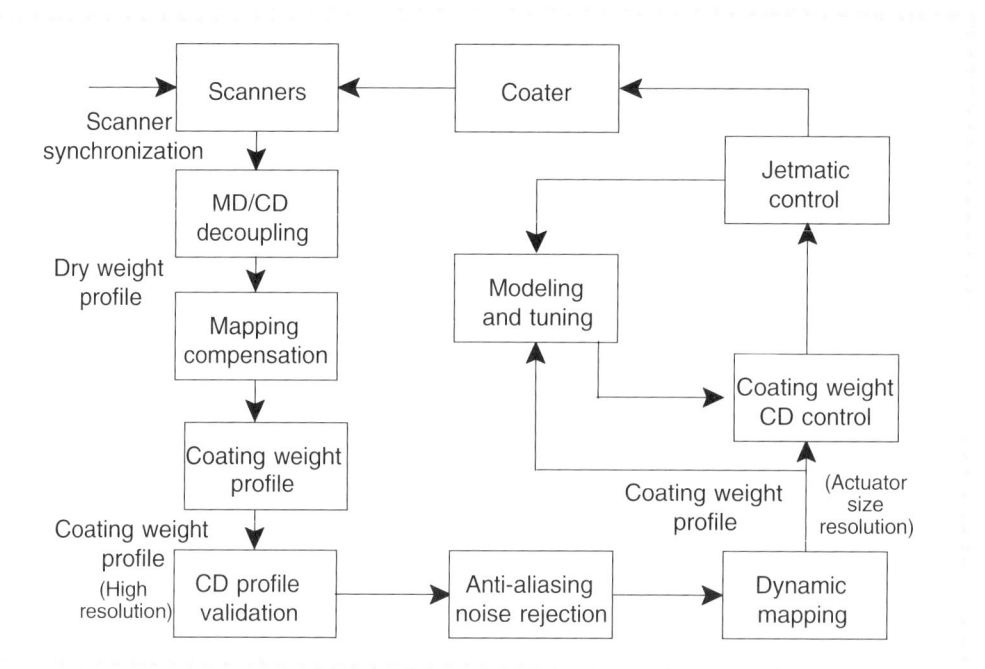

Figure 30. CD coating weight control block diagram[22].

3.1.1 Scanner synchronization

The calculation of the coating weight added by a single station requires two scanners. The location of one is before the coating station, and the location of the other is after the coating station. A transportation delay occurs as the sheet passes through the two scanners. The goal of scanner synchronization is to ensure that sensor data used for the coating weight calculation are measured from the same spot of paper in the web. Scanner synchronization is extremely important for correct coating weight profile[22].

3.1.2 Dynamic mapping

Mapping is an important issue in any CD control, since the moving sheet is often subjected to web wandering and shrinkage. For closely spaced actuators on a wide machine, 0.3% cross direction sheet wander could almost be half the actuator distance. If not compensated, this will influence CD quality significantly[22].

In CD coating weight control, two kinds of mapping exist:

- Upstream dry weight profile to downstream profile (measurement-to-measurement mapping)
- Measurement-to-actuator mapping.

Methods similar to the usual paper machine controls find use.

Figure 31 shows a measurement-to-actuator mapping consisting of two fundamental parts. The first part is a function:

$$Y' = f(X) \tag{4}$$

This relation is also raw mapping. It describes the mapping from each actuator cell to its corresponding measurement cell. The mapping also considers nonlinear sheet shrinkage[22].

The second part in the mapping is a linear compensation relation:

$$Y = \alpha Y' + \beta \tag{5}$$

The slope parameter, α, is usually close to one, and it compensates for the linear sheet shrinkage. The offset parameter, β, is calculated on-line using the center positions of the sheet. It covers sheet wandering[22].

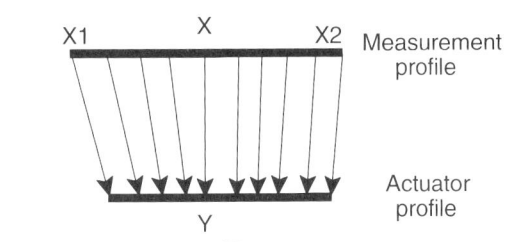

Figure 31. Dynamic mapping[22].

3.1.3 Control and tuning

Coating weight CD control includes decoupling and tuning of the controller gain[22]. Decoupling is necessary since the actuator response shape is usually three times as wide as the actuator zone size in this application. Tuning of the controller gain is necessary because the process varies as the operating conditions change especially due to blade wear and blade changes. The controller gain automatically adjusts to act more aggressively when it detects a bad initial profile. It is restored after the initial transient. The best initial actuator profile can also be automatically saved and reused as a reference for the next blade change[22].

3.1.4 Coater startup

One to three startups in a day is typical in the operation of modern coaters. This requires advanced control strategies. Quality control during the startup is probably the most challenging aspect of coater control.

CHAPTER 8

Coating weight prediction during the speed ramp allows MD control to start as soon as the machine speed ramp begins. The additional coating weight due to the machine speed ramp also brings more water into the system. The moisture control must respond appropriately. The fastest drying units are primarily used during this stage to bring the moisture level to the desired target as soon as possible. This enables the coating weight CD control to start as soon as production speed is reached[22].

3.2 CD moisture

The coating stations use profile control IR radiators to level the moisture profile of the base paper web before coating and for correcting the moisture profile after coating. The profile before coating is often leveled simply by excessively drying the web to a moisture level below 2.5%[23].

Air flotation and infrared (IR) drying have frequent use in addition to conventional steam cylinders to dry moisture carried by the coating material quickly before the next coating station or the reel. The main control problem is to divide the drying capacity between different drying controllers because drying and control capabilities of different actuators differ from each other. Moisture streaks will usually disturb the coating process and cause uneven coating pickup. They will also make the coating weight uneven and distort the moisture profile. The first IR rows after coating are typically base dryers. Approximately 30%–50% of the later drying capacity is for base drying. The remainder is for eliminating streaks[23].

An IR radiator against the last cylinder or after the last cylinder in a free run with reflector is a typical approach for final moisture profile correction[23].

A drying rate model for operator support in coater control is possible[24]. The system uses a dryer configuration consisting of gas infrared dryers (gas IR) and an air dryer. For a one-sided coated sheet, the following processes will occur:

- Heat absorption by the coating layer and the sheet to raise the temperature
- Water drainage from the coating layer to the base sheet
- Evaporation of water from the coating and the base sheet close to the critical temperature
- Heat transfer through convection
- Heat transfer through conduction.

The governing equations in the model are the heat and mass transfer equations. For the gas IR process, radiation and convection dominate. In the air dryer, convection is the only mechanism. Conduction becomes more significant in heavy grades. The outputs of the model are the web temperature and moisture at each dryer location.

Process control in paper mills

The main equations for the model are the following[24]:

$$\frac{dT}{dt} = \frac{1}{C_t C_p}(Q_{IR} + Q_{Convection+} - Q_{Conduction} - Q_{Evaporation}) \quad (6)$$

$$Q_{IR} = \eta Q_{Generated} \quad (7)$$

$$Q_{Convection} = h_{conv} \Delta T_m \quad (8)$$

$$Q_{Conduction} = \frac{1}{L/k}(T_{Surface} - T_{Sheet}) \quad (9)$$

$$Q_{Evaporation} = h_{fg} \dot{m} \quad (10)$$

$$\frac{dm}{dt} = \alpha \ln \frac{P_{Total} - P_{v\_air}}{P_{Total} - P_v} \quad (11)$$

where T is temperature [K]
 C_t coating weight [kg/m^2]
 C_p specific heat [J/kgK]
 h_{conv} convective heat transfer coefficient [W/m^{2}K]
 ΔT_m log mean temperature difference [K]
 L effective length in the lumped conduction equation [m]
 k thermal conductivity [W/m]
 h_{fg} latent heat [W/kg/m^2]
 P_v vapor partial pressure [N/m^2]
 η IR efficiency
 m evaporation rate of water
 α a constant
 Q_{xx} heat transfer by xx mode.

The first equation in the model represents the heat balance in each dryer section per unit area. The energy input terms included depend on the property of the section. IR energy uses the actuator energy rating (190 kW/CD m) and efficiency. Calculation of the convection term uses log mean temperature difference (ΔT_m) and convection heat transfer coefficient between the ambient air and the coated sheet. The conduction term uses a lumped parameter model and the Biot number. Calculation of the heat loss due to evaporation uses the latent heat at a specific surface temperature and mass transfer during the evaporation phase. The last equation is the mass transfer equation known as Stephen's Law. It describes the evaporation rate from the surface.

This model has undergone testing on a mill scale to investigate the effect of energy input changes on the resulting drying rate and the quality of the coated paper.

CHAPTER 8

4 Stock preparation and wet end control

The control of paper quality often uses the measurements in the dry end of the machine: basis weight, moisture, ash content, thickness, etc. The instrumentation level at the wet end is not as sophisticated. Other than level and pressure measurements, only some critical flow rates and pH are measured. Paper quality is, however, defined primarily in stock preparation and wet end processes.

Recent years have seen several new approaches. On-line measurements for pulp consistency, retention, fiber length, pulp drainage, and electrical pulp properties are available and are gaining more attention. They are also improving the control possibilities at the wet end and opening a new window for paper machine operation.

4.1 Stock preparation control

The quality of paper results from controlling such variables as refining degree, pulp blending, and handling of additives[25]. The refiner control resembles the TMP controls presented in Chapter 7. Softwood and hardwood usually undergo refining separately. The operator should be able to change the loading of different refiners in series to use the most suitable refiner discs available.

The control is a multiple stage cascade control where freeness control uses on-line freeness measurement[25]. Freeness deviation changes the setpoint for the specific energy controller. The refiner power and the plate gap control specific energy.

The refiner control requires accurate flow and consistency measurements, or otherwise the measurement and control of the specific energy will not be possible. Consistency control also facilitates the actual freeness control. Fiber length measurement and control improve refiner performance.

Different species have totally different effects on paper properties. The exact recipe used in blending different pulps, recycled fibers, and other raw materials is extremely important[25]. In theory, blending control is easy. Flow rate and consistency measurements and the connected controls should be sufficient. In practice, accurate pulp flow rate and consistency measurements are necessary. To control the blending to fulfill fiber requirements, drainage and fiber length measurements from all fiber flows are also necessary.

4.2 Wet end control

Retention measurements find use today to understand the wet end chemistry and the state of the wet end. Retention is a function of the mechanical filtration effect and colloidal retention effect[25]. Two kinds of retention exist: overall retention and first-pass retention[26]. Overall retention is the ratio of the rate of production at the reel to the rate of all solids added to the system. First pass retention is a measure of the amount of solids remaining in the paper sheet compared to the amount leaving the headbox.

Calculation of retention uses a mass balance of the paper machine assuming that all white water goes to the silo[26]:

Process control in paper mills

$$Ret = \frac{C_{HB}Q_{HB} - C_{WW}Q_{WW}}{C_{HB}Q_{HB}} * 100\% \qquad (12)$$

where Ret is retention [%]
C_{HB} and C_{WW} are consistencies of the furnish in the headbox and white water, [%]
Q_{HB} and Q_{WW} are volumetric flows from the headbox and the wire section [m³/s].

If the white water volumetric flow depends linearly on the flow, then the following is true:

$$Q_{WW} = KQ_{HB} \qquad (13)$$

where K is the coefficient in a simple equation for retention written as follows:

$$Ret = \left(1 - K\frac{C_{WW}}{C_{HB}}\right) * 100\% \qquad (14)$$

One can calculate the retention for different components of papermaking furnish separately using the concentration of a particular component in the headbox and in the white water[26].

A retention measurement system monitors the stability of the paper machine wet end. The information of total solids in the headbox and the white water and filler consistency level gives an early indication of the potential problems in the wet end[25]. White water consistency control also increases the stability of the system. Retention aid addition can control consistency[26].

5 Quality inspection systems

Each step in the production process is a potential contributor of defects. Among the major defects in coated paper are the following[27]:

- Holes and thin spots (These typically occur in paper machines and can be repeating. They can lead to web breaks in coating lines and damage supercalender rolls.)

- Dirt and contaminants that can come with the base sheet form in the coating bath or are deposited during the coating process.

- Streaks that indicate blade marks and blade scratches. (Blade marks are more severe. They are due to particles and debris lodged behind the coating blade. They often lead to a complete absence of coating.)

The papermaking process itself and the system users set requirements for a web inspection system. The system is a stand-alone measurement system without a closed

CHAPTER 8

loop control. The operating staff acts on the information it receives from the inspection system. Web inspection systems should fulfil the requirements in Table 2[28].

Table 2. Requirements for a web inspection system[28].

Requirements	Realization
Detect and process all required data	Single-sensor-architecture with advanced signal processing techniques
Report all defects	Advanced user interface and able to transfer and store defect data in electrical form
Compensate for error sources	Automatic gain compensation pixel-by-pixel
Modular construction	Single-sensor-architecture.
High availability	Automatic compensations and light source with a long life time

A system that consists of a laser (or camera) that scans the web continuously is available[27]. A rotating mirror assembly in the scanner housing sweeps the laser beam across the product up to 6000 times per second. The receiver collects the laser light reflected from the product as Fig. 32 shows[27].

An amplitude threshold detects defects. For example, a hole would allow more high intensity laser light to be reflected from the chrome roller into the receiver. This would produce a sharp positive signal at the CD position in question. A dirt spot would absorb more laser energy and reduce the light entering the receiver. This result would be a negative signal. Setting a proper threshold level allows easy detection of these variations.

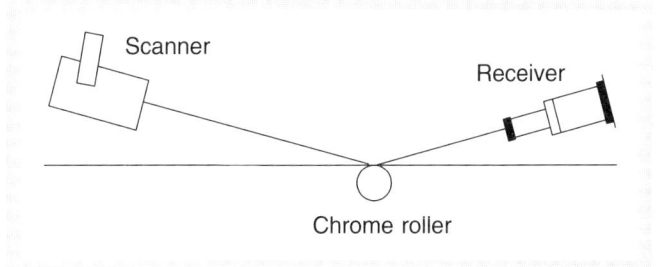

Figure 32. Positioning of the scanner and receiver in the streak detection system[27].

Amplitude threshold works well for larger defects such as holes, dirt spots, and some streaks. Narrow streaks and scratches are much more difficult to detect because the amplitude excursion is small. Storing flaw data from eight successive scans and summing the potential streak data of these successive scans avoids this problem. The technique uses the idea that

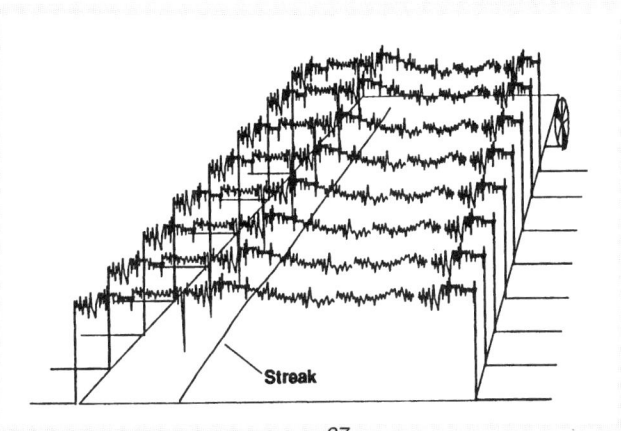

Figure 33. Eight scans of streak profile[27].

180

Process control in paper mills

the flaw signal will show up in the same CD position scan after scan if it is a true streak as Fig. 33 shows[27].

The added signals form a pyramid as Fig. 34 shows[27]. The logic in adding the flaw data is that in the next new scan the pyramid grows by one unit if the flaw is present. If the flaw is missing, the pyramid decreases by two units. The real streak registers when the pyramid grows to eight units.

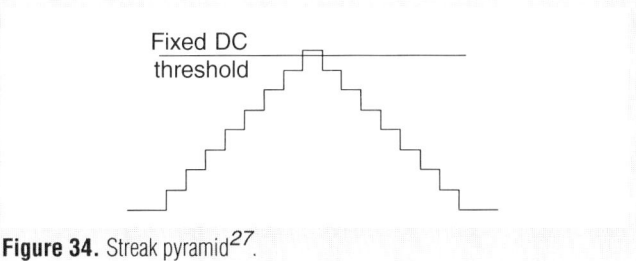

Figure 34. Streak pyramid[27].

Mäkelin[28] describes the web inspection system based on CCD technology. It uses single-sensor-architecture and processing of video signals inside the cameras. Only defect data transfers from the cameras to a data processing system. The minimum detectable hole or spot size of a CCD system is defined by an area seen by one pixel. This is determined by a camera viewing area in the CD direction and by the scanning rate in the MD direction. In scanning systems, this leads to the fact that faster running webs give a larger area seen in the MD direction by a pixel. The resolution depends on the web speed.

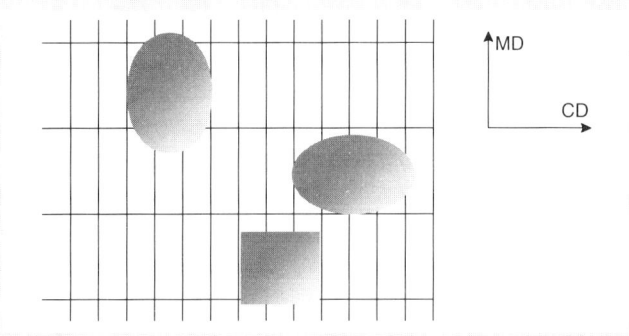

Figure 35. Three different spots giving the same image dimensions[28].

CCD technology makes it possible to view an image or the shape of a defect. Figure 35 illustrates this and also shows how pixel dimensions restrict resolution. The figure shows three different spots with the same dimensions: three pixels in CD and one in MD[28].

Figure 36 shows an application using a web inspection system on an SC production line[28]. Defect

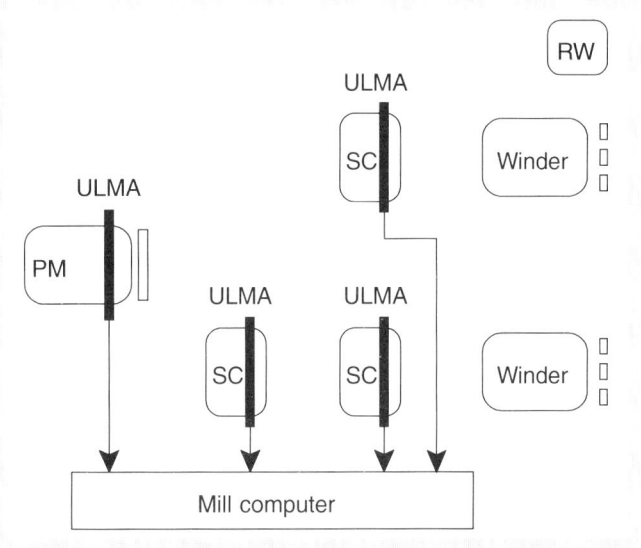

Figure 36. Web inspection system for an SC production line. PM is paper machine, SC is supercalender, and RW is rewinder[28].

data is processed in the mill computer and displayed to the operators using an advanced production and quality management system. The mill computer combines the defect data with the other information it has to generate roll specific defect maps. Defect data from the paper machine is primarily used mainly for paper machine diagnostics and to protect supercalenders from damage caused by several defects. Defect data from supercalenders is used for quality control of the finished product.

6 Grade change automation

6.1 Mass production vs. flexible production

Mass production strives for the highest possible volume and total exploitation of resources by optimizing the output of each operation and minimizing waste production. The importance of big customers is emphasized in information transfer between marketing and production[29].

Problems with mass production are the following:

- Total performance is measured as the sum of each added-value-producing operation
- Unplanned production arises
- Intermediate storage increases
- Production turnaround times are long.

The most ideal situation from the production point of view would be to run only one product continuously so the process would remain stable. This would also avoid waste production due to grade changes. In real life, customers demand many different grades and are unwilling to maintain large stores of their own merely to help a supplier remain on long production runs. For this reason, a plant should be able to change grades quickly and efficiently[30].

In flexible production, the aim is to manufacture large amounts of different products at nearly the unit cost of mass production. One strives for minimizing raw material, intermediate, and final product storage. In this way, reacting quickly to market needs and shortening turnaround times is possible. Production planning links to process control.

Problems with flexible production in the process industries are as follows:

- Continuous changes in process state and process parameters produce waste production and vulnerability to disturbances
- Excessive reduction of intermediate storage may result in disturbances in production output
- Disturbances and risk factors reduce the degree of use.

Process control in paper mills

6.2 Grade change process

During the grade change in a paper mill, many drives and valves that define the process parameters require control. The objective is to change the grade as quickly as possible from one acceptable quality level to another while avoiding breaks. Breaks are particularly harmful to mill production and are very common. When a break occurs, operators must rip away the broken web and insert the head of the roll again into the machine. Grade change weakens the stability of the process and therefore increases the risk of breaks[31]. While a paper machine is running, paper produced between grades that does not conform to quality standards is generally rejected and recycled into the process. When the quality level required for a new grade has been reached, the reelers begin again to accumulate paper. Figure 37 provides a general view of a grade change in a paper mill.

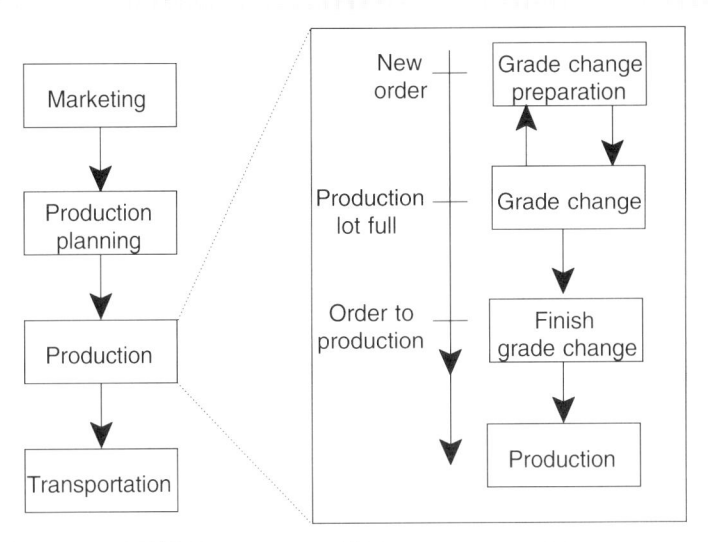

Figure 37. General view of grade change in a paper mill[32].

The grade change proceeds according to the following sequence[29]:

- Normal production: The time, during which production remains within quality margins.
- Preparation for a grade change: Operators' preparation work during normal production before the grade change such as selection of operations required, testing, and fine-tuning.
- Execution of the grade change: The operations selected occur according to the plan unless disturbances or deviations from the normal process conditions exist.
- Termination of the grade change: Controls are readjusted to the normal operation and the success of the grade change is evaluated and reported.

CHAPTER 8

The main activities of the preparation stage are the following[29]:

- Functional mapping of the production line: The system presents in visual form concepts known to the operational and maintenance staff. These are the aims, limitations, interactive relations, and functional requirements of the process.

- Operational plan: Operations during the grade change can occur manually or automatically. The necessary tasks can be examined and altered at the different levels of the control hierarchy as Fig. 38 shows.

- Qualitative validation and planning: The operational plan can be tested qualitatively. Each operation can be given starting, stopping, and support conditions that can be logical or quantitative performance criteria. In addition, the operator can determine the performance sequence of operations or design a control message for a continuing subprocess.

- Scanning of history and operator-experience database: Examination of similar grade changes enables the reuse of information and monitoring of efficiency development in grade change performance.

- Simulation tests: The tools exploit the models used for process control and the operational models made by operators. The system supports the use of an accurate simulator for pre-trial and pre-inspection of the grade change.

The execution stage of the grade change consists of the pre-change operations that must be carried out during a normal run and a transient grade change stage. In the paper machine, for example, changes in basis weight, filler content, and speed cause a shift in the load level of the drying section. Since the dynamics of the drying section are slower than the dynamics of the flow and speed-up processes, the changes in the steam pressure must be done in good time before the grade changes.

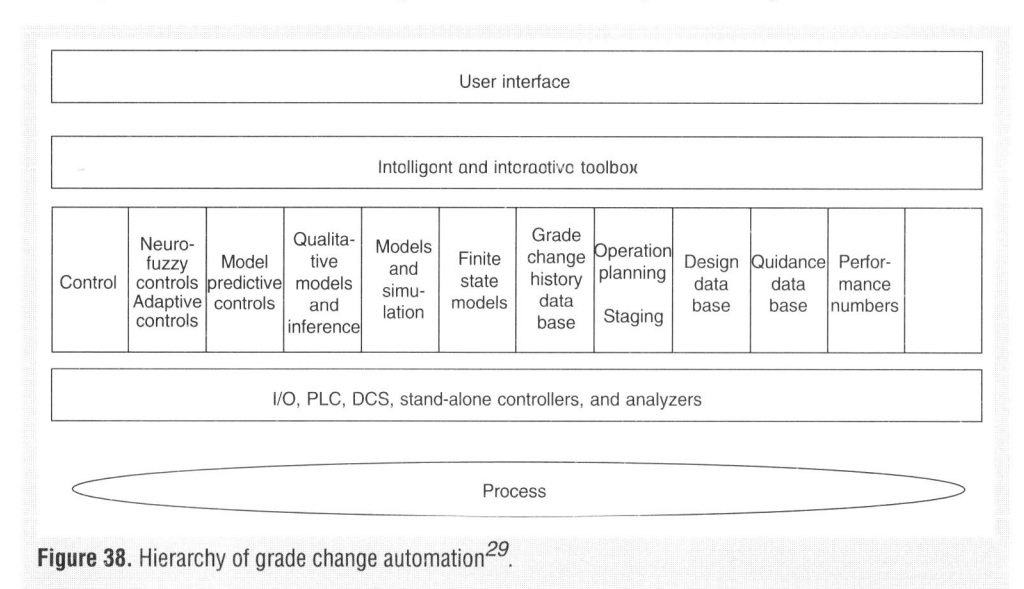

Figure 38. Hierarchy of grade change automation[29].

Process control in paper mills

The selection of an appropriate method for the transient state control is influenced by requirements set for the operation. To obtain additional benefits, more efficient algorithms and calculation tools are required. It must, however, be noticed that more sophisticated methods also need more work both at the planning and implementation stages. Benefits increase in the following order: single-input/single-output (SISO) PID control, multi variable feedforward and feedback control, and model predictive control. Planning the incorporation of neural network technology is beneficial if the process model is very nonlinear or its structure is unknown.

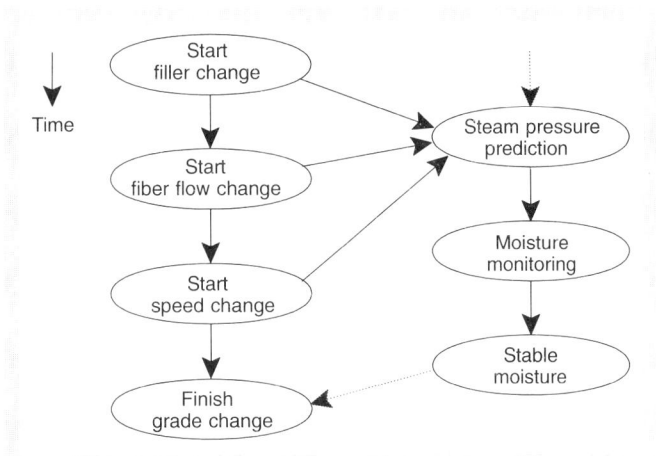

Figure 39. Execution stage of grade change in basis weight[29].

Sequence control is necessary in addition to dynamic control. The sequences handle control timings and disturbance management.

Figure 39 presents the execution stage of the grade change when making a change in basis weight. Figure 40 presents operation of the grade change automation in the case mentioned.

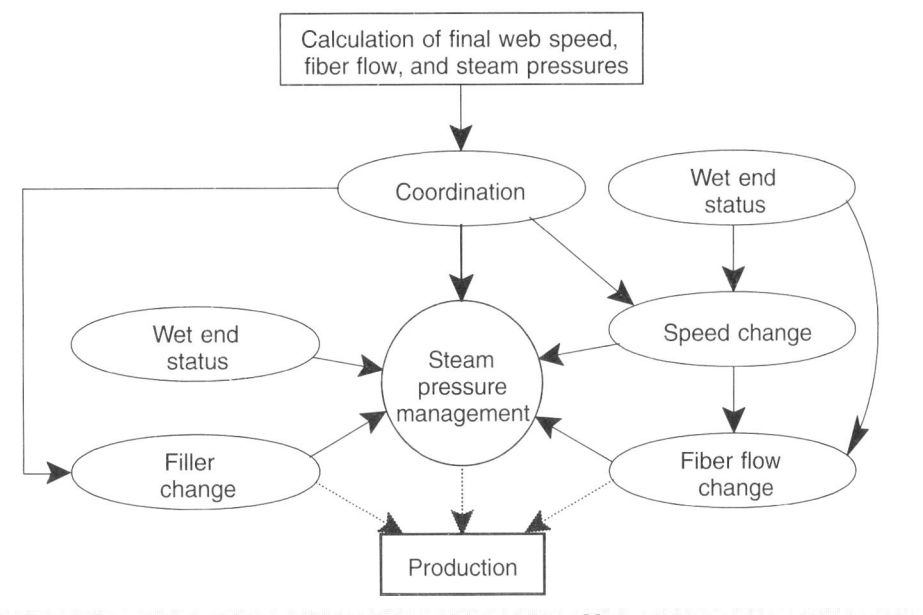

Figure 40. Operations of grade change automation in the case mentioned[29].

CHAPTER 8

Both figures emphasize starting the steam pressure changes before the flow and speed changes begin. Section 6.4 will discuss the possible control actions taken during the grade change in the paper mill and the required methods.

6.3 Time required for grade change

The time required for the grade change in a paper machine denotes the time necessary to shift from one acceptable quality level to another as Fig. 41 shows. Factors influencing this are as follows:

- Mill operations policy including the frequency of grade changes
- Mill structure such as the scale and age of the machinery
- Mill infrastructure including the manpower, experience, and degree of computer integration
- Management's weightings about how operators must be motivated to change grades quickly.

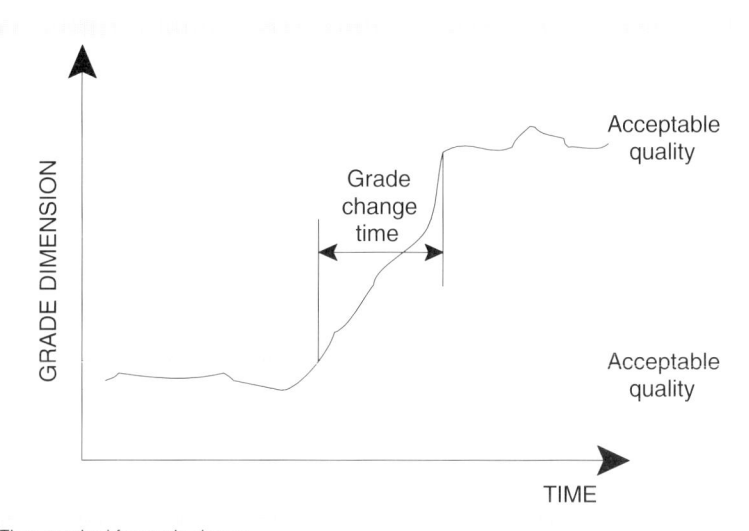

Figure 41. Time required for grade change.

6.3.1 Mill operations policy

Grade change type

The kind of change necessary influences the grade change time. Four possible types occur in paper mills listed in the following order of difficulty:

- Changes of color: Frequently these changes require total washing of the mill to remove all traces of the previous color. Changes of color only apply to mills manufacturing paper of more than one color.

- Furnish changes: The intent can be to increase paper brightness using titanium dioxide or to alter the relation between long and short fibers. Manual adjustments of drive and valve settings are often necessary.
- Changes in basis weight: These changes often require changes in mill parameters such as web speed.
- Caliper changes: These changes require only slight adjustments at the dry end of the paper machine.

The most important types are basis and furnish that differentiate one paper grade from another.

Magnitude of alteration

The time spent in the grade change increases as a function of the magnitude of change. Paper mills strive to plan their production cycles to minimize the "step size" of the change that must be made as Fig. 41 shows.

Frequency of grade changes

Faster grade changes will occur in mills that change grade often. Frequently occurring grade changes are an incentive for a mill to introduce operations to speed grade changes. In addition, increasing experience should accelerate the execution of these changes.

6.3.2 Mill structure

Scale

Although scale probably influences the time spent for a grade change, two possible interactive mechanisms exist. Growth in scale increases the mechanical inertia of the system components. Cleaning containers of pulp traces from a previous grade can be more time consuming, and the acceleration and deceleration of rolls can take longer. These factors increase the time required in a grade change. The pressure on operators to execute grade changes quickly to minimize waste production in a large, expensive mill may shorten the time spent in the grade change.

Technology vintage

Older, lower-level control technology may prevent operators from changing process parameters quickly and reliably. Companies occasionally update motors, valves, and lower level control mechanisms to improve efficiency or increase the possibility to manufacture new papers.

6.3.3 Mill infrastructure

Computer integration

In contrast to lower level computer control that controls drive speeds and valve flows, the intent of computer integration is to coordinate the joint functioning of the mill and the

CHAPTER 8

different systems in the mill environment at a higher level in the hierarchy. A grade change can influence operations over a wide area outside the paper mill such as the integrated pulp mill or the power plant. A grade change usually influences the pulp mill, the groundwood plant, the paper auxiliaries plant, and the coating kitchen. One example is that the pulps required for a new grade do not change in sufficient time. As a result, running the production according to requirements is difficult. Including pulp manufacture in the control strategy can also boost the efficiency of a grade change[32].

Computer integration can handle specific tasks such as quality monitoring and reporting of statistics, A high degree of computer integration may speed grade changes by providing better process control.

Operator experience

The experience gained by operators probably has a significant effect on the time required in a grade change. Experienced operators obviously change grades more quickly and more efficiently, although some exceptions may exist.

6.4 Grade change controls

6.4.1 Alternative approaches

Nearly all paper mills have automation to control the process in the steady state when only one paper grade quality is run. The degree of automation in grade change varies. Automated grade changes operate at the highest levels in the hierarchy of automation and carry out the shift from one grade to another by gradually changing process parameters (ramp changes) according to a preprogrammed sequence. Figure 42 presents the hierarchy of the computer-aided grade change automation.

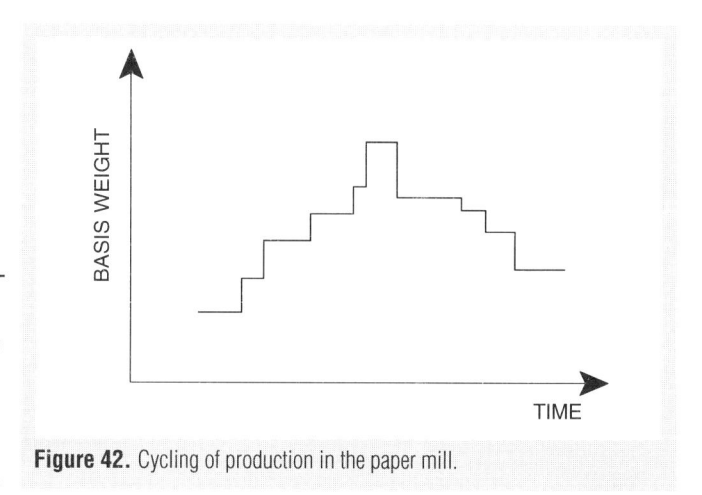

Figure 42. Cycling of production in the paper mill.

The hierarchy consists of the user interface, workbench, tools, and unit controls. The tools are interactive and can use database information and operator experience and intuition. To the operator, the tools appear as an intelligent workbench that can show in visualized form the effect of selected grade change operations on a process. The tools support different operational stages during the grade change.

Depending on the level of automation in the paper mill and also on operator experience in using it, several different alternatives exist for grade change control:

Process control in paper mills

- Manual action: Skillful operators can execute grade changes manually although the level of performance often varies. Grade change control requires adaptation to the actual status of the machine at the appropriate moment and to the properties of the raw materials. In addition, the dynamics of the machine must be known so that required control measures can be planned as a function of time. Grade change therefore involves the kind of control task that automatic control theoretically would be able to perform more quickly and more reliably. Automation of grade change will free the operator for tasks that are not possible to automate. The implementation of grade change automation superior to human performance is difficult. Although grade change automation has often been purchased for paper machine automation systems, it is not necessarily used. This is usually the result of the caution that it requires and its slow speed. In other words, a manual grade change is faster. Speeding automatic control implemented by conventional methods will quickly cause instability problems and oscillations[29, 34].

- Conventional feedback control: Feedback control with no feedforward compensation performs badly. Feedback control reacts to differences between setpoints and measurement values, although grade change requires direct reaction to changes in setpoints and preferably even prediction of future changes in setpoints. In theory, conventional feedback, feedforward, and cross compensation should suffice for implementation of grade change automation. In practice, the system will become too complicated for tuning particularly due to the nonlinearities of the machine.

- Open-loop control with pre-planned ramps: This approach has been applied on paper machines despite certain characteristic shortcomings. With an open control circuit, achieving the desired moisture is difficult. Planning models that would predict the required steam pressures at a sufficiently accurate level for each paper grade irrespective of variations in raw materials and the changes due to new wires and felts is also difficult. With changes in the machine, properties of raw materials, and grades produced, the performance level of an open-loop control gradually weakens.

- Tracking pre-planned trajectories using the feedback controller: This approach combines feedback and open loop control. The ramps mentioned earlier are used with predicted outputs as references for the inputs of feedback controllers. Shortcomings in the approaches described above can decrease to a degree. The negative factor is that it has not been possible to solve problems connected with tuning of the complicated tracking controllers[34].

- Model predictive control: The increase in the calculation power of computers has made it possible to apply model predictive control to real processes. In model predictive control, the controller outputs are calculated using a system model based on minimization of the difference between the predicted and desired behavior of the system. The different model predictive control algorithms differ from each other primarily in terms of the system model used as Table 3 shows and of the cost function needed to minimize the control deviation. The model predictive controller calculates a future output sequence so that the predicted process output follows the desired reference trajectory as closely as possible. Only the first element of the calculated output sequence is used for process control. When the next sample is taken, the same steps are repeated using the latest measurement information, This is the receding horizon principle. Model predictive control is an intuitive and systematic way of implementing and tuning multivariable controllers. Feedforward compensation, dead time compensation, and information on future reference values are characteristic of this approach[34, 35].

Table 3. Some process models used by predictive controllers. Symbol * means that the parameter is a planning parameter of the controller. Symbol Δ denotes a differentiating operator ($\Delta = 1 - q^{-1}$).

Controller	Model	A	B	C	D
DMC	FSR	1	*	1	Δ
PCA	FIR	1	*	1	Δ
MAC	FIR	1	*	1	Δ
GPC	ARIMAX	*	*	*	Δ
EPSAC	ARIMAX	*	*	*	Δ
EHAC*	AROCO	*	*	1	Δ

6.4.2 HMPC (Horizon multi variable predictive control)

The HMPC in Fig. 43 is a model-predictive multi input/multi output (MIMO) controller[36]. It uses process response models to predict future changes in controlled variables.

The HMPC controller also considers disturbance variables using these process variables as feedforward information. The controller predicts the future effects of disturbance variables on controlled variables and initiates compensatory measures to keep the controlled variables within their desired values.

In paper machine applications, a maximum of ten controlled variables, ten manipulated variables, and ten disturbance variables can connect to the HMPC package. Typical controlled variables are basis weight, moisture, and machine speed. Typical manipulated variables are stock flow, steam pressure, and machine speed. The disturbance variables do not belong to the control matrix but do influence the controlled variables. They are therefore treated as uncontrolled, manipulated variables. The refiner power level is one example of disturbance variables.

Properties of an HMPC that influence the performance of grade change control are the following:

Process control in paper mills

- Prediction of steady-state conditions
- Constraint handling
- Active feedback control
- Grade-specific modeling
- Integration into a distributed control system.

HMPC can predict the steady-state responses of controlled variables. This gives the operator an idea of future behavior of the paper machine. Before the grade change, the final settings of manipulated variables are seen. These give the defined changes in controlled variables.

If restrictions are met such as the upper limit of steam pressure, the operator learns this information and can initiate steps before the execution of the grade change. In addition, treatment of restrictions enhances the grade change performance by predicting when the restrictions will be met and by adapting control strategy accordingly. When a restriction is met, a weighting function is employed for machine optimization. The operator defines the weighting values that use the relative importance of the final process values. For example, the controller can automatically reduce the basis weight target as the upper limit of the steam pressure is approached.

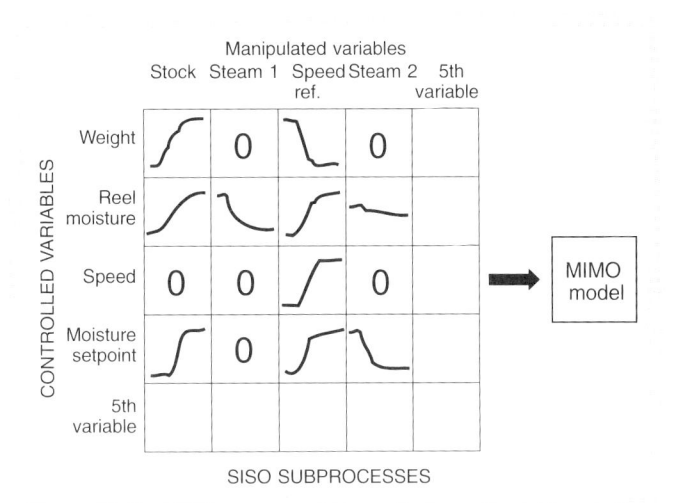

Figure 43. The MIMO process model is used in the multi variable controller. Each grade has its own model. During grade changes, model selection occurs automatically[36].

HMPC feedback properties allow the controller to react directly to process disturbances that occur during a grade change.

HMPC supports grade-specific modeling to ensure optimum control of a machine manufacturing a wide product range. During a grade change, HMPC automatically selects the MIMO model for the grade in question.

Integration into a distributed control system also provides the possibility of ramping other control loops during the grade change.

6.4.3 ADAPLANC (Adaptive planning controller)

ADAPLANC is a variation of the model predictive controller that was developed to solve problems appearing when the conventional model predictive controller is applied in a grade change[34].

CHAPTER 8

The principle behind ADAPLANC is the following. First, a predictive model of the controlled process is constructed. The model equations are used in constructing state and parameter estimators and in an algorithm that defines the controller inputs that give optimum predicted outputs. State estimation guarantees that model and machine states match each other as much as possible. Parameter estimation ensures that the model adapts to changes in raw materials, wires, felts, etc.

In this approach, use of nonlinear models is possible so that mechanistic modeling can be used. In mechanistic modeling, knowledge of machine dimensions can be used as an example. This can reduce the number of tests required for controller planning. The model based on the process structure and the physical laws is more robust and more accurate over a wider operational area than the linear models based on experimental tests used in conventional model predictive controllers.

While the determination of certain parameters required in mechanistic modeling is difficult, they can be estimated from measured data. The use of mechanical modeling allows concentration on the estimation of only the unknown parameters from measured data collected during a normal run.

Mechanistic modeling reveals parameters that probably change during the operation of the machine. An on-line estimation algorithm can therefore be developed solely for these few parameters. This kind of adaptation is safer and more reliable than adaptation using the estimation of all parameters of a big black box model.

Automatic generation of a code makes implementing even such control algorithms that cannot be implemented using standard recompiled procedures or functional blocks easy. Figure 44 shows the implementation of ADAPLANC.

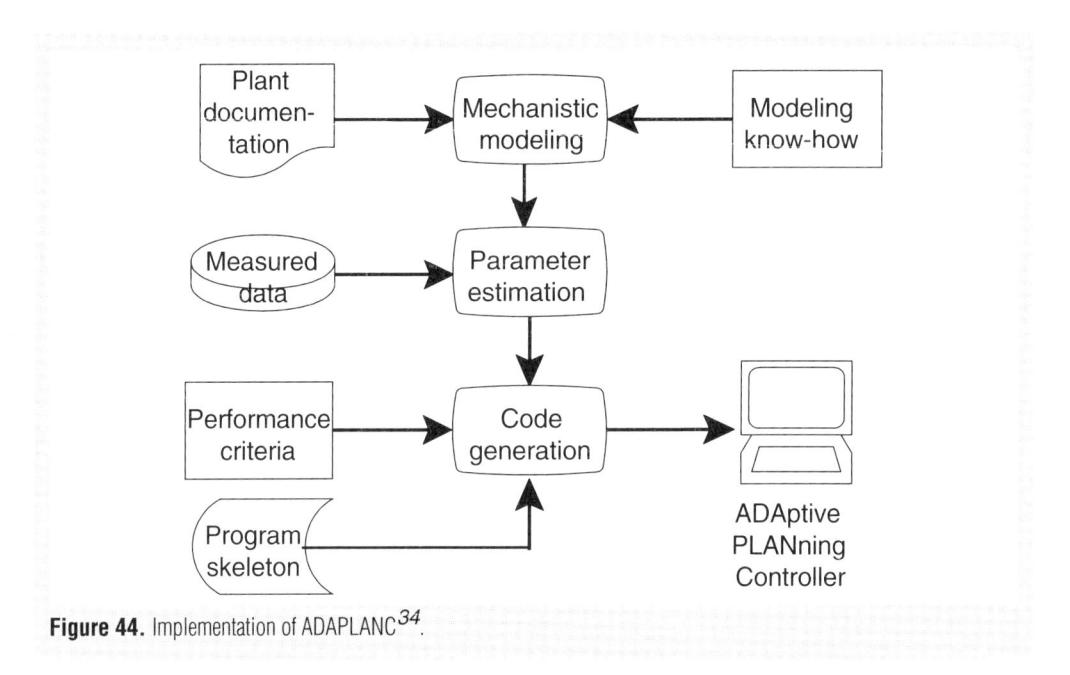

Figure 44. Implementation of ADAPLANC[34].

References

1. Holik, H., 1986 Control Systems Proceedings, SPCI, Stockholm, p. 87.
2. Bialkowski, W. L., 1990 Control Systems Preprints, Finnish Society of Automation, Helsinki, p. 220.
3. Wallace, B. W.and Balakrishnan, R., 1990 Control Systems Preprints, Finnish Society of Automatic Control, Helsinki, p. 278.
4. Brewster, D. B., 1986 Control Systems Proceedings, SPCI, Stockholm, p. 61.
5. Lindeborg,C., 1990 24th EUCEPA Conference Proceedings, SPCI, p. 109.
6. Tran, P., Subbarayan, R., and Chen, S-C., 1997 International CD Symposium, Cross Directional Web Measurements, Controls and Actuators in Paper Machines, Proceedings, Finnish Society of Automation, Helsinki, p. 113.
7. Chen. S-C. and Adler, L., 1990 24th EUCEPA Conference Proceedings, SPCI, Stockholm, p. 77.
8. Wilkinson, A. J. and Hering, A., 1983 PRP 5 Conference Proceedings, International Federation of Control, Antwerp, p. 151.
9. Adams, W. L., 1990 24th EUCEPA Conference Proceedings, SPCI, Stockholm, p. 131.
10. Shakespeare, J. and Kniivilä, J., 1997 International CD Symposium, Cross Directional Web Measurements, Controls and Actuators in Paper Machines, Proceedings, Finnish Society of Automation, Helsinki, p. 86.
11. Heaven, M., Vyse, R., and Steele, T., 1997 International CD Symposium, Cross Directional Web Measurements, Controls and Actuators in Paper Machines, Proceedings, Finnish Society of Automation, Helsinki, p. 100.
12. Kniivilä. J., Shakespeare, J., Korpinen, A., et al., 1995 Automation Proceedings, Finnish Society of Automation, Helsinki, p. 381.
13. Niemi, A., Berndtson, J., and Karine S., 1997 International CD Symposium, Cross Directional Web Measurements, Controls and Actuators in Paper Machines, Proceedings, Finnish Society of Automation, Helsinki, p. 214.
14. Mustonen, H., Nyberg, P., Kniivilä, J., et al., 1994 Control Systems Proceedings, SPCI, Stockholm, p. 172.
15. Kaljunen. T., Parola, M., and Linna, H., 1997 International CD Symposium, Cross Directional Web Measurements, Controls and Actuators in Paper Machines, Proceedings, Finnish Society of Automation, Helsinki, p. 8.
16. Ilvonen, P. and Vierimaa, P., 1990 24th EUCEPA Conference Proceedings, SPCI, p. 151.

CHAPTER 8

17. Åkerlund, K., Happonen, E., and Mustonen, H., 1995 Automation Preprints, Finnish Society of Automation, Helsinki, p. 329.
18. Crotogino, R. and Gendron, S., 1986 Control Systems Proceedings, SPCI, Stockholm, p. 80.
19. Svenka, P. and Brendel, B., 1997 International CD Symposium, Cross Directional Web Measurements, Controls and Actuators in Paper Machines, Proceedings, Finnish Society of Automation, Helsinki, p. 20.
20. Taylor, B. F. and St. James, S. B., 1990 24th EUCEPA Conference Proceedings, SPCI, Stockholm, p. 189.
21. Shapiro, S. I., 1990 24th EUCEPA Conference Proceedings, SPCI, Stockholm, p. 159.
22. Bale, S., Fu, C., Nuyan, S., et al., 1997 International CD Symposium, Cross Directional Web Measurements, Controls and Actuators in Paper Machines, Proceedings, Finnish Society of Automation, Helsinki, p. 92.
23. Peltoniemi, H., 1997 International CD Symposium, Cross Directional Web Measurements, Controls and Actuators in Paper Machines, Proceedings, Finnish Society of Automation, Helsinki, p. 26.
24. Ghofraniha, J., Heaven, M., Lee, L., et al., 1997 International CD Symposium, Cross Directional Web Measurements, Controls and Actuators in Paper Machines, Proceedings, Finnish Society of Automation, Helsinki, p. 107.
25. Kaunonen, A., 1993 XXV Riunione Annuale, ATICELCA, Milan, p. 1.
26. Kaunonen, A., Lehmikangas, K., Nokelainen, J., et al., TAPPI 1991 Papermakers Conference Proceedings, TAPPI PRESS, Atlanta, p. 19.
27. Buts, A., 1990 24th EUCEPA Conference Proceedings, SPCI, Stockholm, p. 177.
28. Mäkelin, L., 1992 ATICELCA Proceedings, SPCI, Bologna, Italy, p. 444.
29. Viitamäki, P., Ventä, O., and Välisuo, H., 1995 Automation Days Preprints, vol I, Finnish Society of Automation, Helsinki, p. 355.
30. Upton, D. M., "Computer Integration and Catastrophic Process Failure in Flexible Production," Harvard Business School, 1994, 16 p.
31. Upton, D. M., "Plant Capabilities for Quick Response Manufacturing," Harvard Business School, 1994. 23 p.
32. Viitamäki, P., 1993 Automation Days Preprints, Finnish Society of Automatic Control, Helsinki, p. 241.
33. Upton D. M., "What really makes factories flexible?" Harvard Business Review, Jul/Aug 1995, p.74.
34. Välisuo, H., Lappalainen, J., Juslin, K., et al., TAPPI 1996 Engineering Conference Proceedings, Book 2, TAPPI PRESS, Atlanta, p. 491.
35. Soeterboek, A. R. M., "Predictive Control: A Unified Approach," Technical University Delft, 1990. 358 p.
36. McQuillin, D. L. and Huizinga, P. W., 1995 1st ECOPAPERTECH, Helsinki, p. 73.

CHAPTER 9

Millwide control

1	**Concepts of millwide control**	**201**
2	**Planning of millwide control**	**202**
2.1	Decision on millwide control system	203
2.2	Mill control analysis	204
2.3	Mill control improvement	206
3	**Implementing millwide automation**	**207**
4	**Benefits of millwide automation**	**210**
4.1	Increased production capacity	212
4.2	Reduction in raw materials and utilities	212
4.3	Mill energy efficiency	212
4.4	Decreased emissions and losses	213
4.5	Better information	213
5	**Millwide information systems in pulp mills**	**213**
5.1	Requirements for millwide control	213
5.2	Information system functions	214
6	**Production scheduling in pulp mills**	**217**
6.1	Introduction	217
6.2	Mill model	217
	6.2.1 Processes	219
	6.2.2 Storage tanks	219
	6.2.3 Rate and grade changes	220
	6.2.4 Planned shutdowns	220
	6.2.5 Random disturbances	221
	6.2.6 Process delays	221
	6.2.7 Bottleneck processes	221
6.3	Optimization problem	222
6.4	Survey of methods	223
	6.4.1 Tamura's time delay algorithm	225
	6.4.2 Heuristic scheduling	227
	6.4.3 Application of expert systems and intelligent methods	229
7	**Millwide control in paper mills**	**230**
7.1	Millwide control functions in paper mills	230
7.2	Order handling	231
7.3	Production scheduling in paper mills	231
7.4	Real time production monitoring	233

CHAPTER 9

7.5	Intelligent quality control systems	234
7.6	Efficiency monitoring	236
7.7	Profitability monitoring	237
8	**Energy management systems**	**238**
8.1	Millwide energy management systems	239
8.2	Boiler load allocation	240
8.3	Turbine load allocation	242
8.4	Purchased power optimization	244
8.5	Compensation of variations in steam consumption	244
8.6	Simulation and optimization functions	245
8.7	Corporate-wide energy management systems	245
8.8	Savings potential	245
	References	246

CHAPTER 9

Kauko Leiviskä

Millwide control

1 Concepts of millwide control

The terms millwide automation, millwide control, computer integrated manufacturing (CIM), and integrated mill systems often have use as synonyms. In the pulp and paper industry, the term production control had early use to describe millwide control functions. Integration is the key concept for millwide control. All these terms refer to integration of process control and distributed automation systems with management information systems. This also means the extension of real time data processing to upper control levels. Sections 1–5 use information from another publication updated to correspond to the situation today[1].

Millwide control has been a topic of discussion for systems and computer people for more than 20 years. It still lacks a unique and well-accepted definition. The system vendors, users, researchers, and consultants all use the term with a slightly different meaning.

The following four early definitions describe some alternative points of view to approach the contents of millwide control:

"Millwide control exists where the authority, responsibility and tools are present, agreed upon and used to plan, coordinate and control daily production toward a measurable optimum[2]."

"The objective of millwide control is to achieve the optimal operation of the mill relative to its orders, costs, inventories and available production capacity[3]."

"Millwide automation is the overall planning, coordination and control of production, quality and energy to maximize the profitability of the mill[4]."

"The concept includes process optimization, but it also incorporates the concepts of participatory management, lowering decision-making to the manufacturing floor, just-in-time operations and total quality control[5]."

A minimum millwide control system should include the following functions[1]:

- Coordination and control of primary processes and equipment
- Management and supervision of operations
- Forecasting and simulation by using models and algorithms to prepare for upcoming situations
- Necessary data acquisition and reporting functions.

CHAPTER 9

Taking the view of a mill manager, the millwide control system should provide answers to the following types of questions[1]:

- How well am I running the mill now?
- Will I meet my budget and other goals?
- Am I profitable (and how much), or will I lose money?
- What will the situation be tomorrow, next week, and next month?
- What are the existing or future problems?
- How can I adapt to changing, dynamic market situations?
- What is the "best" alternative?
- When should we start the next maintenance shutdown?

In the pulp and paper industry today, many systems that might be millwide control systems are available. These systems differ primarily in their functions and in their hardware and software realizations. Hierarchical structures describe the functions of such systems[6,7]. Using experience today, one can say that the millwide control system is a planning and coordination tool for the mill management and operational staff at different levels of organization. It is especially for those staff members who handle the daily operation of the mill. A millwide control system collects, records, processes, and uses information from the entire mill. It does not use information only from one department or even from one area. Information comes from field instrumentation, distributed control systems (DCS), process management systems, laboratory, administrative data processing systems, and manual inputs. Data transfer obviously plays an essential role in the design of millwide control systems as does combining and displaying data from different sources.

Millwide document management and control could also come under the broad category of millwide control. These documents include piping and instrumentation diagrams, electrical documents, startup procedures, and equipment repair part lists. These documents can be provided and updated on-line through the mill network using on-line documentation tools such as hypermedia programs, Internet browsers, etc.

2 Planning of millwide control

Planning millwide control is a stepwise process that continues even during installation. It therefore provides a mill automation plan with the time and resource schedule for the system installation. The following stages generally occur[8]:

- Decision on millwide control system
- Mill control analysis
- Analysis of possibilities to improve mill control
- Cost and benefit analysis.

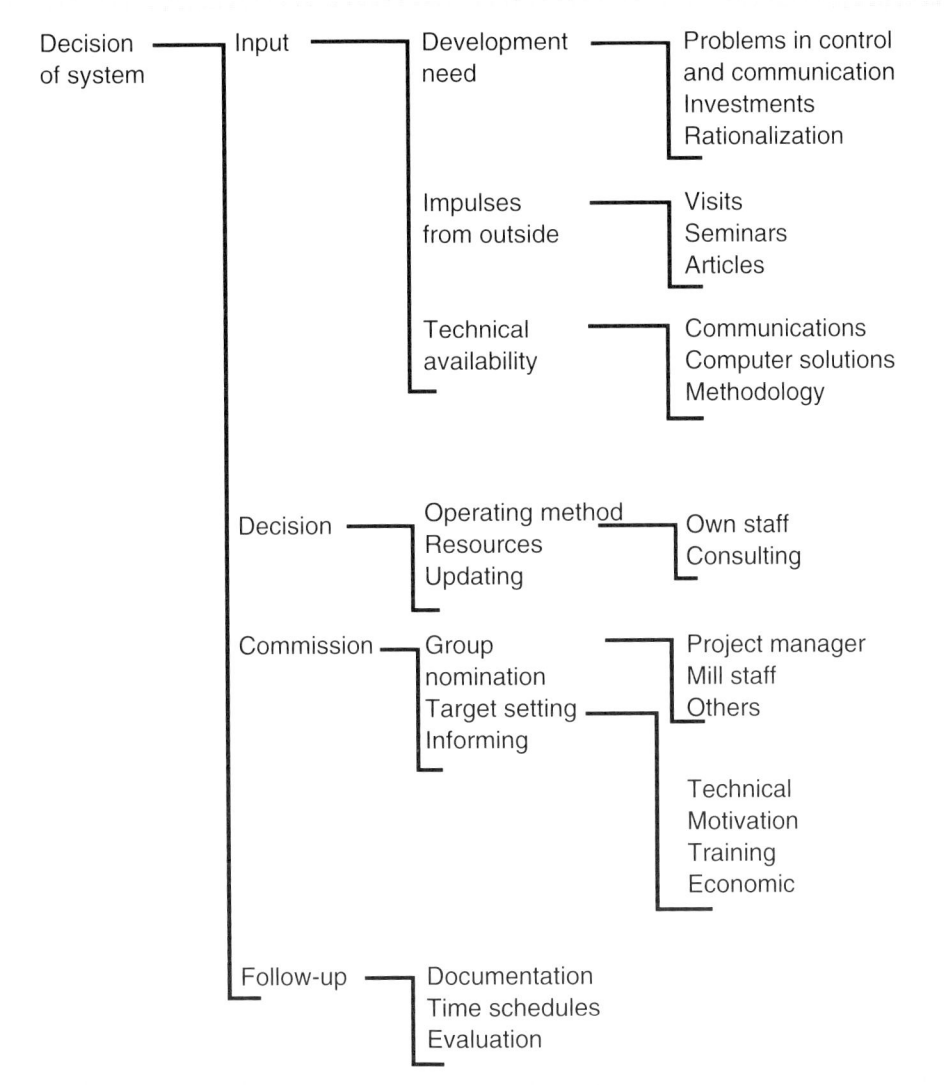

```
Decision ——— Input ——————— Development ——— Problems in control
of system                    need              and communication
                                               Investments
                                               Rationalization

                             Impulses ———————— Visits
                             from outside      Seminars
                                               Articles

                             Technical ——————— Communications
                             availability      Computer solutions
                                               Methodology

           Decision ———————— Operating method ——— Own staff
                             Resources             Consulting
                             Updating

           Commission ——————— Group ——————————— Project manager
                              nomination        Mill staff
                              Target setting ——— Others
                              Informing

                                                Technical
                                                Motivation
                                                Training
                                                Economic

           Follow-up ———————— Documentation
                              Time schedules
                              Evaluation
```

Figure 1. Topics to consider when making a decision on a millwide control system.

A methodology for planning a millwide information and control system should include the concepts of strategic planning from the top-down approach and should emphasize tactical, evolutionary implementation using the bottom-up approach[5].

2.1 Decision on millwide control system

The installation of a millwide control system in new mills is a necessity today. In existing mills, considerations to invest in millwide automation usually originate from problems in mill control and data communication and their effects on the profitability of production as Fig. 1 shows. Opinions on these problems and their causes and on their avoidance

CHAPTER 9

often vary considerably. A thorough analysis is therefore necessary as a first step. This requires a multi disciplinary team of people involved in the mill operations and the development of the mill control systems to clarify the situation in the mill. This group should consist of specialists in the following:

- Production management
- Process engineering and operations
- Control and systems engineering.

This team should primarily consist of mill staff, but a consulting engineering company can also be a component[9]. A millwide "champion" is necessary[5]. This is a person who is willing to accept the risks associated with the millwide project implementation and work as an advocate and leader.

Top management must make the decision on millwide automation because it influences the operation of the entire mill. The system must fit within the overall strategy and emphasize the critical elements of mill operations. The project should also have support from upper management. Forming an executive steering committee can assure this. The committee mission is to design the process, develop guidelines, measure progress, and assist with implementation[5].

2.2 Mill control analysis

The planning of the millwide control system should begin with documentation of the existing mill control and all the components that influence the millwide control system. Fig. 2 shows that the analysis involves the following[8]:

- Organization analysis: What organizational levels will be the primary users of the system? What are their needs and expectations? Is today's structure the best from the mill control point of view, or will the system installation result in changes in organizational structure? How is the responsibility of mill operations divided between different organizational groups and on which level? In what time span are decisions made?

- Area analysis: What parts of the mill will be included in the millwide system? How are they connected physically and from the mill control point of view? Is the entire mill network to be installed simultaneously or will it use the connection of separate "islands of automation?"

- Control system analysis: The existing mill automation is the starting point of millwide control discussions. The existing and proposed control and instrumentation systems require documentation and critical analysis of their condition. This documentation also helps to estimate the database of the millwide system. Besides equipment descriptions, it must also include lists of measurements and control loops included in the systems.

- Controllability analysis: Mill controllability can be studied in many ways. In process analysis, the nominal, maximum, and minimum capacities of the mill pro-

Millwide control

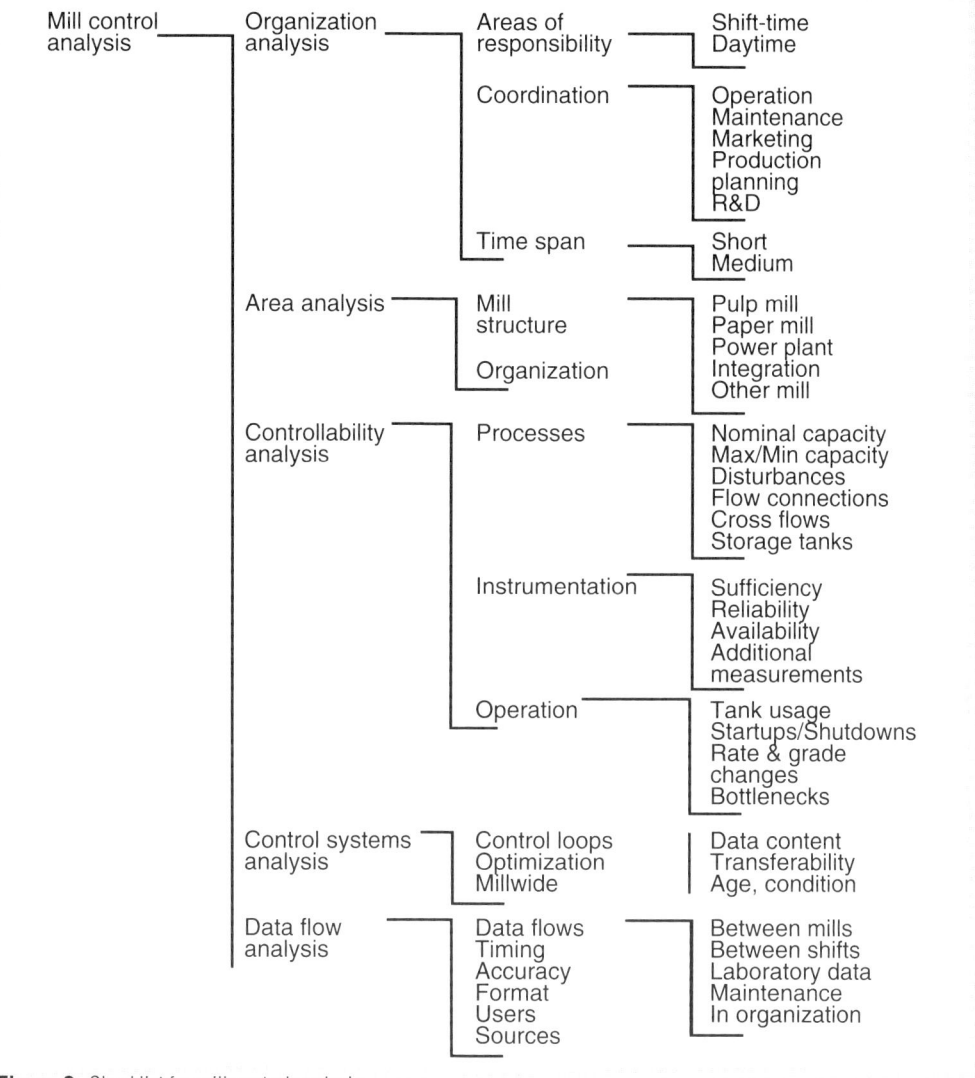

```
Mill control ─── Organization ─── Areas of          ─── Shift-time
analysis         analysis         responsibility        Daytime

                                  Coordination     ─── Operation
                                                       Maintenance
                                                       Marketing
                                                       Production
                                                       planning
                                                       R&D

                                  Time span        ─── Short
                                                       Medium

                 Area analysis ── Mill             ─── Pulp mill
                                  structure            Paper mill
                                                       Power plant
                                  Organization         Integration
                                                       Other mill

                 Controllability ─ Processes       ─── Nominal capacity
                 analysis                              Max/Min capacity
                                                       Disturbances
                                                       Flow connections
                                                       Cross flows
                                                       Storage tanks

                                  Instrumentation  ─── Sufficiency
                                                       Reliability
                                                       Availability
                                                       Additional
                                                       measurements

                                  Operation        ─── Tank usage
                                                       Startups/Shutdowns
                                                       Rate & grade
                                                       changes
                                                       Bottlenecks

                 Control systems ─ Control loops   ─── Data content
                 analysis          Optimization        Transferability
                                   Millwide            Age, condition

                 Data flow      ── Data flows      ─── Between mills
                 analysis          Timing              Between shifts
                                   Accuracy            Laboratory data
                                   Format              Maintenance
                                   Users               In organization
                                   Sources
```

Figure 2. Checklist for mill control analysis.

cesses are collected with the process connections and possibilities to change these connections. Mass and energy balances of the processes are generated. In instrumentation analysis, the reliability and availability of existing instrumentation is defined. This analysis also shows the need for additional measurements and points where redundant information is necessary. Operational analysis studies the principles of mill operations and determines the reasons to operate it this way. Determining better modes of operation and possibilities to reach them is also valuable. Operational analysis requires historical data on disturbances and stability diagrams. Besides the normal mode of operation, information on startup and shutdown situations, quality changes, and other special situations is also necessary. Bottleneck processes and situations also require identification.

- Data flow analysis: Data flows between control rooms of different processes, between the shifts, and between different levels and branches of the organization require analysis. This analysis should include data flows between automation systems and manual systems. This will clarify the mill database and needs for data transfer and report generation. Normal operation and exceptional situations require inclusion because the importance of some key data can be emphasized and the route of its transfer can be different during disturbances.

In mill and control analysis, interviews, inquiries, follow-up of the mill operation, data collection, analysis, and discussions with mill people are all equally important. They also form the basic data for the benefit calculations considered later.

2.3 Mill control improvement

Mill control improvement means that the planning, decision making, and monitoring functions associated with production have better speed, accuracy, and timeliness. This leads to better use of all mill data. When planning these improvements, the following possibilities of Fig. 3 exist[8]:

- Changes in processes and operational strategies: This means eliminating bottleneck processes, choosing consistent strategy, and finding compensating capacities.

- Changes in organizational structure: This can mean detailing responsibilities and diminishing differences between shifts with better data communication and improved monitoring of the processes and common rules of judgement and training. Millwide systems also improve cooperation between different branches in the organization.

- Changes in data communications and usage:

    - Improved communication leads to more efficient manual procedures with more consistent routines, improved reporting, additional measurements, and consistent operation during disturbances. It also means centralized control rooms, improved use of process control systems, and more efficient planning meetings and training.

    - A computerized millwide system handles data acquisition, data processing, and reporting functions. The system can also include a planning system with models and algorithms and a simulation system. Hardware consists of mill computers and links to other computers or distributed systems forming a network. Besides the necessary software tools for millwide functions, the role of data communication software is important today.

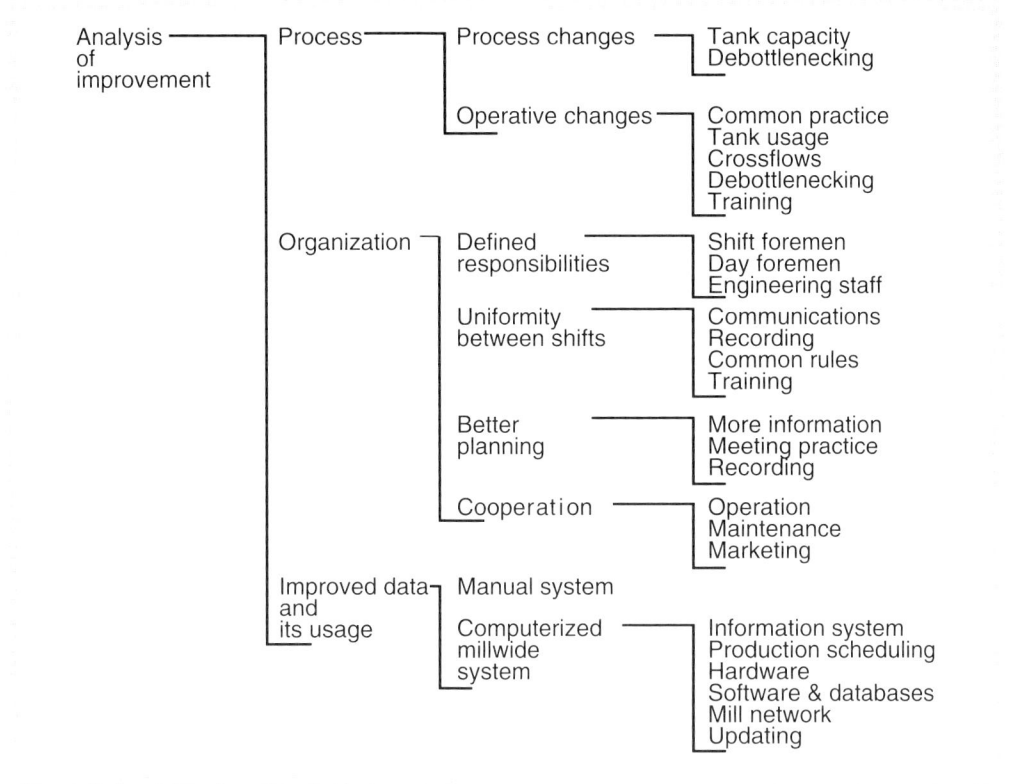

```
Analysis ──┬── Process ──┬── Process changes ──┬── Tank capacity
of         │             │                     └── Debottlenecking
improvement│             │
           │             └── Operative changes ──┬── Common practice
           │                                     ├── Tank usage
           │                                     ├── Crossflows
           │                                     ├── Debottlenecking
           │                                     └── Training
           │
           ├── Organization ──┬── Defined          ──── Shift foremen
           │                  │   responsibilities     Day foremen
           │                  │                        Engineering staff
           │                  │
           │                  ├── Uniformity       ──── Communications
           │                  │   between shifts        Recording
           │                  │                         Common rules
           │                  │                         Training
           │                  │
           │                  ├── Better           ──── More information
           │                  │   planning              Meeting practice
           │                  │                         Recording
           │                  │
           │                  └── Cooperation      ──── Operation
           │                                            Maintenance
           │                                            Marketing
           │
           └── Improved data ──┬── Manual system
               and             │
               its usage       └── Computerized    ──── Information system
                                   millwide             Production scheduling
                                   system               Hardware
                                                        Software & databases
                                                        Mill network
                                                        Updating
```

Figure 3. Possibilities for mill control improvement.

Another publication presents one systematic approach for planning millwide systems[8]. Fadum[10] reports the use of a structured analysis approach. Thomas[9] has stressed the role of top management in planning millwide automation and the importance of data analysis in actual planning. Pujol[11] sees two distinct roles for mill management in millwide projects. One is to staff the project team, and the other is to decide how to handle the changes caused by the millwide system (reassignment of personnel, training, etc.). Van Haren[5] sees ownership as an important management issue. By letting people who will use the system design the display screens as an example, their motivation to use the system increases.

3 Implementing millwide automation

A millwide system project is essentially an automation project. Because of the extent of millwide automation and the various functions it must handle, hardware today is inevitably some kind of local area network where sufficient memory and computer capabilities are available. The system design follows hierarchical system principles. Systems usually use a multi vendor environment, and computers of different age connect together.

CHAPTER 9

```
Computerized──┬─ Information ──┬─ Data acquisition ──┬─ Manual inputs
millwide      │  system        │                     └─ Measurements
system        │                ├─ Data handling ─────┬─ Initial data
              │                │                     │  Calculations
              │                │                     └─ Tranformations
              │                └─ Reporting ─────────┬─ Status reports
              │                                      │  Periodic reports
              │                                      │  Cost reports
              │                                      │  Disturbance reports
              │                                      │  Trends
              │                                      └─ Alarms
              ├─ Production ───┬─ Modelling ─────────┬─ Model area
              │  scheduling    │                     │  Variables
              │                │                     │  Initial values
              │                │                     │  Parameters
              │                │                     └─ Updating
              │                ├─ Optimum ───────────┬─ Initial data
              │                │  schedules          └─ Results
              │                └─ Simulation ────────┬─ Tank levels
              │                                      │  Species/grades
              │                                      │  Energy estimates
              │                                      └─ Raw material
              │                                         estimates
              │                   Follow-up ─────────── Alarms
              ├─ Hardware ──┬─ Measurements ────────── DCS
              │             │  Data links              Mill networks
              │             │  Computers               Office networks
              │             │  Work stations           Outside connections
              │             └─ Personal computers
              ├─ Software ──┬─ Operating systems ──┬─ Reports
              │             │  Basic software      │  Calculations
              │             │  Applications        │  Datalinks
              │             └─ Special software    └─ Statistics
              └─ Updating      Service contracts
                               Staff
                               Training
```

Figure 4. Implementing millwide automation.

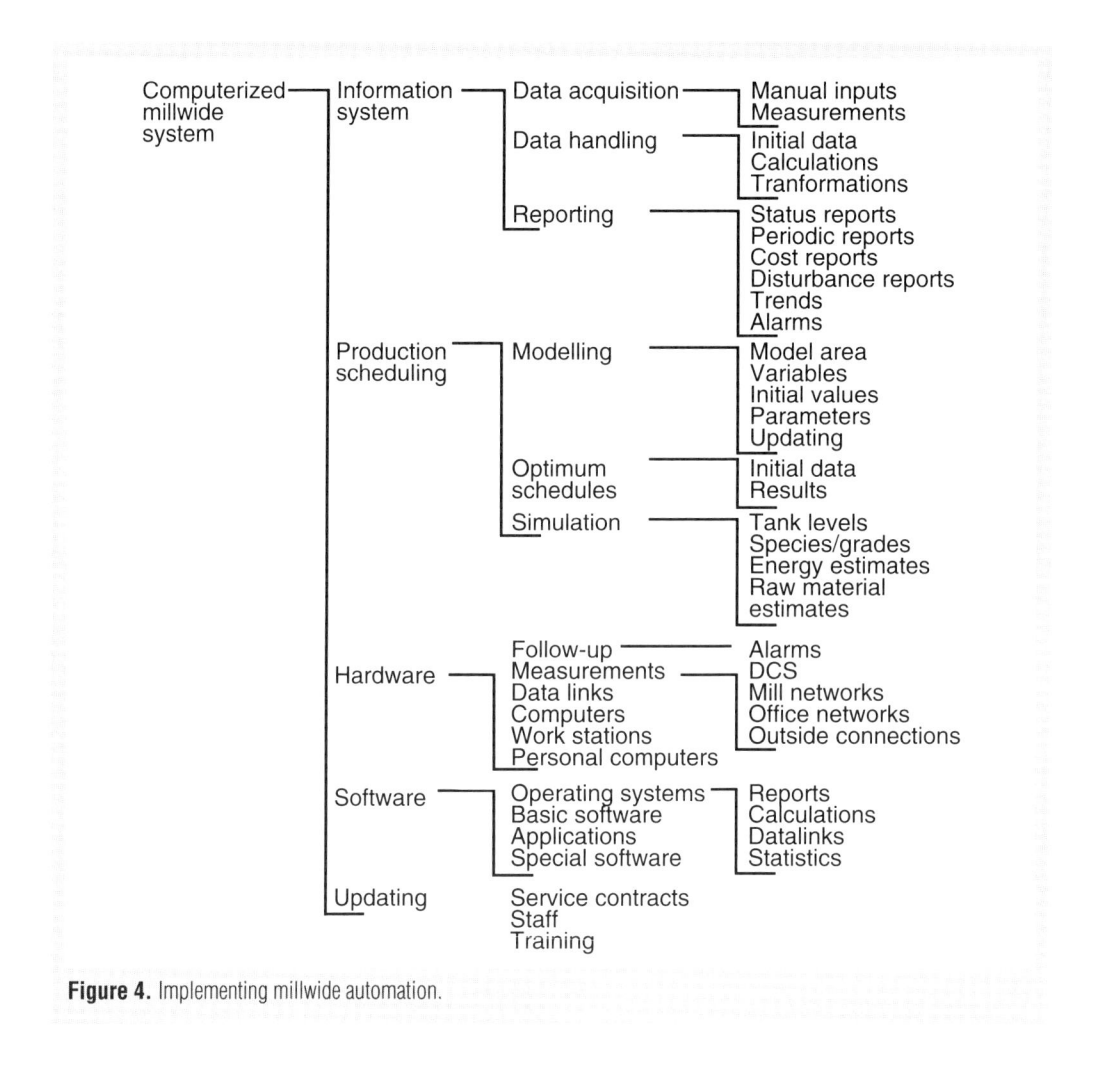

Figure 4 shows some important points for implementing millwide automation.

Because the use and the users of a millwide system vary considerably, special care is necessary for user friendly and interactive operator communication. Different software tools are also necessary during normal operation. Their use must be easy to learn.

Connecting different computers and automation systems of different age causes problems. Standardization work is still necessary to solve this.

In connecting millwide functions, an important feature is that the absolute accuracy of the primary information, i.e., some basic measurements, is stringent. This problem usually arises with measurements of mill production and product quality. For example, all consumption data is usually on a production basis.

The same accuracy is also necessary from material and energy balances and other models used in the millwide systems. Some model correction and validation capabilities are therefore necessary in the millwide systems.

Implementation of millwide automation also influences mill management and organization. Managers will realize better profit-oriented decision making tools with more timeliness that save their time for other tasks[12]. One suggestion is to create a new, distinct organization branch to handle millwide systems[13]. A separate job such as a production planner may be necessary for using the systems. At the very least, one must redefine the role of existing organizational groups.

Common critical elements of the implementation plan are the following[5]:

 - Engineering and design
 - Equipment selection and purchase
 - Construction
 - Software development
 - Training programs.

The same reference[5] summarizes the different costs in a millwide project as follows:

 - Hardware: 30%–40%
 - Software: 30%–40%
 - Management, testing, and specifications: 20%–40%.

Parallel projecting is recommended instead of the serial approach. The major decisions to be made are technology oriented. In a bottom-up implementation approach, some of these decisions are as follows[5]:

 - Evaluation of current systems
 - Selection of controls and automation systems
 - Segmentation of areas and cells
 - Communication standards
 - Database structure and size
 - Data access and limitations.

In all these decisions, the long projection time causes some problems. Managing a millwide project plan requires strong leadership. "Islands of automation" consisting of distributed control systems, programmable logic controller (PLC) systems, and "single window" systems for paper machines can form natural milestones in building millwide systems. Careful planning is necessary to make all these pieces fit together[5].

CHAPTER 9

4 Benefits of millwide automation

Listing and imagining the potential benefits and improvements of a computerized millwide production coordinating and control system is easy. Providing these estimates in numbers for a specific mill is difficult and complicated. Top management who allocates money for investments needs more than guesswork for decision making.

The most obvious benefits and savings are as follows [14]:

- Production rate will increase, and the product quality will improve.
- Consumption of raw materials per produced ton will decrease, and the energy efficiency of the mill will improve.
- Emissions and other environmental loading will decrease.
- The reporting system will be more effective. The real-time information from production efficiency and economy will be particularly important.

Actual numbers are difficult to find in the literature. Some suggestions are that most benefits of millwide control are not possible to estimate so no efforts should occur. Millwide control opportunities are mill specific. A mill with several energy sources may be able to trade off one fuel for another. A mill with several pulp sources will find pulp utilization to be profitable. A mill with several machines will usually find more opportunity from scheduling than a mill with one machine. A mill with many customers will have more need for a customer specific quality database than a mill with few customers. Users should understand and work through their mill specific economics in order to have a reasonable estimate of the benefits of millwide control to any specific mill.

The following data are from published information:

- Payback time for a production control system in a pulp and paper mill will be about two years [15].
- Increased production capacity is 3%–5% [16].
- Decrease in staff is six people, increased capacity is 2900 tons/ year, decreased oil consumption is 1000 tons/ year, and the decrease in electric power consumption is 600 MWh/year [17].
- Production increase is 7.9%, and energy savings are 4.9% from an energy balance [18].

In a major investigation concerning user opinions and needs, Table 1 shows a listing of the most important benefits of millwide systems [1].

Millwide control

Table 1. Benefits of millwide control systems[1]

Benefit	Percentage of answers
Better use of equipment and capacity	61
Better control of costs and effectiveness	58
Better decision making	55
More accurate and timely information	47
Increased production	45
Decreased losses and risks	37
Easier and more flexible planning	18
Saving in personnel	16
Increased safety in operation	11
Reduction of time loss	3
Better customer service	3
Better inventory management	3

Fadum[19] recently estimated the sources of benefits in millwide applications in paper, pulp, and power. Table 2 shows his results.

Table 2. Potential benefits of information systems in the pulp and paper industry[19].

	Production	Quality	Cost	Environment	Benefit
Paper	Improve prod. flex grade change	Product quality and process quality variations	Grade cost and cost accounting	Sewer losses	Maximum cost effective production and improved customer service
Pulp	Pulp mill scheduling	Process quality tracking	Process cost monitoring	Sewer losses	Maximum pulp production with reduced quality variations at lowest cost
Power	Energy management scheduling	Steam and electric quality	Energy accounting	Air and water monitoring	Most cost effective supply of energy while reducing environmental impact
Benefits	1 to 15% production increase	5 to 25% reduction in quality losses	1 to 10% cost reduction	Reduce loads by 0 to 50%	Optimum manufacturing efficiency

The following chapters are updated from an earlier paper[14].

CHAPTER 9

4.1 Increased production capacity

The key points in maximizing production capacity are optimum loading of the bottleneck department, minimizing rate and grade changes, and optimum use of buffer storage. Mathematical methods for solving this coordination problem have received extensive study. Calculation of the theoretical potential for increase in production capacity uses estimation of the theoretical maximum capacity of the bottleneck process and comparison with the present situation. The amount of this theoretical margin achievable in each mill depends on the following:

- Reasons for capacity losses such as eliminating lags or breaks in information exchange between departments
- Size of buffer storage and cross connections and modes of operation when parallel lines exist
- Nominal capacities of different departments.

Besides mathematical methods for solving the coordination problem, a computerized millwide control system also offers monitoring tools to increase short-term production capacity. For example, production rate deviations from the planned target (optimum) will be immediately noticed and a corrective action will result. This will mean avoiding any unnecessary decrease of capacity and change in production rate. The system also watches the load in all departments, follows typical key variables (measured or calculated) for predicting and preventing disturbances, and gives information about any maintenance necessary.

4.2 Reduction in raw materials and utilities

Consumption of raw materials and utilities per produced ton will typically be smallest with stationary operation at nominal load. A good process control will further minimize this consumption. If a disturbance or a production rate or grade change occurs, the new stationary state occurs only after long delays that might be hours or days.

This will also mean that specific consumptions can be higher for very long periods compared with a typical stationary operation. Quality of the final product can also suffer from changes. Minimizing the production rate and other changes will also decrease losses due to higher specific consumption.

4.3 Mill energy efficiency

Improved coordination between different departments and especially between power plant and production departments will give remarkable energy savings. The most important factors here include:

- Decrease of steam pressure changes that are dangerous for the boilers
- Decrease in exceeding the purchased power limit
- Optimum use of purchased power

- Decrease in disturbances at the swing boiler
- Stabilization of steam consumption to give improvements in total thermal efficiency of the power plant and increased reliability of the unit.

Energy management systems normally include many of these factors as discussed in Chapter 9.8. Having good coordination and exchange of information with the production departments is still important. This means real time links between the production planning system and the energy management system. A systematic mill study will give estimates for the above savings.

4.4 Decreased emissions and losses

An unavoidable increase in losses and emissions due to production rate and grade changes exists. Minimizing these factors with a millwide planning and coordinating system will also minimize their losses. A direct mill study will be necessary to estimate the monetary values for these savings and improvements.

4.5 Better information

An effective real time reporting and information system is necessary to make savings possible. Some direct savings due to computerized real time reporting are possible. For example, manual reporting will be minimized, and savings in labor costs will occur. This point depends highly on existing reporting and current mill practice.

5 Millwide information systems in pulp mills

5.1 Requirements for millwide control

General developments have also changed millwide applications in pulp mills by making data acquisition, processing, storing, and presenting more efficient. The backbone of the system and its data communications is the local area network. Several requirements it must fulfil are the following[20]:

- Meet international standards
- Avoid disturbances
- Support the selected hardware and be open for future solutions
- Support expansion
- Allow outside connections (telecommunications, external networks, and intra- and Internet)
- Allow vendor support and services
- Show field-proven application.

In pulp mills, the millwide information system is primarily a tool for the production management staff to improve monitoring and control of mill operation and product quality[20]. The system collects and stores process and laboratory data, upgrades this data, and distributes it timely and in proper format as printed reports and video displays. This is the basis for decision making for short-term operations and for long-term planning[18]. The system also connects with other data systems in the mill such as administration, maintenance, process management, etc.

The users and uses of mill information systems vary considerably. Different people in a mill organization also set varying requirements for the millwide information system[20]:

- Process and laboratory data must be collected into a common data base
- Data is stored systematically
- Data is upgraded to reports and displays
- Data distribution is done effectively to the proper persons at the proper time
- Manual reporting is replaced by a computerized reporting system and redundancies are avoided.

Computerized mill information systems will lead to improved monitoring of production and quality. They will also facilitate the early identification of process disturbances and their reasons. Data processing is efficient, and data is available in different forms. This makes the entire decision making process in pulp mills more effective. Easy connection with other systems and flexible possibilities to expand the system in the future are also important factors.

5.2 Information system functions

Depending on the case, the pulp mill information system can include production, energy, quality, and environmental monitoring activities and reports given to operators, supervisors, and maintenance and laboratory personnel[18].

Many ways to divide the overall information system into smaller functional subsystems are possible[18]:

- Status reports give an overview of each sub-process or the whole mill. They include the most important production, quality, and consumption data. Trends of production rates and storage tank levels are also displayed. The time interval is variable so that a longer history or a more detailed view of the latest development of data is available depending on the situation. Status reports also include so-called media reports for the present situation and the history for steam; electricity; process water; hot, warm, and waste waters; and emissions and chemical balances. They facilitate continuous mill operation and help the staff visualize an overview of mill status.

- Periodical reports give a summary of data for a certain period (day or month) in question. They include cumulative values of production and consumption data, specific consumption, mean values and standard deviations of quality data, etc. They help evaluate mill performance during a certain period and the analysis of mill operations.

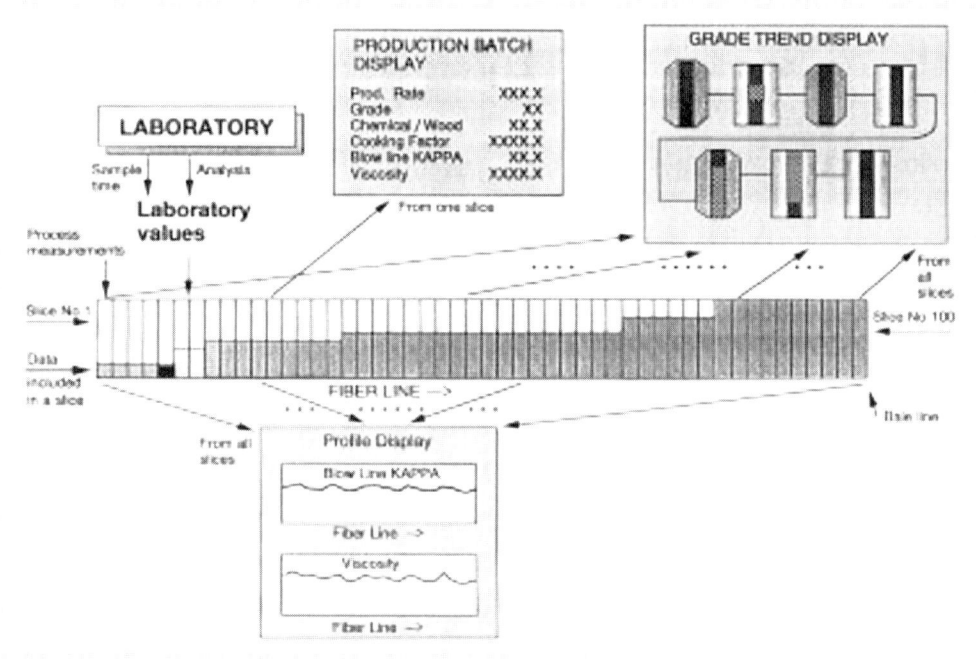

Figure 5. Principle of pulp quality monitoring (courtesy of ABB).

- Environmental reports primarily use laboratory values. They include status, daily, and monthly reports. They must fulfil the regulatory requirements and make possible mill control from the environmental viewpoint.

- Grade reports usually show all necessary data from the fiber line during a period for producing a certain grade. The grade report is finalized after the grade change, and the results from the earlier grade are printed or displayed. The report shows averages, cumulative values, and maximum and minimum values.

- The laboratory data system must make laboratory data recording, laboratory data entry, and laboratory reports possible. The system collects data from several sources: different laboratories, process measurements, and special analyzers. Data is presented in status, trend, and periodic reports. Average and cumulative values and standard deviations with maximum and minimum values are available for the user.

- Pulp quality monitoring allows following a production lot such as one cook from the digester house to the bale line and connecting different kinds of production and quality information to this lot. This serves quality control purposes and helps clarify reasons for possible deviations. Pulp quality fingerprint (PQF) records the production history of a specific unit or "slice" of pulp as it proceeds through the fiber line[21]. This usually uses a slice of 5–7 tons. The information is time stamped to record retention and passage times. Figure 5 shows the system principles[21].

CHAPTER 9

- Monitoring the operation registers operation times and shutdowns of different processes and devices. The results are shown in daily and monthly summary reports including cumulative shutdown times and number of shutdowns for a certain period. These numbers describe the mill efficiency in a process by process format.

- Cost reporting includes cost factors and a summary of production costs in different production departments of the mill. Specific consumption and unit prices are displayed. The report presents the material and time costs separately.

- Maintenance and troubleshooting management uses disturbance registration. It tries to record the reasons for disturbances and shutdowns and report the production losses. This serves the operation and maintenance staffs.

The millwide information system relies on process measurements and real time information. Reporting today should also use distributed data bases, office networks, and personal data processing.

Another publication[22] describes a fiber line information system. According to Fig. 6, this has four levels of development from process data to production line information. The authors think that many mills have now reached level 2, "Mill data," but some have advanced further.

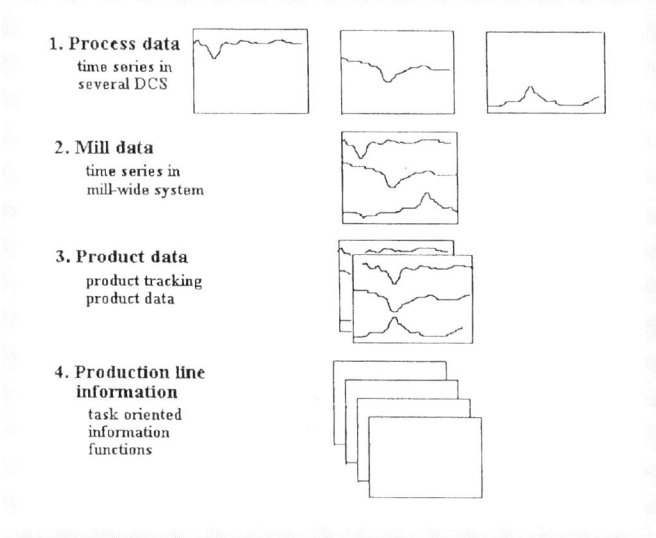

Figure 6. Four levels of development from process data to production line information[22].

The concept of "Fiber Line Information Presentations" (FLIPs) includes at least the following functions[22]:

- Pulp data information includes kappa number and brightness levels along the fiber line. All variables are indicated according to their process position.

- Last shift survey includes the same data as before but for the latest shift of the operator with the shifts before and after. This allows the operator to survey consequences of his actions that have not been visible during the time of his shift.

- Forecast shows what will happen to pulp still in the production line before it exits from the dryer. This helps operators at the beginning of a shift.
- Residence and passage times show the moments when a certain bulk of pulp will reach separate process positions.
- Costs and quality summarizes some key variables for a specific bulk of pulp all along the production line. Costs are given relative to the costs of reference treatment for the pulp grade in question or as absolute costs.
- Processing profile displays parts of costs and quality data.
- Process analysis supports the operator in investigating the relationships of process variables for improved operation, training, fault detection, etc.
- Calibration support continuously checks the data quality of the pulp flow and consistency measurements.

Fiber tracking is the backbone of the FLIPs system. It includes flow pattern models, on-line parameter identification, delay time and mixing calculations, "resampling" of time series to pulp data and required corrections, and calibrations.

6 Production scheduling in pulp mills

6.1 Introduction

Pulp mills are very complex systems considering their total control. They consist of processes with storage tanks between them[23–26]. Shutdowns and disturbances in one process propagate and influence the operation of other processes. This may lead to production losses and unnecessary changes in production rates that cause additional quality disturbances. These can sometimes increase the environmental load of the mill. Production rate changes may also cause undesired wear on process equipment. The propagation of disturbances can lead to enforced shutdowns of processes due to full or empty storage tanks. The startup of the process after this kind of disturbance can also be problematic. Obviously, all these factors can cause considerable economic losses.

Improved production scheduling reduces these losses. First, this means determination of production rates for different processes so that full use of the storage tank capacity of the mill is possible. Second, the system must be able to help the staff continuously with decision making. This is especially important when disturbances and unexpected situations occur. This point is usually neglected in theoretical studies that concentrate primarily on the first problem, but its importance from a practical point of view is significant.

6.2 Mill model

For production scheduling purposes, the pulp mill is usually described as a system of processes and storage tanks between them possibly with a steam balance as Fig. 7 shows. The model flows are for pulp, liquor, and steam. Concentration of chemicals is assumed constant. The dynamics of individual processes are usually neglected, and the

CHAPTER 9

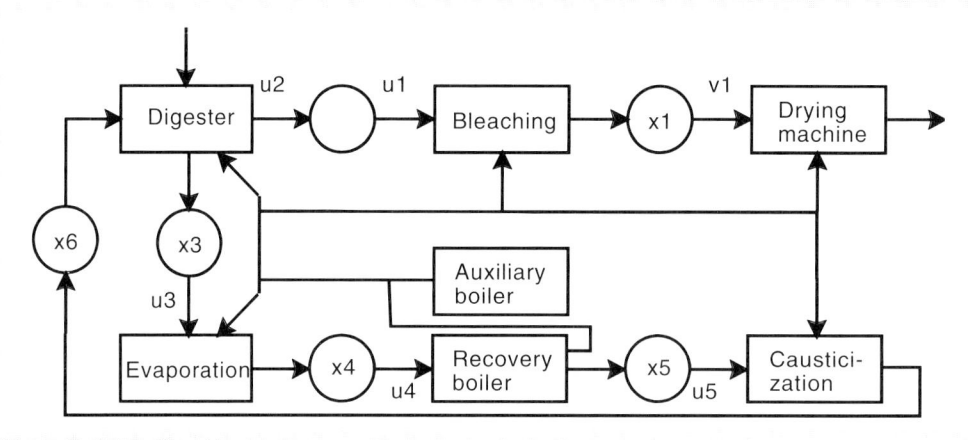

Figure 7. Pulp mill model.

model only includes storage dynamics. One important definition in the scheduling models is the concept of transfer ratio. This is the ratio between flows around each process. It is usually assumed constant during scheduling. In practice, this means that processes always use raw materials and prepare products in the same ratio when producing the same quality.

The modeling usually uses common state space notations. The amount of material in each storage tank describes the state of the system by the state vector, $x(t)$. The production rates of the processes are control variables forming the control vector, $u(t)$. A given pulp production is usually the deterministic disturbance variable, $v(t)$. The following vector-matrix differential equation describes the relationship between the variables:

$$\frac{dx}{dt} = Bu(t) + Cv(t) \tag{1}$$

where B and C are coefficient matrices describing the relationships between the model flows (transfer ratio).

Since most storage tanks have only one input flow and one output flow, most elements in B and C matrices equal zero.

If a steam balance is included in scheduling, an additional variable describing the steam development in the auxiliary boiler is necessary. This is a scalar variable denoted by S. Steam balance is the following:

$$S(t) = Du(t) + Ev(t) \tag{2}$$

Note that the right-hand side of the equation includes consumption and generation terms. The capacity limits of tanks and processes constrain the variables in the model as follows:

$$x^{Min} \leq x(t) \leq x^{Max} \tag{3}$$

$$u^{Min} \leq u(t) \leq u^{Max} \tag{4}$$

$$S^{Min} \leq S(t) \leq S^{Max} \tag{5}$$

Due to the fact that scheduling concerns long times, no complete and complicated process models are necessary. If all the small storage tanks are part of the model, the system dimensions increase to make this difficult to handle. These tanks also have no meaning from the control point of view. Combining small storage tanks usually simplifies the model. Another alternative is to eliminate the small tanks by "lumping" the processes around the tank together as Fig. 8 shows.

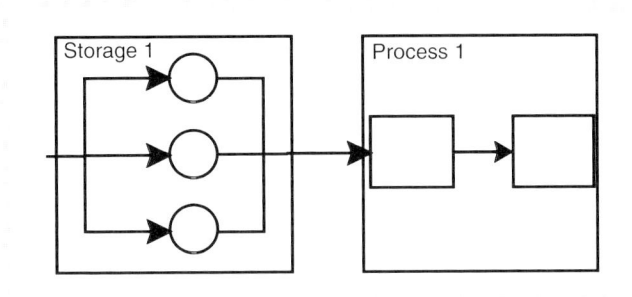

Figure 8. Definition of tanks and processes.

6.2.1 Processes

The following sections are a modification and updating of another publication[23].

Disturbances such as rate or grade changes, planned shutdowns or temporary restrictions, and random disturbances that can cause a complete shutdown or merely limit the production to some extent always influence the operation of process departments. Scheduling usually uses these phenomena by considering the constrained nature of control variables (See Eqs. 3–5.)

Because the flows around each process are assumed proportional, the selection of the control variables is arbitrary from the mathematical point of view. One flow entering or leaving the process is selected to describe the production rate of the process as the control variable. In practice, the control variables should be selected respective to the mill practice and measurement possibilities. In pulp mill processes, the incoming flow is usually measured and selected as the production rate.

Production rates of fiber line processes are usually given in tons/a.d. ton of pulp per hour. Correspondingly, the production rates of the chemical recovery processes are in cubic meters of the chemical in question per hour. For the evaporation plant, tons of evaporated water per hour, for the batch digesters, cooks/day, for the recovery boiler, tons of dry solids burned per hour are sometimes used as a measure of production rate.

6.2.2 Storage tanks

The importance of storage tanks in controlling the mill is common. By keeping the proper level in tanks, one avoids the propagation of disturbances from one process to the neighboring one and to the entire mill. This leads to fewer production rate changes

CHAPTER 9

and shutdowns. It also means smaller production losses as discussed earlier. If the optimum tank level is maintained, a longer time margin in disturbance situations is also available.

In optimization methods, difficulties arise from weighting the costs of storage levels against the costs of production rate changes. This problem is usually due to excessively tight problem definition. In practice, exact maintenance of optimum storage levels is not necessarily the most important thing. Instead, the storage level should remain inside the optimum or recommended region. When the level goes outside this area, a control action must begin. Each tank must also have some certain allowable operational limits. Tank limits should therefore be applied according to Fig. 9.

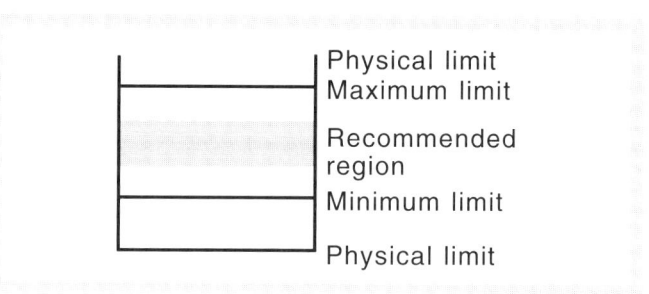

Figure 9. Different regions of the storage tank.

In the model, the storage tank or the amount of material inside the tank is in tons or in cubic meters depending on the case. Unit conversions are used, if necessary.

6.2.3 Rate and grade changes

In addition to possible production losses, production rate changes may also cause quality variations and decrease the efficiency of the process. Energy losses and incorrect dosing of chemicals may also occur. A production rate change may be due to the production plan for the sake of coordination, or it can occur because of filling or emptying the storage tank before or after the process. Process technology also causes limitations in the rate at which the production can change without endangering the equipment. Most processes can be shutdown almost immediately, but a startup or rate increase always takes some time. Scheduling considers this by setting a constraint to the speed of rate changes or by considering the production losses during startup as Fig. 10 shows.

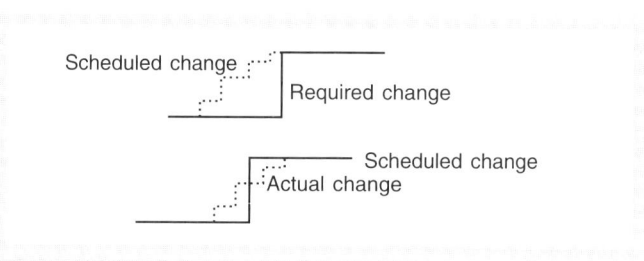

Figure 10. Rate changes in production scheduling.

6.2.4 Planned shutdowns

Planned shutdowns are disturbances scheduled in advance. This provides sufficient time to make preparations to compensate for their effect. The compensation is carried out by applying proper levels to the storage tanks before and after the process department to be stopped, i.e., emptying the tank before the process and filling the tank after it. Scheduling programs do this by increasing the production rate of the shutdown process before the shutdown by the amount that is lost during the shutdown.

6.2.5 Random disturbances

Disturbances that leave only a short time or no time to prepare any compensation for their effect add uncertainty to production scheduling. Increasing the size of storage tanks can meet this uncertainty, but this also increases investment costs considerably.

Scheduling considers random disturbances so that a reserve capacity in tanks can compensate for the effect of the disturbance. The size of this reserve capacity depends on the statistical behavior of the processes around the tank; their tendency for breakdowns, the duration of disturbances, the built-in capacities, bottleneck processes, etc. The efficiency of this approach also depends on possibilities to control the tanks, i.e., how they can remain inside the allowable limits.

6.2.6 Process delays

From the production control viewpoint, process dynamics can be neglected. Compared with the storage dynamics, time constants of the pulp mill processes are usually so small that they do not require consideration. Process delays can have great significance in some situations.

Two different kind of process delays exist in pulp mills: pure transport delays that describe the passage of material through the processes and delays that describe how fast a production rate change proceeds through the process. No correlation necessarily exists between these delays. In continuous digesters where long transport delays are typical, rate change occurs almost without any delay, as an example. To determine the necessity of delays from the scheduling viewpoint, the strategies and the method of rate changes require careful study.

Transport delays are important when considering the processing of a grade or species change through the production line. One must note that the production rates of processes that have long transport delays (digester, bleach plant, and recovery boiler) must remain as smooth as possible. Rate changes cause quality impairment that is especially difficult to offset in these processes because of the long transport delays.

6.2.7 Bottleneck processes

A bottleneck process is one whose capacity restricts the production of the entire mill. The production capacity of a process depends on its nominal capacity and on the disturbances influencing its operation. The bottleneck can therefore be permanent or temporary. Various process departments can also become bottlenecks due to changes in operating conditions such as species changes, machinery breakdowns, and lack of raw materials, chemicals, or energy.

Determination of temporarily varying bottlenecks is a very important function in the production scheduling system. Optimization approaches effectively handle bottlenecks caused by lack of built-in capacity or a planned shutdown. They are not so flexible to deal with abrupt disturbances and bottlenecks. If some machine component suddenly

CHAPTER 9

breaks or an unpredicted lack of raw materials or utilities disturbs the process operation, the operator needs to learn quickly how long he can continue the production of the remainder of the mill. He also wants to have an answer to questions about mill control if he cannot prevent the propagation of the disturbance in question to other processes. This requires flexible simulation and heuristic reasoning or an expert system type of approach.

The production capacity of a longer-term bottleneck process requires maximization to optimize mill production capacity. This calls for a specific control strategy for the storage tanks, that assures complete use of tanks in all situations. The strategy depends heavily on the reliability of process operation. It can be formulated using repair schedules and disturbance statistics. Figure 11 shows some cases.

The first case shows a built-in bottleneck. The production of the bottleneck process should be maximized in all conditions. One way is to keep the tank level before the process high so enough raw material is always available. The second case shows a process that tends to become a bottleneck because of disturbances. To assure operation of other processes, the tank level before this process should be low, and it should be high after it.

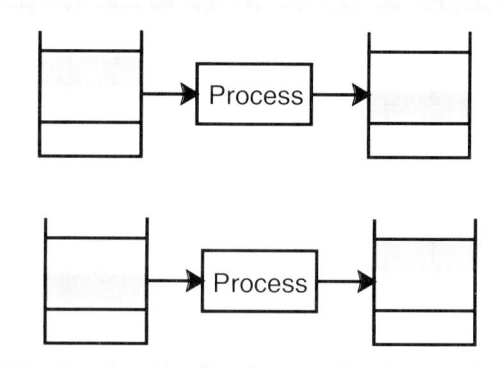

Figure 11. Control strategies for bottleneck processes.

6.3 Optimization problem

Setting targets and constraints for an optimization problem in a qualitative way is very easy. In some cases, putting the items in exact mathematical form is more difficult. In the following, the list of usual targets includes short comments for accomplishing them in most scheduling methods[24, 25].

- Planned pulp production must occur in due time or the amount of pulp must be maximized. Accomplishing the first target sets pulp production as a deterministic disturbance in the model. This means that one potential control variable is tied leading to uncontrollable models in some cases. A more "heuristic" method is therefore necessary to achieve more flexibility. Using a soft constraint in fuzzy optimization can improve this. Maximizing the pulp production requires an objective function that includes pulp production as an independent variable. Heuristic methods can also do this by trying different pulp production levels and selecting the highest one.

- Storage capacity between the processes must be totally used before production rate changes are allowed. This requires an objective function to set penalties to the rate changes until the storage tanks go inside their limits. This is the most usual type of objective function used in scheduling methods. It usually has a quadratic form:

$$J = \frac{1}{2}\left\{ \|x(K) - x^o(K)\|^2_{Q(K)} + \sum_{k=0}^{K-1} [\|x(k) - x^o(k)\|^2_{Q(k)} + \|u(k) - u^o(k)\|^2_{R(k)}] \right\} \quad (6)$$

Variables denoted by a superscript ^{o} are reference values for levels and production rates. The sensitivity of this objective function to variables x and u is tuned by selecting the weight factors Q and R. Note that a different weight must usually be set to the level at the end of scheduling, $x(K)$.

- Planned shutdowns must be part of the production schedules. This usually occurs by using the constrained nature of the model and setting the upper constraint of the production rate of some process to zero during the defined interval. The actual planning of shutdowns is quite difficult to implement in production scheduling because it requires very heuristic methods.

- Unplanned disturbances must be handled. Setting the storage limits in a proper way (See the previous section.) compensates for the effects of stochastic disturbances. An earlier comment noted that these situations require their own interactive planning tools. Only few systems have these.

- Dealing with process and tank constraints must be possible. This is accomplished by using the model constraints shown in the previous section. This requires constrained optimization methods.

- Possibilities to change the target levels of the storage tanks such as for preparing a planned shutdown must exist. Higher weights are therefore set for the tank levels at the end of a scheduling period.

- An energy balance must be part of the scheduling. Different possibilities to do this vary from simple balance calculations to allocation of energy resources.

6.4 Survey of methods

Pulp mill production scheduling has been an object for active scientific research for more than 25 years. Bengt Pettersson did pioneering work in Sweden at the end of the 1960s. He studied mathematical methods for production scheduling at the Billeruds AB Gruvön mill[27,28].

Pettersson studied simulation and optimization methods. Simulation guaranteed short solution times, but the manual work associated with problem specification and testing of different production schedules was too extensive. The sub-optimality of the solution also caused problems. Some attempts were made to systemize the simulations by defining standard functions for production rates and combining them using combinatorial methods.

Pettersson also studied the application of a standard Linear Programming (LP) package for solution of the scheduling problem. The magnitude of the LP matrix was about 750 rows and 400 columns. In the test runs, this problem was too extensive for solution by the process computer available then. Formulating a proper objective function to weight the costs of production rate changes against the divergence from the desired target levels of the tanks was also difficult.

For production scheduling in the Gruvön mill, Pettersson developed an algorithm using the successive solution of many smaller LP problems (about 50 rows and 40 columns) defined and linked together by the maximum principle. Installed in 1969, this was probably the first scheduling application.

A systems theoretical approach for this problem was introduced in Finland[29]. A simulation technique solved the problem of stochastic optimization of storage sizes and corresponding control strategy of production units. The research was a cooperative program of Helsinki University of Technology, Ekono, and Oy Wilh. Schauman Ab Pietarsaari Mills. The results were implemented in industrial practice by sizing the storage tanks to give improved coordination of processes because of efficient use of the buffering capacity of the entire mill[29-32].

The same authors first solved the problem of production scheduling and energy management as a whole[31,32]. They used dynamic mathematical programming to include energy generation, distribution, and price variables in the net profit function. They also compared the net profit function with the one Pettersson used that minimized the number of production rate changes. For the second case, the comparison showed that the net profit of the mill decreased compared with the net profit maximization. This proved that the objective function should preferably be the linear-quadratic type. In the middle of the 1970s, the Pulp and Paper Combinate in Stamboliisky, Bulgaria, tested the same approach.

Swedish Forest Products Research Laboratory (STFI) studied an efficient network flow algorithm (PNET) originally developed for electrical energy distribution networks. This group of algorithms has the advantage of small computer memory demands and short execution time. Both factors were important for process computer applications available then. The software was installed at the Iggesunds Mills and the Skutskär Mills in Sweden and at the Kemi Oy Mill in Finland. Considerable literature concerning this application exists[17, 33, 34]. A similar system is marketed today as part of a pulp mill production control system (PMPC)[21, 35].

In France, Centre Technique du Papier (CTP) in cooperation with Laboratoire d'Automatique de Grenoble and Cellulose d'Aquitaine studied a scheduling method during the 1970s[36]. In 1985, CTP installed a system called Optimill at the Papeteries de Gascogne Mimizan mill[37, 38]. This system includes functions to create the process configuration interactively, collect process information, identify the mill model, simulate the control strategy given by the user, report possible bottlenecks, and perform optimization using Tamura's algorithm.

Balaz[39] developed a method that combined heuristic and optimization approaches. This is the method of nonsynchronous production scheduling. This method allows nonsynchronous process output changes, and it looks for the best time structure of control.

Skopin[40] presented a method for optimum production control of a bleached pulp line. The method was used to evaluate the optimum tank levels in unplanned shutdowns, determination of schedules for a shutdown process, and transition to a specific state of the mill after an unplanned shutdown.

The University of Oulu studied a group of hierarchical control algorithms[24, 41–44]. Tamura's time delay algorithm was specially modified according to the specific features of the pulp mill production scheduling problem. The advantages of this method are its user-oriented, simple solution routine, modest computer memory requirements, moderate execution time, and consistency with the hierarchical structure of integrated automation systems. Information on further development of this approach is available[45].

Parallel to these studies, the University of Oulu also studied heuristic simulation and network flow algorithms[46]. Heuristic simulation offered a flexible approach for solving production scheduling problems despite its sub-optimality. It offered qualitatively satisfactory results in most cases. It does require considerable programming work in putting the heuristic rules in the system and case-by-case modification seemed complex. To avoid these disadvantages, a test system that used expert system framework for rules development was built[47]. Work also involved a simulation and heuristic based method[48].

In summary, the methods and algorithms developed since the late 1960s for solving the pulp mill scheduling problem fall into six groups:

– Mathematical programming (LP)

– Optimal control

– Hierarchical multi level optimization

– Network flow algorithms

– Simulation with heuristic methods

– Expert systems and intelligent methods.

Golemanov has shown[29, 49] that the first three methods have a common base in the systems theory. In the following discussion, two examples provide more details.

6.4.1 Tamura's time delay algorithm

Tamura's time delay algorithm is part of the goal coordination algorithms. In these algorithms, the optimization problem is broken into smaller sub-problems solved with simpler methods. The division occurs according to state variables, control variables, or both. The algorithm works with a discrete model. The above mentioned mill model must first be discretized. It can also deal with the constraints in state and control variables. The quadratic objective function (Eq. 6) is suitable for the algorithm. A specific feature is that it can handle process delays according to the equations below[50]:

CHAPTER 9

$$x(k+1) = A_o(k) + A_1 x(k-1) + \ldots + A_\theta x(k-\theta)$$
$$+ B_o u(k) + B_1 u(k-1) + \ldots + B_\theta u(k-\theta) \tag{7}$$

$$x(0) = x_o \tag{8}$$

In solving the pulp mill scheduling problem, the division occurs according to index k. Instead of the original minimization problem, its dual problem is solved:

$$\underset{x,u}{Min}\ J = \underset{p}{Max}\ M(p) \tag{9}$$

This leads to a two-level algorithm where the lowest level solves the optimum values for x and u for the constant p, and the values for p are upgraded on the higher level. This requires some optimum search algorithm as Fig. 12 shows. It usually uses a gradient search algorithm. The gradient corresponds to the above state equation. More detailed solution information is available[24, 50].

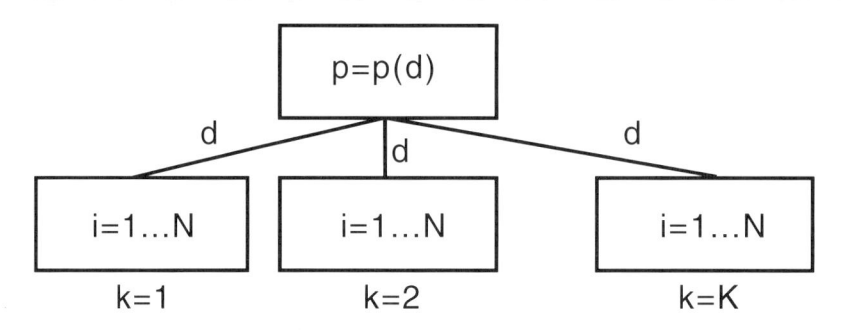

Figure 12. Tamura's time delay algorithm as a two-level description.

This algorithm has the following advantages:

- Time delays are included without increasing the number of state variables. This minimizes calculation times.
- It easily handles the constraints given as inequalities.
- The objective function only has terms that have a clear physical meaning.

Test results are available[24], and some comments are the following:

- Compared with other approaches, the algorithm is rapid and has low memory size requirements. With modern computers, this fact is not as important.
- Changes in weight factors influence operation. This determines the processes used in compensating the disturbances or the storage tanks where the levels change more frequently. In practice, this tuning requires good knowledge of the algorithm or good user interface. Note also that the weight factors do not have any physical meaning.

Millwide control

- The algorithm itself is easy and simple. Upgrading to a standard program package is possible.
- The method requires updating procedures for the mill model or at least the active monitoring of the model performance and manual changes when necessary.
- The method suffers from the tight definition of the optimization problem. The mathematical formulation does not leave room for larger changes in mill operation practices. This disadvantage is common to all optimization approaches.
- Energy balance is not directly part of the optimization, but it is a constraint.
- The algorithm minimizes deviations in the state and control variables. In some cases, maximizing the production would be beneficial, but this is difficult with this approach.

6.4.2 Heuristic scheduling

As an example of heuristic scheduling, consider an actual industrial installation. The installation was divided into two stages with an extensive millwide information system built first[51] and the scheduling system later[48].

The scheduling system calculates the production schedules of nine pulp mill processes. The number of storage tanks is eight. Calculations start from the existing tank levels and production rates at the beginning of the scheduling period.

After the initialization and modification stages, schedule calculations proceed according to Fig. 13[48]:

- Reference schedules are calculated for each production rate variable defined in the mill model specification and used as inputs to the simulation routine. As initial data, simulation needs production rates of all processes for all scheduling intervals, initial state of storage tanks, transfer ratios of processes, length of the scheduling interval, and length of the entire scheduling period.

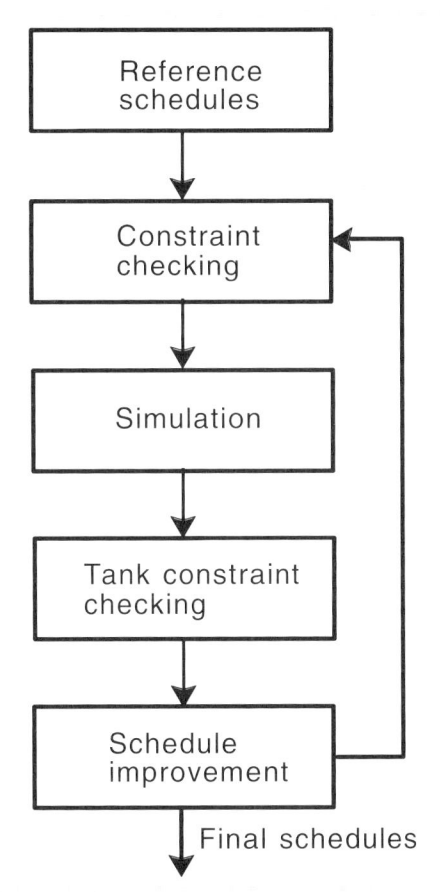

Figure 13. The proceeding of production scheduling calculations[48].

CHAPTER 9

Using this information, storage trajectories are calculated as a function of time.

- Checking ensures that reference schedules fulfil the capacity restrictions. If not, a bottleneck situation is reported.
- In the simulation, tank level trajectories respective to the prevailing schedules are calculated using a mill model.
- The user can modify the schedules manually and do simulations with these modified schedules.
- Checking is necessary to see if calculated tank levels fulfil the restrictions. If they do, results are given to the user who can accept or modify them.
- If the tank level restrictions are not fulfilled, the schedules are iteratively improved.

Schedule improvement in Fig. 14 uses a set of heuristic rules. A short summary of its targets are as follows[48]:

- Constraints set for scheduling must be kept (capacity constraints of tanks and processes, rate change speed limits, etc.)
- Number of production rate changes must be minimal, and the necessary ones should occur in a logical, coordinated way
- In each case, bottleneck processes, if any, should be detected and reported to the user.

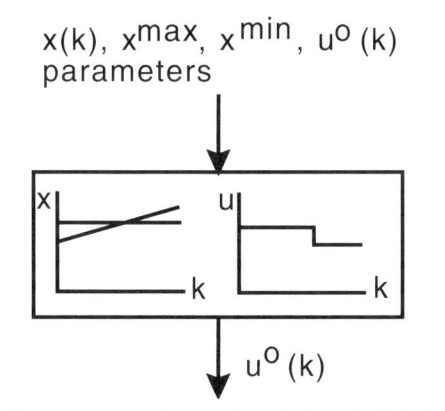

Figure 14. Principle of schedule improvement[48]. The figure shows a situation where a tank level constraint is violated, and because of it the production rate of the process before the tank will be decreased. This is notified as a change in the reference production schedule of the process.

The basic ideas behind the heuristics are the following[48]:

- No control action occurs as long as the tanks remain in an allowable region.
- If a tank level constraint is violated, the control action is first subjected to the process that naturally controls the tank. Most often this is the downstream process. If its capacity is not sufficient or it cannot be used for other reasons such as a shutdown, the other process is selected. Control actions consider the capacity limits of processes and rate change speed limits.
- If the rate changes are not sufficient to keep the tank under control, a process shutdown is suggested.

- After the tank in question is put under control, the effect of respective control actions to other tanks is tested by simulation. If some other tank limitations are violated, similar control actions are repeated.

Although the basic ideas seem simple, accounting for all possible situations and their combinations leads to a complex set of heuristic rules.

6.4.3 Application of expert systems and intelligent methods

As mentioned in Chapter 3, expert systems and intelligent methods are sometimes combined with conventional algorithmic approaches. The result is a sub-optimal or feasible schedule or presentation of several schedules for operator selection. One problem here is to guarantee the optimum solution. Simulations assure feasibility.

Expert systems and intelligent methods give versatile and efficient tools for the formulation of production scheduling problems[26, 47, 52, 53]. Compared with optimization approaches, their main advantages are the following:

- Optimization is not limited to a single objective function. Depending on the case and the selected mill control strategy, the objective function can be changed.
- The qualitative factors in scheduling are easy to implement. The opinions and adapted operation strategies of the mill staff can be considered and compared with the optimum strategy. This also has training benefits.
- Changing the mill control strategy and the mill model used in scheduling is easy.
- Stochastic and uncertain features are easier to consider.

Production scheduling applications suffer from the common expert system disadvantages:

- Knowledge acquisition problems: Generating a consistent and correct rule base and keeping it updated for a long time when the mill and its operational strategy are changing is difficult. Some kind of adaptation is necessary.
- The optimum solution is impossible to guarantee. Feasibility might even be a problem. The priorities in the solution approach might also change in different situations. By using different strategies, the "optimum" solution is probably neglected in the early stages of reasoning.

Fuzzy logic would be a suitable tool to improve the rule-based systems[26]. By using linguistic variables and fuzzy presentations, the vagueness that is inheritant in this kind of production system can be handled. For example, scheduling can include it as soft constraints. In practice, the constraints given to storage tanks are seldom strict, but they can be broken in some cases. This is also true for the objective function. Another advantage is in the inference system. In using heuristics or a conventional rule base, a danger exists that a feasible solution is overruled in the early stages of using heuristic

CHAPTER 9

rules. This depends very much on the order of preference in which the scheduling rules are handled. Using fuzzy logic representation can avoid this.

Genetic algorithms and evolutionary computing lend themselves to production scheduling problems[54]. This makes solution of a multi objective problem possible.

7 Millwide control in paper mills

7.1 Millwide control functions in paper mills

The following discussion is a modification and updating of another publication[55]. According to Fig. 15, the millwide control system in a paper mill includes the following monitoring and planning functions:

- Order handling, i.e., the receiving and acceptance of customer orders
- Formulation of the long-term basic production schedule of the mill usually for a period of one month
- Production scheduling using the basic production schedule usually for a period of one week. This also includes the allocation of machines for different kinds of products.
- Real-time monitoring of production and realization of the production schedule, and order status
- Quality monitoring
- Efficiency monitoring
- Profitability monitoring

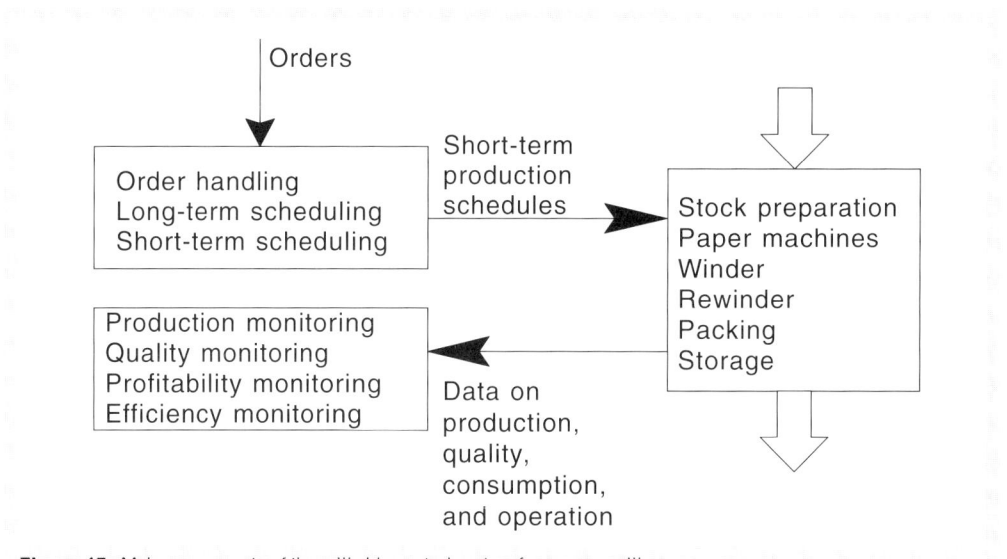

Figure 15. Main components of the millwide control system for a paper mill.

Millwide control

For design purposes, the millwide control system usually has two parts: the information system and the scheduling system. The millwide information system collects, processes, and distributes data that is needed or created at the operative control level of the mill. A millwide scheduling system forms a tool for the operator who is responsible for the collection and scheduling of orders.

7.2 Order handling

Order information received by the millwide control system consists of information on new orders, changes in existing orders, cancellation of orders, delivery orders, and invoice information.

Each order requires consideration of the following information:

- Order code
- Information on delivery times, customer, etc.
- Product and quality standards
- Roll and machine roll information
- Delivery information
- Invoice information.

7.3 Production scheduling in paper mills

Mill profitability will increase with the following:

- Increasing sales amount
- Increasing sales prices
- Reducing fixed costs
- Reducing variable costs.

With the production scheduling system for a paper mill, one can influence the first and last points above. Minimizing trim losses and grade change costs can reduce variable costs. Optimization of trim losses and grade changes also results in immediate savings in manufacturing time for an order group. In this case, the capacity increases, and the sales amount can increase if the markets are not restricting.

In production scheduling, the following factors require consideration: trim losses, grade change times, paper machine costs, inventory costs, late delivery costs, planned shutdowns, capacity limitations, availability of raw materials, and machinery condition[55].

Figure 16 shows a flow chart describing the phases of paper mill production scheduling and real-time production monitoring[55]. The short-term production scheduling uses the basic production schedule for one month to form a frame for scheduling. This frame essentially tells the order in which to produce different paper grades to decrease grade change costs. In principle, each grade is produced at least once per month. The first criterion in scheduling is the delivery time of the customer order.

CHAPTER 9

Orders for delivery within 4 weeks are then arranged according to brightness. Orders of the same brightness are further arranged according to basis weight and dye. Then orders belonging to the same grade are further grouped and combined so the detailed production schedule for the paper machines and the winders makes minimization of trim loss possible.

Another publication[56] discusses paper production scheduling in a Japanese mill. In this case, the large number of pulp types (18), different paper runs (more than two hundred per month), and products make the scheduling difficult. Planning for monthly scheduling occurs near the end of each month. The weekly paper production schedule covers eight days.

The paper production scheduling system has three sub-functions:

- Product group scheduling that assembles products similar in quality and sharing approximately the same delivery period. Product grouping uses delivery time, amount of demand, and mutual arrangement rules between products.

- Individual scheduling assigns particular products into their respective product group.

- Balancing of scheduling improves the schedule considering the total demand of each kind of pulp and associated energy consumption. This occurs at 30 min intervals for the entire duration of the schedule.

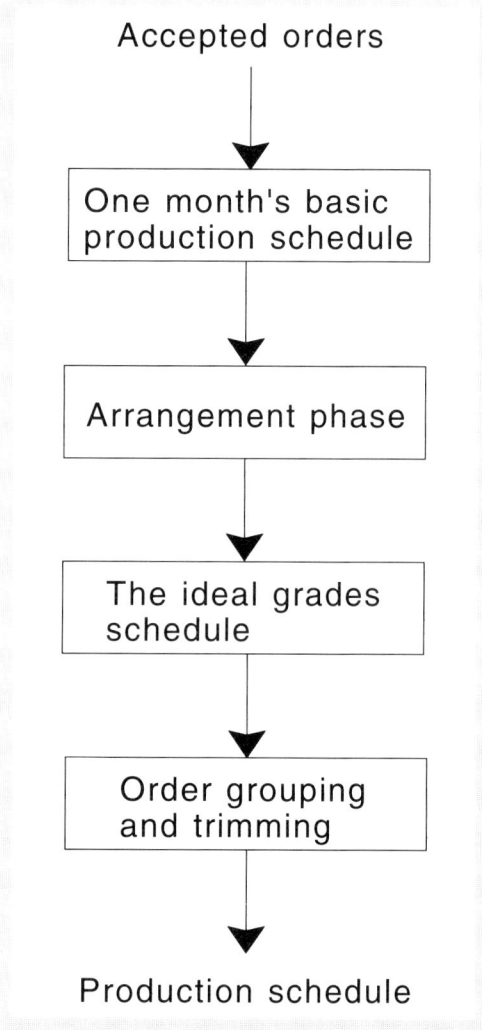

Figure 16. Principles of paper mill production scheduling.

An expert system realized the scheduling. Collection of the expert rules used close examination of manual scheduling. Opposite to operations research methods, expert systems provide the user with several feasible schedules. Human judgement selects the best schedule from four candidate schedules as Fig. 17 shows.

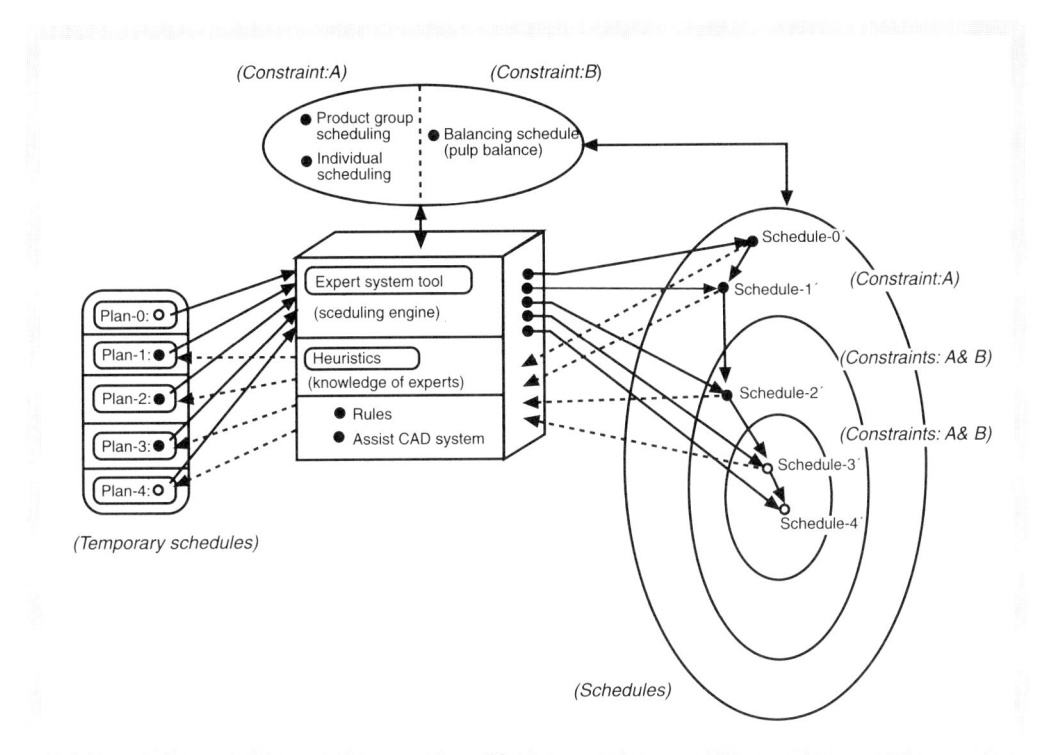

Figure 17. Principle of the schedule making process[56].

7.4 Real time production monitoring

The monitoring system serves two functions. It gives the operator information on how to realize the scheduled production. It also gives information on the order status to improve customer service. The operator can certainly make modifications in the schedules if major disturbances or lags occur using the information received from the monitoring system. These functions occur in a very flexible way when augmenting monitoring and scheduling in the same system.

Information is collected for shift, daily, and monthly production reports. Rejected rolls and order status information are registered and reported. Delivered products are reported by grades. The production data is also used in profitability and efficiency monitoring, budgeting, and production scheduling.

Order status monitoring tries to produce real-time information on the progress of orders through the production line. It therefore supports the sales division and customer service. The progress of a single order through different points in the production line is registered, and the operator also receives alarms and instructions from the system. Registration of the progress of orders uses introduction of a special order status code updated as the order progresses. For example, updating of the order status code might occur after each of the following actions[55]:

- Order receipt
- Order acceptance into the long-term schedule
- Order scheduled and trimmed
- Production of order begins in the paper machine
- Interruption in order production
- Order produced in the paper machine
- Order produced in the winder
- Order packed
- Order delivered.

Recording of actual production rates of the paper machines and winders occurs simultaneously. Monitoring can optimize production according to the orders. Divergence in realization of the orders according to the schedules and produced amounts therefore requires recording.

7.5 Intelligent quality control systems

Expert systems and neural networks have had use in quality control systems in paper mills. The primary objectives of a paper quality expert system are the following[57]:

- Secure the quality of paper
- Minimize variations between shifts
- Reduce production costs
- Support operators
- Provide a flexible simulation tool
- Use existing knowledge
- Train new staff members.

The basic functions of one system[57] are to collect real time process and quality data, evaluate the measured quality against customer specifications, recommend necessary corrective actions, and simulate fulfilling of these actions as Fig. 18 shows.

The system is rule-based. Its knowledge base consists of natural language rules and mathematical formulas and functions. The quality model represents the process knowledge as a quality matrix. Rules come from mill experts and known relationships between control and quality variables.

Several workstations can operate with the same knowledge. One could be in the paper machine, and the other could be in the paper laboratory. Figure 19 shows the operation environment.

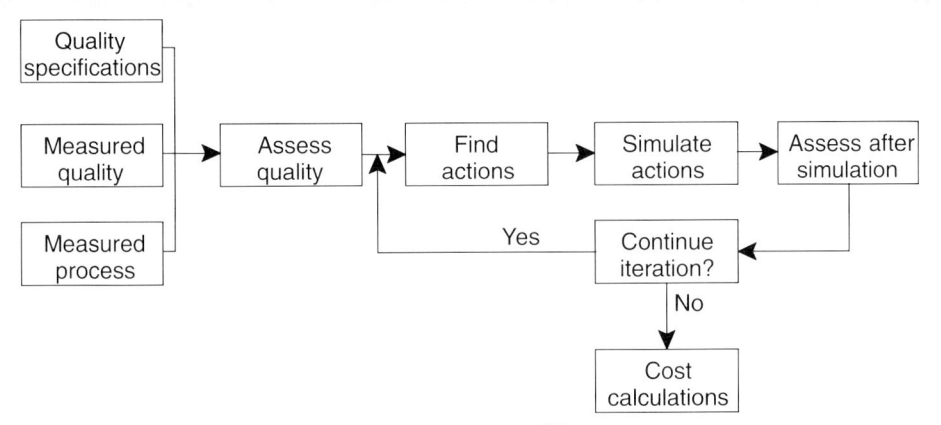

Figure 18. Main functions of the paper quality expert system[57].

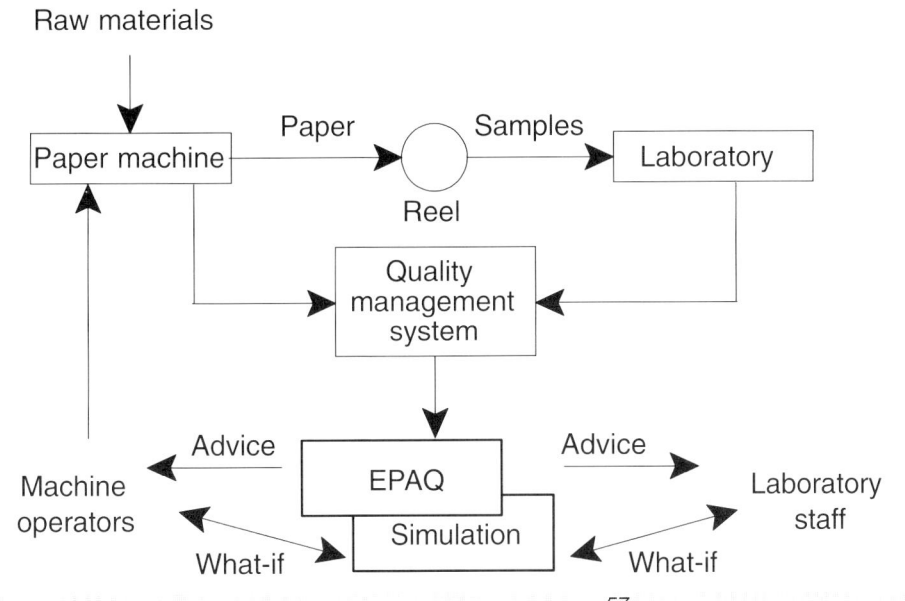

Figure 19. Operation environment of the paper quality expert system[57].

Q-OPT[58] is a neural network kernel used in the quality control of paper machines. Figure 20 shows its main functions:

- Real time prediction of output values using on-line measurements
- Optimization
- Sensitivity analysis
- Statistical analysis.

CHAPTER 9

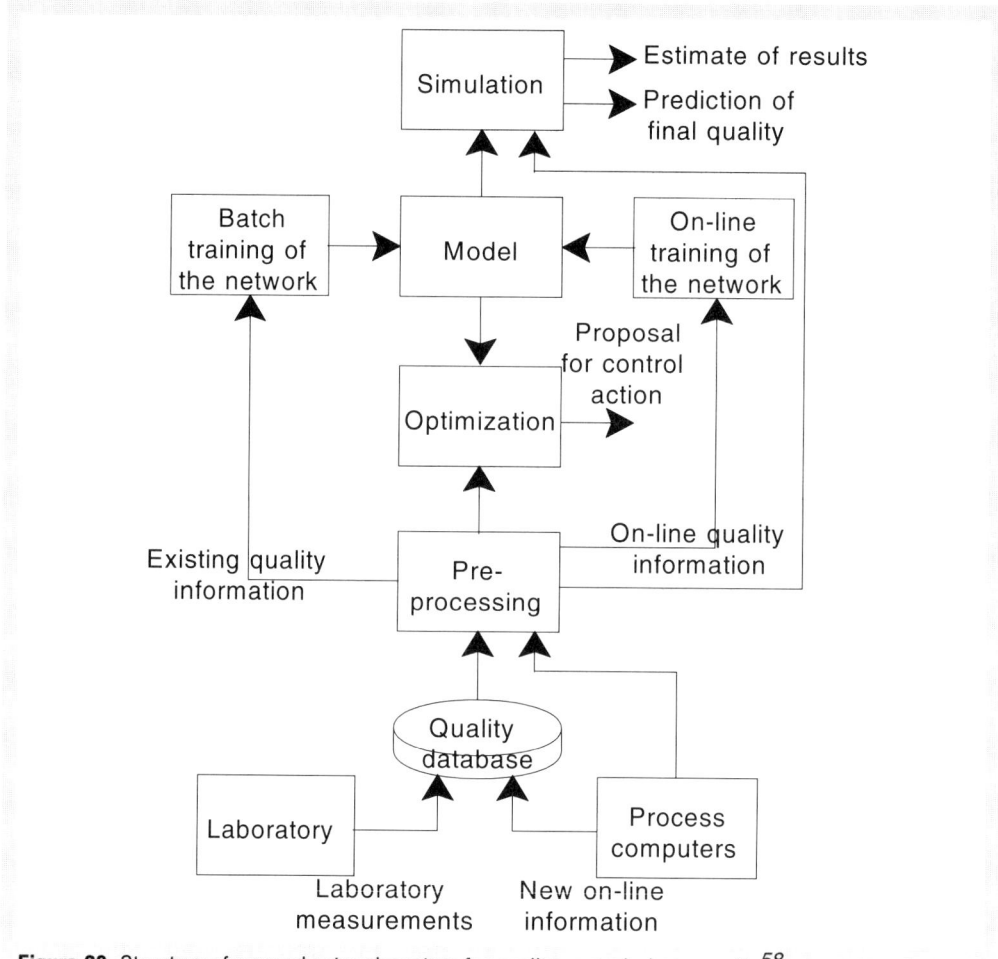

Figure 20. Structure of a neural network system for quality control of paper mills[58].

7.6 Efficiency monitoring

Efficiency monitoring uses production and operation data such as shutdown times and break times of the paper machine, breaks in the winder, lost time in all the paper mill sections, total production time, amount of production, grade changes, and variables including machine speed, width of paper, etc.

Using this information, the total efficiency of the paper machine can be expressed in terms of specific efficiency numbers. Figure 21 describes the definition of production efficiency and time efficiency factors. These factors indicate the efficiency of operation, production, length use, width use, and approval of products. Effect numbers monitor operation, productivity, and production effects[51-63].

Millwide control

```
Planned
downtime                      Length
Lost                          loss (X)
time (A)                      Width
                              loss (Y)
                              Reject (Z)

Actual                        Net
production                    production
time (C)                      (PN)
```

B=A+C
Available
production time

Time efficiency
=100(B-A)/B %

Gross production
PG=PN+X+Y+Z

Operating efficiency
=PG/B tons/h
Production efficiency
=PN/C

Figure 21. Definition of efficiency coefficients.

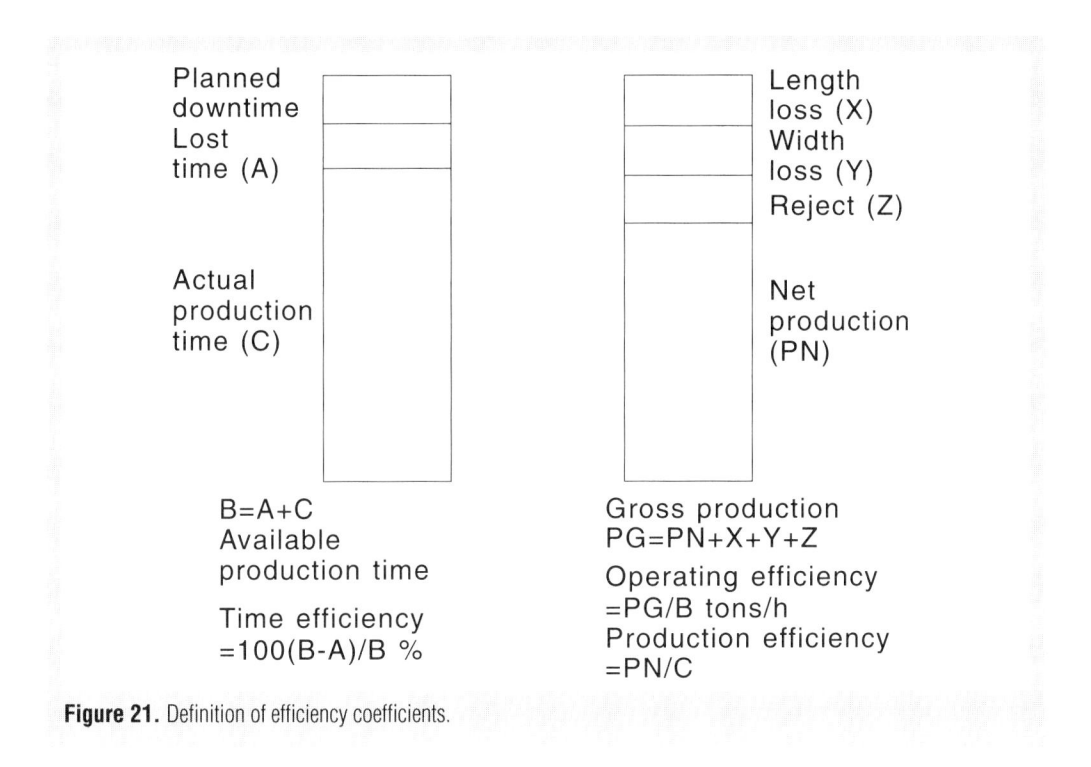

7.7 Profitability monitoring

Raw material consumption in the paper machine is measured as fiber components and stock additives dosed to the mixing chest. When the gross amount of production, amount of rejected rolls, and rejects and prices of raw materials per weight unit are known, calculation of raw material costs for each product unit is possible. Consumption of steam and electric power is measured. Calculation of energy costs per product unit uses energy prices. Since consumption data depends on grade, calculating costs for each grade is required. After each grade change, the millwide computer system collects and saves raw material and other cost information totals for further processing.

Profitability of each order group is calculated by using its cost structure. The sales price (the price received from the customer) must cover delivery and production costs and a sufficient part of a paper mill's common costs as Fig. 22 shows.

```
                    ▼  Delivery costs:
                       freight, forwarding, insurance, etc.
Sales price         ▶  Immediate production
                       costs
                    ▲  Profit
```

Figure 22. Factors influencing sales price of paper.

Only the immediate variable costs are components of the monitoring of profitability of paper grades manufactured at a mill. This includes the costs of raw materials, operation, energy, and packaging. Monitoring of other costs is necessary in draft budget making when considering actions for next year for product groups, market areas, capacity, and customers.

8 Energy management systems

In many industries, energy is an important factor from the actual production viewpoint and for cost and availability. In some cases, this is a critical resource. Variations in energy prices have required more efficient energy use. At least the following three approaches are available for this purpose:

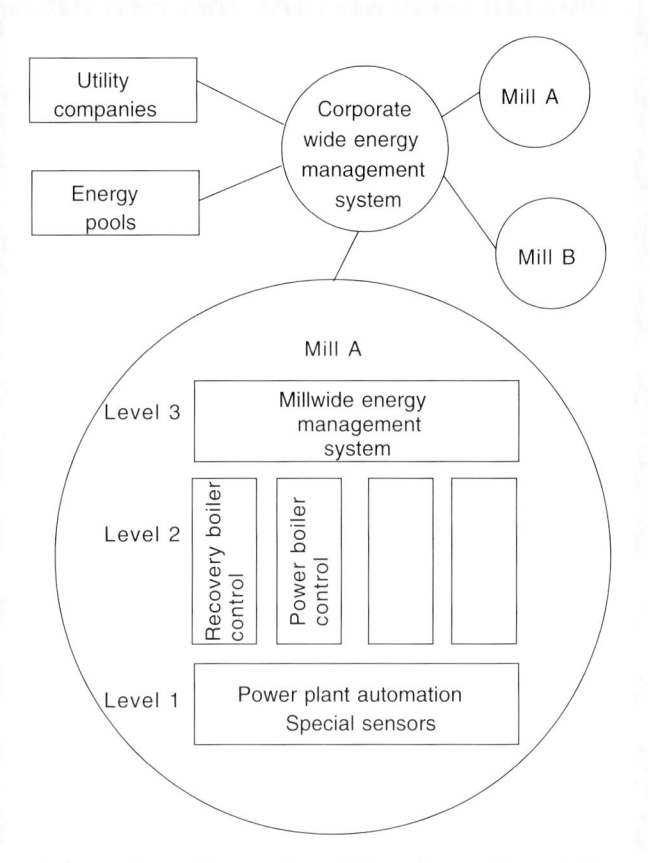

Figure 23. Structure of energy management systems in the pulp and paper industry[64].

- Selecting the products so they use less energy
- Designing processes that use less energy or energy that is more readily available or cheaper
- Improving process instrumentation, control, and management to give more constant and lower energy consumption.

Energy management can have four hierarchical levels as Fig. 23 shows:

- Stabilizing controls that use basic instrumentation and feedback controls
- Optimization and optimum control of the processes that handle energy production (power plant processes)
- Millwide control and optimization of consumption and generation of energy
- Corporate-wide energy management.

These functions try to minimize total energy costs of the mill simultaneously with assuring availability of energy. Disturbances are handled by some compensating functions and corporate-wide coordination of energy production and consumption that is primarily responsible for power and fuel acquisition.

The targets and need for energy management vary depending on the energy consumption structure and the method of acquisition. The following discussion gives some basic principles.

Energy management functions are simple when either electricity or steam dominates the consumption structure and the remaining utility use is minor. If the electricity consumption dominates and no appreciable internal generation exists, the main concern of the energy management system is purchase of electricity at the lowest possible cost.

If the mill primarily uses steam generated in its internal power plant, the most important energy management function is minimizing the production costs of the steam and balancing the steam production and consumption during disturbances. The difficulties encountered depend on the structure of the power plant: the number of boilers and fuels. If only one boiler or fuel is available, energy savings can result from optimum control of the boiler and use of external compensation capacity during disturbances. With many boilers and fuels, the optimum load allocation of boilers provides the most important source for cost savings.

Energy management becomes complicated in mills that have their own steam and electricity generation or purchase electricity from external sources. This is true where consumption of different forms of energy depends heavily on the operation strategy selected. These are the situations in most integrated pulp and paper mills. The saving potential in these cases is remarkable.

8.1 Millwide energy management systems

The main functions of the millwide energy management system are the following:

- Optimization of energy consumption and acquisition
- Coordination of production and consumption of energy
- Management of disturbances
- Reporting.

These functions have further division into the following functional modules:

- Boiler load allocation
- Turbine load allocation
- Compensation of variations in steam consumption
- Purchased power optimization
- Optimization of pressure levels in user systems
- On-line simulation and optimization.

CHAPTER 9

In the control hierarchy of the mill, energy management is typically the upper level activity. In many cases, it can also handle on-line control of some process variables.

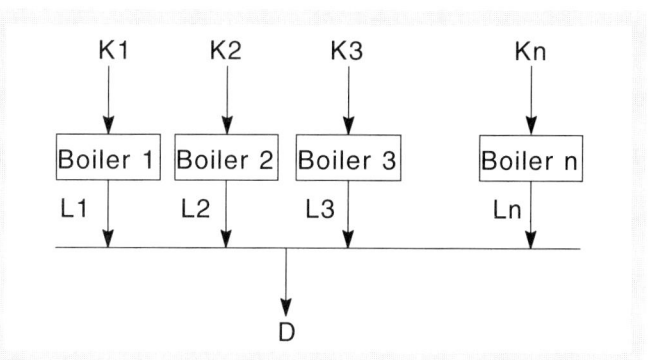

Figure 24. Example of power plant with several boilers feeding the same steam header.

8.2 Boiler load allocation

Boiler load allocation tries to divide generation of the required steam between boilers so that generation costs are minimum. Two kinds of boilers exist:

- Process boilers whose production rate depends on the mill operation. They have a limited potential for energy management. Recovery boilers belong to this category.

- Primary boilers that use purchased fuels and whose production rates require optimization.

For boiler load allocation, the boiler with multiple fuels is usually considered as separate boilers: one boiler for each fuel as Fig. 24 shows. The performance characteristics are defined separately for each fuel. The objective function for optimization is the total cost function of steam production as follows[65–74]:

$$Min \sum C_i = C_1 + C_2 + \ldots C_n \tag{10}$$

where n is the total number of boilers and the required amount of steam is taken as the constraint for the optimization:

$$\sum L_i = D \tag{11}$$

where C_i is the cost of steam production for a boiler [monetary unit/h]
L_i the steam generation of the boiler [kg/h]
D the steam requirement of the processes [kg/h].

The quadratic performance curve estimates the performance of each boiler:

$$e(L) = aL^2 + bL + c \tag{12}$$

where $e(L)$ is the boiler performance [%]
a, b and c are the coefficients determined experimentally.

The boiler load usually has a minimum value that differs from zero and a certain maximum value. Figure 25 shows some examples of efficiency curves[69].

The cost of the steam generation for each boiler is now a function of the boiler load:

Efficiency curves

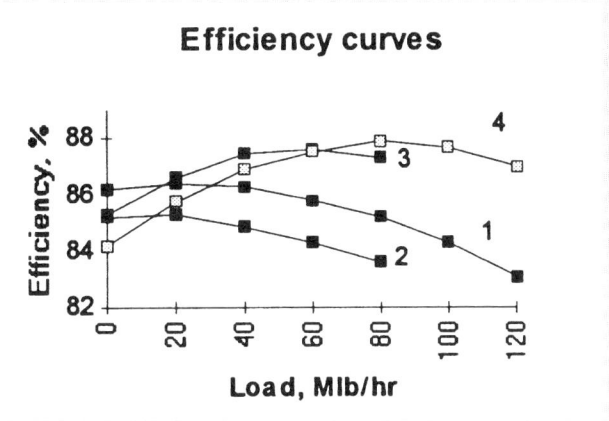

Figure 25. Examples of boiler efficiency curves[69].

$$c_i(L_i) = \frac{K_i(f)L_i}{e(L_i)} \quad (13)$$

where $K_i(f)$ is the fuel cost per kg of steam.

The following equation gives the value for the fuel cost:

$$K_i(f) = \frac{K_i p_i(f)}{h_i(f)} \quad (14)$$

where
h_i is the enthalpy change in the boiler [kJ/kg]
$p_i(f)$ the fuel price per kg of fuel
$h_i(f)$ the heat value of the fuel in question [kJ/kg].

Most optimization methods involve incremental costs. This is the tangent of the cost function at a certain point according to the following equation:

$$\Delta c_i(L_i) = \frac{dc_i(L_i)}{dL_i} \quad (15)$$

where $\Delta c(L)$ is the incremental cost for the boiler load per kg of steam.

The boiler load allocation uses the cost function developed for each boiler (actually for each fuel) and the incremental costs calculated for different boiler loads. Several methods exist, but they usually use successive iterations. One possibility is to start from equal percentile loads for each boiler to fulfil the total steam requirement. The load is then taken from the boiler that has the largest incremental cost to the boiler that has the lowest cost. This continues until the incremental costs of each boiler become equal, or the load constraints are met.

8.3 Turbine load allocation

Turbine load allocation closely resembles the boiler load allocation especially in situations where all turbines feed from the common high pressure header, all turbines are condensing turbines, or the turbines produce steam to the same low pressure header[71]. In practice, the situation is often much more complex because the same power plant can have several high, medium, and low pressure levels and condensing and back pressure turbines fed from different high pressure headers. These facts with reductions and auxiliary condensers easily lead to mill specific optimization models.

Turbine load allocation is easy when the turbines generate internal, cheaper electricity while simultaneously fulfilling the steam requirements. In these cases, the only optimizing function is to minimize the costs for power generation. Two approaches are available[71]:

- Optimization uses the power requirement. Energy costs are minimized on the need for internal generation.

- Optimization uses the steam requirements. With certain steam production, the internal power generation is maximized.

In practice, some facts make the optimization more complex:

- Power is sometimes purchased at a price lower than that for power generated by the mill's own condensing turbines and auxiliary condenser. This can even decrease the price of power generated by the mill's own back pressure turbines.

- During disturbances, the steam balance may drift to an area that is not beneficial for power generation. The steam requirements then increase, but the power requirement remains the same. In these situations, the entire power plant operation must be directed to a steady-state.

Turbine models are linear. For example, the following equation gives the power generation of the back pressure turbine in Fig. 26:

$$G = a_1 + a_2 f_2 + a_3 f_3 \tag{16}$$

where f_2 is the flow of the medium pressure steam [kg/s]
 f_3 the flow of the back pressure steam [kg/s]
 a_1, a_2 and a_3 are coefficients determined from turbine characteristics
 G is the power generation of the turbine [MW].

Other discussion presents more detailed models[65, 70, 71].
The turbine material balance is the following:

$$f_1 - f_2 - f_3 = 0 \tag{17}$$

where f_1 is the input throttle steam flow [kg/s].

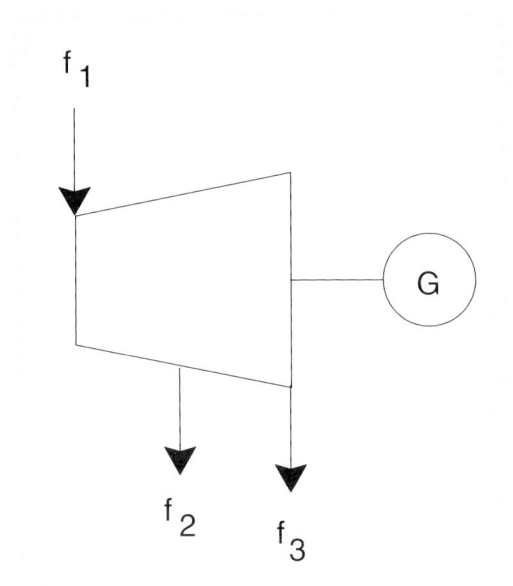

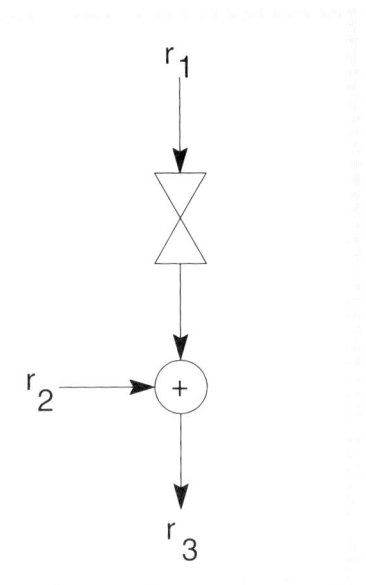

Figure 26. Turbine model.

Figure 27. Reduction model.

All flows are constrained. Turbine load allocation also requires the material and energy balances of reduction stations, material balances of steam headers, and steam requirements of the mill processes. Figure 27 shows the material balance for the reduction as follows:

$$r_1 + r_2 - r_3 = 0 \tag{18}$$

where r_1 is the input flow of steam [kg/s]
r_2 the flow of the cooling water [kg/s]
r_3 the output flow of steam [kg/s].

The energy balance is written with the help of corresponding enthalpies as follows:

$$h_1 r_1 + h_2 r_2 - h_3 r_3 = 0 \tag{19}$$

Because the turbine models are linear, linear programming can be used. The objective function is the following:

$$Max\ G = G_1 + G_2 + \ldots + G_n \tag{20}$$

where G_i denotes the steam generation of each turbine [MW].

CHAPTER 9

Turbine load allocation can also use iterative methods as the boiler load allocation. Practical constraints are easier to consider, but the solution may be sub-optimal compared with the solution from linear programming. One possibility is to calculate costs for alternative flow routes of steam and choose the routing with minimum costs that fulfil power and steam requirements.

8.4 Purchased power optimization

In addition to internal power generation, pulp and paper mills purchase power from utility companies or from other mills of the same company. The purchased power optimization balances the internal generation and the power need momentarily so the overall cost of power is minimum. Purchase contracts vary from mill to mill, but peak consumption in purchased power usually leads to increased costs. The purchasing strategy and the division of power purchase between different sources define the optimum operation point. Complicated contracts with continuous variation in the power need usually mean that computerized methods for purchased power optimization are beneficial from an economic sense.

The most usual way to monitor purchased power is tie-line control. This means cumulative monitoring of the power purchase in 1–3 min intervals during one hour. Using the calculated consumption, an estimate is calculated describing the situation at the end of the period. An alarm sounds if the estimate shows that the consumption limit will be exceeded at the end of the period.

In optimization, monitoring by itself is not sufficient. Reacting to deviations using some corrective actions must be possible. This happens by increasing the internal power generation by an amount corresponding to the deviation from the purchased power constraint. Here the energy balance of the power plant is the limiting factor. Another possibility is to decrease power consumption by closing processes that use excessive power and that can be stopped with minimum disturbance to production. These compensating processes vary from mill to mill.

Fauconnier[75] has reported about US$1 million annual savings resulting from use of the peak load controller. This means only five weeks payback time for the system.

8.5 Compensation of variations in steam consumption

Optimum operation of the power plant is usually calculated for a time span of one day. Dynamic load variations and disturbances are compensated using internal or external compensating processes before these disturbances influence the power plant operation. If this compensation is done successively, the use of expensive control fuels will be minimum.

Internal compensating processes include the following:

- Increase in the level of feed water tanks when steam consumption decreases
- Use of auxiliary condenser in a similar situation
- Use of a district heat network or the steam accumulator
- Control of primary boilers.

External compensating processes vary from case to case. Processes, that use considerable steam and whose operation is easy to change are used. When an excess of steam exists, the production of these processes increases. In the opposite situation, their production decreases.

The energy management system follows the capacity of these compensators and their ability to react quickly to the disturbances. Using this information, the priorities of different compensators change in different disturbances. This makes the most economical use of the compensating processes possible and assures that the mill production is also considered when compensating is done.

8.6 Simulation and optimization functions

In large, complicated power plants, collecting the separate functions listed above for a global model of the whole mill that includes simulation and optimization functions is beneficial. This model is a design tool for the staff responsible for coordination of energy production and consumption. The model operates on-line or off-line.

The power plant model is useful for testing the economy of different operating strategies and their effects on energy costs at the daily level. The model starts from the existing situation. The user changes the initial values (steam or power requirements) or model parameters (fuel prices or constraints), and the model calculates the new operating point by minimizing the energy costs.

8.7 Corporate-wide energy management systems

Purchased power monitoring and control is an important task of the corporate-wide energy management system[18]. This function uses the purchased power estimate on a 15–60 minutes time scale. The estimate is calculated according to the current rate of power consumption and internal generation.

Corporate-wide planning functions require simulation ability. The operation alternatives of several power plants should be simulated under normal and exceptional conditions. The system must be able to resolve the impact of current and predicted changes on energy consumption and production at mill and corporate levels.

8.8 Savings potential

The benefits of energy management systems depend on the mill in question. The actual benefits result only after detailed mill studies. Table 3 presents a summary of some applications[66].

Table 3. Benefits of energy management systems[66].

Function	Savings in 1000 FIM
Boiler load allocation	200–600
Turbine load allocation	100–300
Steam leveling	420–1500
Optimization of pressure levels	200–300
Purchased power optimization	100–800
Boiler control	800–3000

CHAPTER 9

References

1. Leiviskä, K. and Uronen, P., in Systems and Control Encyclopedia. Supplementary Volume 1 (M. G. Singh, Ed.), Pergamon Press, Oxford, 1988, pp. 467–472.

2. Miller. J P., TAPPI 1983 Annual Meeting Preprints, TAPPI PRESS, Atlanta, p. 101.

3. Trapp, P. R., Tappi J. 67(3):40(1984).

4. Fadum, O., 1983 ISA/CPPA Control Conference Preprints, CPPA, Montreal, p. 112.

5. Van Haren, C. R. and Brown, G. R., Tappi J. 72(2):65(1989).

6. Uronen, P. and Williams, T. J., "Hierarchical computer control in the pulp and paper industry," Report 111, Purdue Laboratory for Applied Industrial Control. West Lafayette, 1984, pp. 93–102.

7. Williams, T. J., Computers in Industry 8(2):233(1987).

8. Leiviskä, K., Poikela, T., and Rautanen L., "Stages in preplanning of production control system," Report 85, University of Oulu, Department of Process Engineering, Oulu, pp. 40–55.

9. Thomas, W. D., Pulp Paper 61(4):112(1987).

10. Fadum, O., Tappi J. 68(6):44(1985).

11. Pujol, J. B. Tappi J. 67(12):56(1984).

12. Bareiss, R A., Tappi J. 67(9):138(1984).

13. Wallace, M. D., Tappi J. 66(7):27(1983).

14. Uronen, P., Leiviskä, K., and Poikela, T., Pulp Paper Can. 85(9):T231(1984).

15. Edlund, S. G. and Einarsson, L., 1982 EUCEPA Symposium on Process Control Proceedings, SPCI, Stockholm.

16. Eriksson, L., 1978 7th World Congress Proceedings, IFAC, Pergamon Press, Helsinki, p. 213.

17. Ranki, J., Tappi 66(6):115(1981).

18. Sutinen, R., Niemi, T., and Leiviskä, K., 1990 Control Systems International Conference on Instrumentation and Automation in Pulp and Paper Industry Preprints, Finnish Society of Automatic Control, Helsinki, p. 330.

19. Fadum, O., Pulp Paper 70(4):59(1996).

20. Vuojolainen, T., 1990 Control Systems International Conference on Instrumentation and Automation in Pulp and Paper Industry Preprints, Finnish Society of Automatic Control, Helsinki, p. 320.

21. Fadum, O., Pulp Paper 69(3):47)1995).

22. Lundqvist, S. -O. and Kubulnieks, E., 1994 Control Systems Preprints, SPCI, Stockholm, p. 60.

23. Leiviskä, K., 1982 Nordic Workshop on Models for the Forest Sector Proceedings, Research Report 1982:1, University of Umeå, Umeå, p. 293.

24. Leiviskä, K., "Short term production scheduling of the pulp mill," Doctoral Thesis, Acta Polytechnica Scandinavica, Ma 36. Helsinki, Finland, 1982.

25. Leiviskä, K., 1990 Control Systems, International Conference on Instrumentation and Automation in Pulp and Paper Industry Preprints, Finnish Society of Automatic Control, Helsinki, p. 304.

26. Leiviskä, K., CESA '96 Multiconference, Symposium on Control, Optimization and Supervision Proceedings, IMACS, Lille, 1996, vol. 2, p. 1246.

27. Pettersson, B., Pulp Paper Mag. Can. 71(5):59(1970)

28. Pettersson, B., Tappi 52(11):2155(1969).

29. Golemanov, L. A., "Systems theoretical approach in the projecting and control of industrial production systems," Doctoral Thesis, EKONO-publication no 113, 1972.

30. Golemanov, L. A. and Blomberg, H., Tappi 56(1):109(1973).

31. Golemanov, L. A. and Koivula, E., Paperi ja Puu 55(2):53(1973).

32. Golemanov, L. A. and Koivula, E., 1974 Symposium on Optimization Methods (applied aspects) Preprints, IFAC, Pergamon Press, Varna, 1974.

33. Edlund, S. G. and Johansson, R., 1977 International Symposium on Process Control Preprints, CPPA, Montreal, p. 36.

34. Edlund, S. G. and Kallmen, C., 1974 British Paper and Board Industry Federation International Symposium Preprints, BPBI, Kent, p. 11.

35. Fasth, C., ABB Review (7):29(1991).

36. Chalaye, G. and Foulard, C., 1976 3rd IFAC PRP Conference Proceedings, Pergamon Press, Brussels, p.109.

37. Ruiz, J., Muratore, E., Ayral, A., et al., 1986 6th PRP Conference Proceedings, Pergamon Press, p. 205.

38. Ruiz, J., Muratore, E., Ayral, A., et al., 1990 Control Systems International Conference on Instrumentation and Automation in Pulp and Paper Industry Preprints, Finnish Society of Automatic Control, Helsinki, p. 343.

39. Balaz, J., "Selected problems of pulp production control," Parts I and II, Pulp and Paper Research Institute, Bratislava, 1985, 1986.

CHAPTER 9

40. Skopin, I. I. and Saffonova, M. R., 1983 5th PRP Conference Proceedings, IFAC, Antverp, p. 99.

41. Leiviskä, K. and Uronen, P., 1979 IFAC/IFORS Symposium Preprints, Pergamon Press, Oxford, p. 25.

42. Leiviskä, K. and Uronen, P., 1979 2nd IFAC/IFORS Symposium on Optimization Methods - Applied Aspects Preprints, Pergamon Press, Oxford, p. 291.

43. Leiviskä, K. and Uronen, P., "Hierarchical control of an integrated pulp and paper mill," Report No 113, PLAIC, Purdue University, West Lafayette, 1979, pp. 47–72.

44. Leiviskä, K.and Uronen, P., Preprints of the 4th PRP Conference, Pergamon Press, Oxford, p. 151.

45. Monteiro, P. P. and Dourado, C. A., Information and Decision Technol. 18:241(1992).

46. Leiviskä, K., Komokallio, H., Aurasmaa, H., et al., Large Scale Systems 3(1):13(1982).

47. Leiviskä, K., Huttunen, R., and Uronen, P., "Pulp mill production scheduling with an Expert System," Paper presented at the Universidade de Coimbra, Portugal, April 1987.

48. Leiviskä, K. and Niemi, T., 1988 Pulp and Paper Symposium Paskov '88 Ostrava, p. 22.

49. Golemanov, L. A. and Leiviskä, K., 1986 6th PRP Conference Preprints, IFAC, Pergamon Press, Akron, p.184.

50. Singh, M. G., Dynamical Hierarchical Control, North Holland, Amsterdam, 1977, pp. 26–36.

51. Leiviskä, K. and Niemi, T. 1986 Pulp and Paper Symposium 'Paskov 86, Ostrava, p. 14.

52. Dourado, C. A., 1992 CIM in Process and Manufacturing Industries Workshop Proceedings, Pergamon Press, Oxford, p. 139.

53. Dourado, A. and Santos, A., IEEE International Conference on Systems, Man and Cybernetics Proceedings, La Touquet, France, p. 731.

54. Santos, A. and Dourado, A., 1997 15th World Congress on Scientific Computation, Modelling and Applied Mathematics Proceedings, Berlin, vol. 4, p. 43.

55. Salo, R., Leiviskä, K., and Jutila, E., 1982 APMS "82" IFIP Working Conference Preprints, Bordeaux, France, p. 35.

56. Mori, Y., Nishimura, M., Chujima, T., et al., TAPPI J. 80(11):158(1997).

57. Åhman, T., 1990 Control Systems, International Conference on Instrumentation and Automation in Pulp and Paper Industry Preprints, Finnish Society of Automatic Control, Helsinki, p. 185.

58. Anon., Q–OPT Process Modelling System. Available [Online{ <http://www.kareltek.fi/taipale/q–opt.htm> [March 20, 1990.]

59. Gleason, M., Pulp Paper 69(7):79(1995).

60. Mardon, J., Vyse, R., and Ely, D., Pulp Paper Can. 92(12):87(1991),

61. Mardon, J. and Vyse, R., Pulp Paper Can. 83(10):65(1982).

62. Reese, R., Tappi J. 75(9):129(1992).

63. Säfvenblad, S. and Wigren, B., 1986 Control Systems Preprints, SPCI, Stockholm, p. 26.

64. Muukari, P. and Nettamo, K., Pulp Paper 60(3):192(1986).

65. Aarnio, S., Tarvainen, H. J., and Tinnis V., Tappi 63(5):73(1980).

66. Alahuhta, T. and Muukari, P., Sähkö 57(6):10(1984).

67. Andreassen, O. and Olsen, T. O., Modelling, Identification and Control 4(2):107(1983).

68. Baldus, R. F. and Edwards, L. L., Tappi 61(9):77(1978).

69. Cho, C. H., Instrumentation Technol. 25(10):55(1978).

70. Cho, C. H. and Blevins, T. L., Tappi 63(6):91(1980).

71. Kaya, A. and Keys, M. A., Automatica 19(2):111(1983).

72. Lasslett, N. F. and Filmer, P. J., Appita 35(2):131(1981).

73. Leffler, N., Tappi 6(9):37(1978).

74. Suh, S. and Olsen, T. O., 1980 PRP 4 Automation Preprints, Pergamon Press, Oxford, p. 381.

75. Fauconnier, W. R., Pulp Paper Can. 94(11):T365(1993).

CHAPTER 9

CHAPTER 10

Process analysis, modeling, and simulation

1	**Introduction**	**249**
2	**Process analysis**	**250**
2.1	Process as a black box	250
2.2	Steady-state and dynamic analysis	250
2.3	Systematic experimental design	251
2.4	Statistical variables	252
2.5	Statistical testing	255
2.6	Correlation methods	257
2.7	Regression analysis	259
2.8	Test signals and dynamic models	261
	2.8.1 Step tests	262
	2.8.2 Impulse tests	262
	2.8.3 Random signals	263
	2.8.4 Other dynamic models	264
2.9	On-line process analysis in the pulp and paper industry	264
3	**Models and modeling**	**269**
3.1	Model classification	269
3.2	Selection of model type	271
3.3	Model scope	272
3.4	Model construction	272
3.5	Solution of model equations	273
3.6	Model validation and use	273
4	**Properties of flow sheet simulators**	**274**
4.1	Definitions	274
4.2	Structure of flow sheet simulators	275
4.3	Example of simulation	275
4.4	Module library	279
4.5	Physical and thermodynamic data	279
4.6	Design and optimization functions	280
4.7	Dynamic flow sheet simulators	280
5	**Flow sheet simulators in the pulp and paper industry**	**280**
5.1	GEMS	280
5.2	MASSBAL	281
5.3	Other flow sheet simulators	282
6	**Intelligent methods and simulation**	**283**

CHAPTER 10

7	**Application viewpoints**	**285**
7.1	Process flow-sheeting	286
7.2	Comparison of process and mill operational alternatives	286
7.3	Simulation in process control	287
7.4	Simulation in production planning and control	287
7.5	Training	287
8	**Problems in modeling and simulation in pulp and paper mills**	**289**
8.1	Adjustment of models	289
8.2	Lack of suitable models	290
8.3	Documentation	290
8.4	Presentation of results	290
	References	291

CHAPTER 10

Kauko Leiviskä

Process analysis, modeling, and simulation

1 Introduction

Gaining insight into process and control operations requires data, information, and knowledge. Depending on the particular case, several possibilities exist. This chapter considers three: process analysis, modeling, and simulation.

Process analysis deals with measured data from the process. It primarily uses statistical methods to clarify the quality of data and the information included in it. This must relate to process knowledge and know-how, or the methods have no use. Many methods have use to analyze a process in operation. They essentially belong to the category of experimental modeling and vary from straightforward statistical analysis, correlation methods, and regression analysis to more complicated methods of on-line process identification. To be effective, these methods require systematic experimental design. Process analysis usually has use off-line after collecting the necessary data, but on-line applications also exist. Process analysis tools also have automatic use for process status and condition monitoring.

Several areas of engineering use mathematical models. In research and development, mathematical models have use in studies of process internal phenomena such as flow, mixing, reactions, heat and mass transfer, etc. The models provide a thorough understanding about what is actually happening inside the process. In product design for the process industries, different kind of models define the effects of variables on product quality and amount. Process optimization also occurs often in the early stages of a design project.

Simulation is the process of building a mathematical model for a system and using that model for a systematic investigation. Process simulation systems are program packages for the simulation of one separate process or an entire mill with standard tools. Flow-sheeting is a term generally used with specific computer programs to solve steady-state material and energy balances for design and cost estimation. The word simulation then does not have restriction to solution of model equations but encompasses the entire process of model construction, model validation, and model use.

In process design, modeling and simulation methods have use today to study alternatives for process equipment and connections, optimize process operation, and

CHAPTER 10

discover the best ways to use raw materials and energy. Equipment sizing also uses models and simulation. In control engineering, models and simulation can determine control strategies for the process.

Simulation is also an interesting tool for process operation. It has use in disturbance and alarm analysis, startup and shutdown planning, real time control and optimization, and operator training.

2 Process analysis

2.1 Process as a black box

Process analysis can proceed in several stages. One possibility to describe process behavior is to calculate different statistical values that are typical for the process in question. These values show average behavior

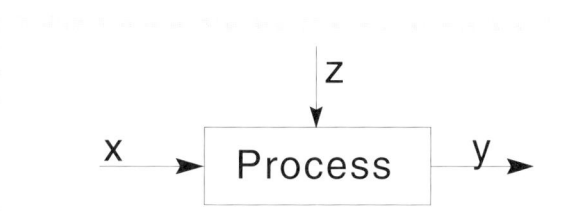

Figure 1. Process as a black box.

and variance around the average, correlation of the measured signal with itself or some other signals, etc. This approach examines the process as the black box of Fig. 1. Note that the knowledge of process behavior and mechanisms influencing it also help interpret the results of statistical process analysis using data.

In the above figure, x denotes the vector of input variables or control variables that can vary in a controlled manner and have use as independent variables in experimental models. For example, temperatures, flow rates, and retention times fall in this category. Output variables, y, depend on input variables. They usually describe the product quality and amount. Disturbance variables, z, influence the process operation. They are uncontrolled but measurable. Input variables compensate their effect on output variables. Examples of disturbance variables are raw material properties and outside temperatures not under active control. In process analysis, we are interested in dependencies between input and output variables and the effect of disturbance variables on process operation. Several possibilities are available to analyze these dependencies and build models that describe the process phenomena.

2.2 Steady-state and dynamic analysis

If process dynamics is not necessary, steady-state analysis and models are sufficient. Two primary approaches are possible:

- Data analysis uses mathematical models on the normal operation data. Correlation methods, parameter estimation, trend analysis, etc, are applicable. The application of intelligent methods for data mining is also becoming common. Usually, sufficient variations in process operation exist to guarantee a good starting point for analysis. The risk is that some important variable can remain completely outside the analysis or some dynamic transients can disturb the data.

- Input variables change according to a systematic plan. This also defines the validity of the model in the experimental design stage. The results are more reliable and represent the process behavior. The disadvantage is that disturbances must be introduced to the process.

In steady-state analysis and modeling, the process must reach the steady-state after each change. Otherwise, misleading results will occur. Experiment design must fit the problem and guarantee a suitable range of process variables.

Dynamic models and analysis are primarily necessary when control or control design is in question. Then the needs of control methods require consideration. Two kinds of models are possible: transfer functions for classical control methods and state space models for modern control methods.

The usual practice in dynamic modeling is to make changes in one input variable and record the response. Section 2.8 discusses these methods. This requires data acquisition during the entire experiment. Measurement of successive steady states is insufficient. This obviously gives a larger amount of data compared with steady-state analysis.

Another possibility is use of normal process variations and on-line identification methods or correlation analysis. In practice, several variables change simultaneously. The effects of a certain variable are then difficult to define.

2.3 Systematic experimental design

All industrial experiments use comparison. For example, we may want to compare the performance of a new process with an older one, define the improvement in product quality due to a change in process, learn the effects of new raw materials and catalysts, calculate savings of a new automation system, etc. Experiments aid the decision process. They must provide data and results that are useful in a reasonable way. Raw test results seldom provide a basis for decision. They usually require interpretation using statistical analysis. Applying systematic methods in experimental design and data processing further facilitates this analysis.

If we want to define the effect of several input variables on a certain output variable, we can change one input at a time and measure its effects after the process has attained a new steady state. This approach takes considerable time and provides no possibilities to determine possible interconnections between variables. These interconnections are very common. Their influence requires consideration in process optimization and control.

Consider a case in which we want to test three variables. To have statistically significant results, each test is repeated four times. To determine only the linear effects in the steady state, each variable uses two levels. This requires 32 successive experiments to test all possible combinations.

In this case, even 32 tests tell nothing about interconnections between variables. This is a common problem in experimental design. Systematic methods simultaneously minimizing the number of experiments can avoid the problem.

Experimental design applies matrix representations. Examples of Hadamard Matrix Design[1] and Taguchi's Orthogonal Arrays[2] are available.

Trial number	\multicolumn{7}{c}{Factors}
	A	B	C	D	E	F	G
1	0	0	0	0	0	0	0
2	0	0	0	1	1	1	1
3	0	1	1	0	0	1	1
4	0	1	1	1	1	0	0
5	1	0	1	0	1	0	1
6	1	0	1	1	0	1	0
7	1	1	0	0	1	1	0
8	1	1	0	1	0	0	1

Figure 2. Example of matrix design[2].

Figure 2 shows an example of Orthogonal Arrays[2].

The array shows seven 2-level factors denoted by zeros and ones. They can denote two different levels of continuous variables (high and low) or two different values for discrete variables (two catalysts and two processing methods or the situation before and after a process change). The array has 8 rows and 7 columns. Columns indicate tested factors, and rows show trial conditions. Each column contains four zeros and four ones. The columns are orthogonal or balanced because the number of level combinations equals the number of columns[2].

With this array, an experimental design for seven variables can use only eight experiments. The experimenter may use different designators for the columns, but the eight trials will cover all combinations independent of the column definition[2]. This assures consistency of a design done by different experimenters and leads to a simplified, standardized factorial design.

Examining the results from orthogonal array experiments uses the analysis of variance (ANOVA) method. This method tells how much of the variation each influencing factor has contributed. By studying the primary effects of each factor, one can define the general trends of influencing factors. Another approach that is better for multiple runs is the use of signal to noise ratio (S/N) to determine the optimum set of operating conditions[2].

2.4 Statistical variables

The following text introduces the most common statistical variables used to describe the characteristics of a group of observations. Most statistical program packages and spreadsheet programs can calculate them. Although the variables may seem too simple and straightforward for any use in detailed process analysis, they do provide answers to several questions:

- Has the process operated inside its normal operation range?
- Have any major disturbances occurred in the process?

- Are some larger variations occurring that have no explanation in process operation and are due to measurement problems?
- Do sufficient variations exist in the process for analysis and modeling?

Assume we have k samples with n repetitions for each sample and $N=nk$ is the size of the entire population. Then the sample mean is the following:

$$\bar{x} = \frac{\sum_{i=1}^{n} x_i}{n} \qquad (1)$$

The mean of the whole population is as follows:

$$\bar{\mu} = \frac{\sum_{i=1}^{N} x_i}{N} \qquad (2)$$

The grand mean is the average of all the sample means. It provides a good estimate of the population mean:

$$\bar{X} = \frac{\sum_{j=1}^{k} \bar{x}_j}{k} \qquad (3)$$

When arranging the measurements in order of magnitude, the median is the value of the middle item. This applies if the number of measurements is odd. If the number is even, the median is the average of the two middle values.

Mode is the most commonly occurring value in a group of measurements. When arranging the data into a frequency distribution, the middle point is the modal value.

Range is the difference between the highest and lowest observations. It is a simple way to describe variations in a variable:

$$\bar{R}_i = x_{Hi} - x_{Lo} \qquad (4)$$

If we have k samples with n repetitions for each sample, the mean range is as follows:

$$\bar{R} = \frac{\sum_{i=1}^{k} R_i}{k} \qquad (5)$$

The sample variance in the above case is then the following:

$$\sigma^2 = \frac{\sum (x - \bar{x})^2}{n} \qquad (6)$$

CHAPTER 10

The standard deviation is the square root of the variance as follows:

$$\sigma = \sqrt{\sigma^2} \qquad (7)$$

The following equation provides a calculation of the estimated standard deviation. In this equation, the sum of squared deviations is divided by $n-1$ instead of n to correct the bias error that could occur especially with small sample sizes:

$$s = \sqrt{\frac{\sum_{i=1}^{n}(x_i - \bar{x})^2}{n-1}} \qquad (8)$$

The meaning of standard deviation has an easy explanation by assuming that the measurements follow a normal distribution. This is usually true with continuous variables. The standard deviation estimates the spread of measured values. This defines the width of the bell shaped curve in Fig. 3.

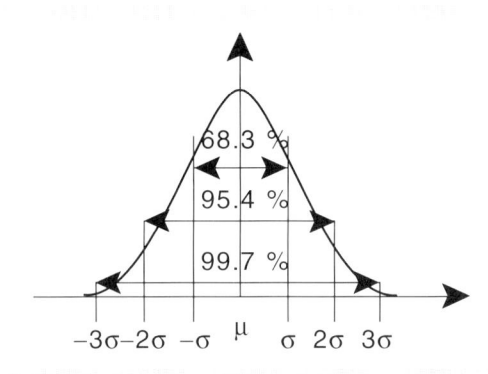

Figure 3. The bell shaped normal distribution.

Figure 3 shows that 99.7% of all values occur inside the area of 6σ symmetrically around the mean value. This also means the following:

- 2.5% of the values lie above m + 1.96σ
- 2.5% of the values lie below m - 1.96σ
- 0.2% of the values lie outside m ± 3.09σ.

Chapter 3.9 of this book used this fact in defining the values for upper and lower warning and action limits for the mean control charts.

When using several repetitions of a sample, the standard error of means defined in the following equation is sometimes useful instead of sample standard deviation:

$$\sigma_{\bar{x}} = \frac{\sigma}{\sqrt{n}} \qquad (9)$$

If each sample has four repetitions, the standard error of means is half of the corresponding standard deviation. The reason for its use is that it is more sensitive to changes in process and can also detect changes more easily.

2.5 Statistical testing

Accurately measuring the properties and distributions of variables is usually impossible. Test data covers only a small part of all populations in question. Using this partial information, we want to define the characteristic values of the entire population. We also want to define simultaneously what risks are present when we make our decision using this partial information for product quality as an example.

Solution of these problems uses statistical testing. Usually, the mean and the variation (often as variance) of the variable are tested. Several distributions are available, but selection of the correct one depends on the problem setting. If we have two groups of data such as two sets of kappa values produced in different process conditions and we want to test if their means are similar, we assume that we know the variance of kappa spread and use the normal distribution. If the variance of the entire population is unknown, the only possibility is to use sample variances and t-distribution. If more than two data sets require analysis, analysis of variance is used.

If we want to compare the variance of the data set to some pre-specified value, we use x^2-distribution. To compare the variances of two separate data sets, F-test is used.

The density function of the normal distribution is as follows:

$$f(x) = \left(\frac{1}{2\pi\sigma}\right)^{\frac{1}{2}} \exp[-(x-\mu)^2/(2\sigma^2)] \tag{10}$$

The test value of the distribution usually has the following definition:

$$z = \frac{(x-\mu)}{\sigma} \tag{11}$$

where x is the value of the variable to be tested
        μ the known mean value with the following:

$$f(z) = (1/2\pi)^{\frac{1}{2}} \exp[-z^2/2] \tag{12}$$

The use of normal distribution in testing two data sets uses the confidence interval. Assume that we want to compare the mean values of two data sets, A and B, and we know their means, standard deviations, and the corresponding sizes of the data sets (number of observations). If the mean value of data set B falls into the following confidence interval calculated using the properties of data for A, both means are identical:

$$confidence\ interval = \mu_A \pm \frac{U_p \sigma}{\sqrt{N_A}} \tag{13}$$

where U_P is the value of the sum function of the normal distribution respective to the user-defined probability.

CHAPTER 10

Tabulations of these values are available in statistics textbooks. Usually, one selects 95% or 99% probability; it means that the risk of drawing faulty conclusions from the test is 5% or 1%, respectively. The sum function of the normal distribution is as follows:

$$P(z) = \int_{-\infty}^{\infty} f(z)dz \tag{14}$$

One can interpret this as the area of the bell shaped curve of the normal distribution.

If one uses normal distribution in testing big data sets, the standard deviation calculated from the data can replace the standard deviation of the entire population. If data sets are small, this is not possible because the estimate is unreliable. In such cases, t-distribution is useful. Its test value has the following definition:

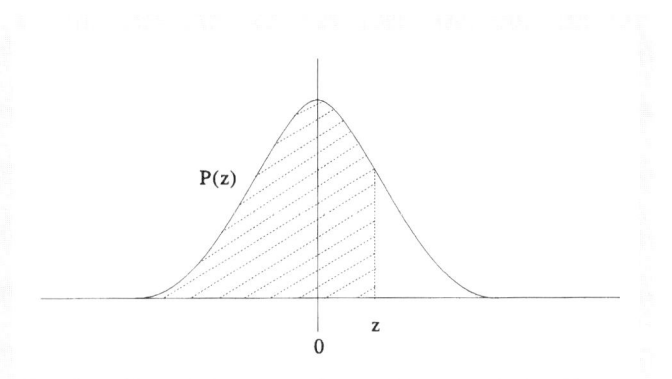

Figure 4. Graphical interpretation of the normal distribution sum function.

$$t = \frac{|\mu_A - \mu_B|}{\sqrt{\frac{s_A^2}{N_A} + \frac{s_B^2}{N_B}}} \tag{15}$$

where A and B are the data sets for comparison.

The values of the sum function of t-distribution are tabulated as a function of their degrees of freedom, $n-1$. The use follows that of normal distribution.

χ^2 distribution is useful to test variance of a data set against some predetermined or tolerance value. Its test value has the following definition:

$$\chi^2 = \frac{s^2(n-1)}{\sigma^2} \tag{16}$$

where s is the standard deviation calculated from the data set
σ the value against which to compare it.

Testing follows the same procedure as in previous cases.

F-distribution is useful in regression and variance analysis. It tests the equality of two variances using the following test value:

$$F(df_1, df_2) \approx \frac{s_1}{s_2} \tag{17}$$

2.6 Correlation methods

Autocorrelation describes the correlation between momentary values of the same signal. It can be calculated for all stationary, continuous, and ergodic signals. The autocorrelation function has the following definition:

$$\phi_{xx}(\tau) = \lim_{t \to \infty} (1/2T) \int_{-T}^{T} x(t)x(t+\tau)dt \qquad (18)$$

where τ is the delay
$-T - T$ the considered time interval.

The autocorrelation function has the following properties:

$$\phi_{xx}(\tau) = \phi_{xx}(-\tau) \qquad (19)$$

$\phi_{xx}\tau$ does not depend on t $\qquad (20)$

$$\phi_{xx}\tau \leq \phi_{xx}(0) \qquad (21)$$

$$\phi_{xx}(0) = \overline{x^2(t)} \qquad (22)$$

$$\phi_{xx}(\infty) = \overline{x(t)}^2 \qquad (23)$$

Figure 5 presents the principal form of the autocorrelation function for a stationary stochastic signal. It also visualizes the above notations.

The autocorrelation function tells how the signal repeats itself during time. It also indicates the cyclic behavior of the process. Pure stochastic variations make the autocorrelation zero with all delays but zero as Fig. 6 shows. The value of the autocorrelation function usually decreases with increasing time. With faster variations, the autocorrelation function rapidly approaches small values. Studying the performance of control loops can use this property.

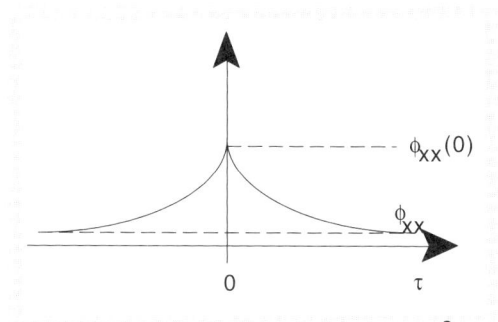

Figure 5. General form of autocorrelation function[3].

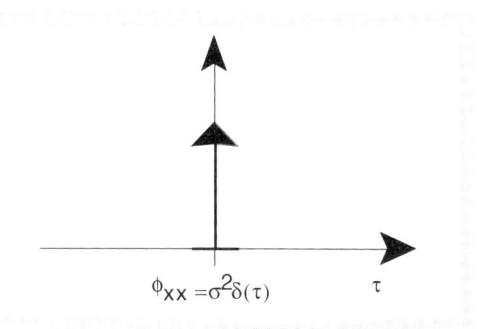

Figure 6. Autocorrelation function of white noise.

CHAPTER 10

The crosscorrelation function between two signals has the following definition:

$$\phi_{xy}(\tau) = \lim_{t \to \infty} (1/2T) \int_{-T}^{T} x(t)y(t+\tau)dt \tag{24}$$

The value of crosscorrelation function varies between -1 and 1. A higher value denotes higher correlation, and the sign tells the direction of interaction between the variables. The maximum of the crosscorrelation function is not necessarily at $t = 0$. The crosscorrelation of two independent signals is zero.

In addition to defining correlated or non-correlated signals, the crosscorrelation function can define the delays between two variables. The location of the peak in the crosscorrelation function tells the timely distance of two variables. If the pure time lag is in question, this peak approaches -1 or +1. If some time constant is also included, the location of the peak is no longer so easy to find, and the method easily overestimates the time lag.

Power spectrum or power spectral density is a way to analyze signals using the frequency domain. The peak in the power spectrum shows the frequencies that have considerable disturbances. If the variations are periodic, the power spectrum has only one narrow peak. If only stochastic variations occur, the power spectrum has a more even distribution over different frequencies.

The area under the power spectrum between any two values is proportional to the variance of the signal on that frequency band. The area of the entire spectral curve is proportional to the signal variance.

Power spectrum defines the correlation between different signals. If two or more signals show significant variations or peaks at the same frequency, they depend on each other. One may be the reason for another, or they may have a common reason. Power spectrum is also useful in the performance indication of control systems by comparing the spectra before and after the control and comparing the variances in two cases.

The Fourier transform usually defines the power spectrum. From the autocorrelation function, we obtain the following:

$$\phi_{xx}(j\omega) = \int_{-\infty}^{\infty} \phi_{xx}(\tau) e^{-j\omega\tau} d\tau \tag{25}$$

Correspondingly, we obtain from the crosscorrelation:

$$\phi_{xy}(j\omega) = \int_{-\infty}^{\infty} \phi_{xy}(\tau) e^{-j\omega\tau} d\tau \tag{26}$$

Correlation analysis can also facilitate and complement statistical process control. The following text is a continuation of the example in Chapter 3.9[4]. Figure 7 shows the autocorrelation function for the bale line viscosity data shown earlier in Fig. 3.33. It shows some cyclical behavior in viscosity, but it does not provide any explanation. Figures 8 and 9 show the crosscorrelation of viscosity to two other variables: ClO_2 charge in Fig. 8 and unbleached kappa in Fig. 9. The first correlation is obviously erroneous due to bad data or correlated input signals. The real reason for viscosity changes is probably unbleached kappa numbers as Fig. 9 shows. Note also the long delays connected to both variables. This is evidently due to long retention times in the bleach plant and the pulp storage tanks.

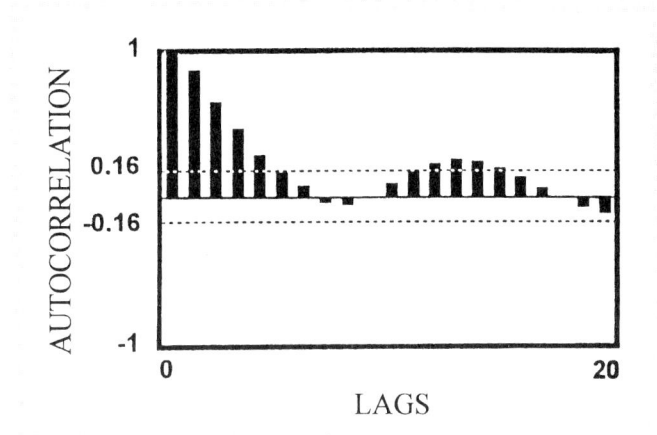

Figure 7. Viscosity autocorrelation[4].

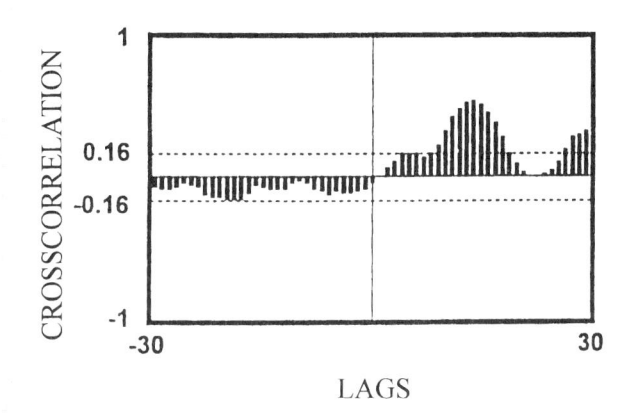

Figure 8. Crosscorrelation of D100 ClO_2 charge with viscosity[4].

2.7 Regression analysis

Linear regression analysis has wide use in process analysis and experimental modeling. It is a good tool, but its application requires good experiments and knowledge of the user. A basic requirement is linearity of the model:

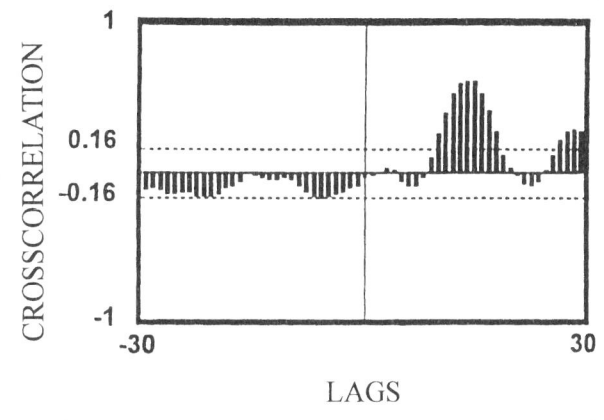

Figure 9. Crosscorrelation of unbleached kappa number with viscosity[4].

CHAPTER 10

$$y = \beta_o + \beta_1 x_1 + \beta_2 x_2 + \ldots + \beta_m x_m + \varepsilon \tag{27}$$

$$y = \beta_o + \sum_{j=1}^{m} \beta_j x_j + \varepsilon \tag{28}$$

The dependent and independent variables in the above equations need not be actual process variables. One can use different scaling and transformations. The beta coefficients are the model parameters, and epsilon is the random error.

The regression analysis proceeds in principle as follows:

- Model parameters are estimated using the method of least squares by minimizing the error between the model and actual measurements.

- The statistical significance of the model is tested using different statistical values based on model properties.

- If necessary, the model structure is changed, and parameter estimation is repeated. Changing usually means model simplification. The procedure undergoes repeating until the model performance is satisfactory.

Regression analysis assumes that random errors, epsilon, connected to different observations are independent and follow normal distribution. If the errors are not independent but correlate with each other, the estimates for parameters designated beta can be erroneous. If the errors do not follow normal distribution, the statistical tests for model significance will give invalid results.

Regression analysis usually has use in building steady-state models. This case fulfills the necessary requirements. Regression analysis is also applicable in dynamic modeling, if the models can be written as difference equations and the errors do not correlate with each other. This requirement is not easy to meet. The least squares method used in regression analysis assumes that random errors occur only in dependent variables, and independent variables are due to them. This is not the actual case. No cure for this problem and the errors caused by it are allowable.

The usual way to evaluate the performance of the regression model is to calculate the correlation coefficient between the model and the original measurements. A value close to one for the correlation coefficient indicates a better model. Another technique used in a regression analysis program is to calculate F-test values. Higher F means a more reliable model. The F-test also has use when estimating if a new variable improves a model.

Reliability of regression analysis depends on the observations. A model can never be better than the data behind it. For regression analysis, one must therefore check the data to find probable deficiencies connected with it.

This is possible using the correlation matrix that most regression analysis programs calculate or by calculating the correlation between the variables in question by some other method. Correlation coefficient is the measure of linear dependence between two variables. If the correlation between two input variables is close to one, this

Process analysis, modeling, and simulation

is a warning that the results might be erroneous. Because the input variables correlate, one of them is usually eliminated during analysis. Elimination is a random selection that may lead to discarding an important control variable while leaving a less important one in the model. Parameter errors can also result.

2.8 Test signals and dynamic models

In dynamical modeling, different test signals play a significant role in experimental design. The most usual test signals are the following:

- Step
- Impulse
- Ramp
- Sine function
- Random signals such as pseudo-random binary sequence (PRBS).

These signals determine the transfer function of the process under study. The choice of the test signal influences the speed and accuracy of dynamic modeling.

Choosing a large amplitude for the test signal usually produces a good model. Excessively high amplitude can lead to nonlinear models. Excessively low amplitude is also dangerous because the process noise disturbs the output signal too much. These factors usually lead to a compromise between model accuracy and practical limitations for testing.

Test signals usually only apply to one variable at a time. With step, impulse, and ramp signals, a change should occur in both directions. This helps identify any possible nonlinear nature of the process. The changes should also occur successively. This means the first change is upward. Sufficient time passes for the response to stabilize. Then the next change occurs downward. Figure 10 shows one possible strategy. Applying the test signal only when the process is in a steady state is essential, or erroneous models will result.

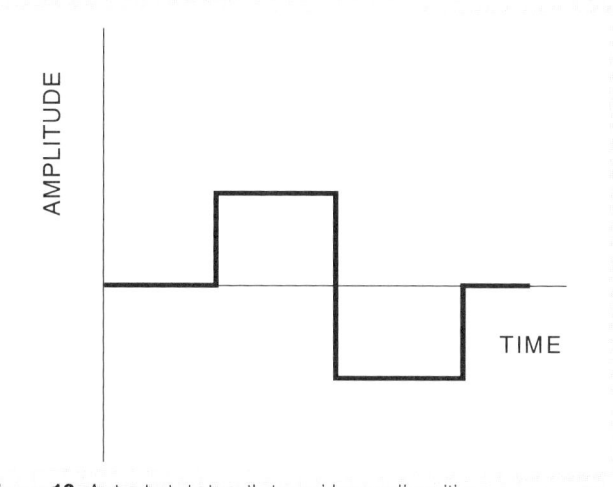

Figure 10. A step test strategy that considers nonlinearities.

CHAPTER 10

2.8.1 Step tests

In a step test, input variables change stepwise. Mathematically, the step function has the following form:

$$f(t) = \begin{cases} A \text{ when } t \geq 0 \\ 0 \text{ when } t < 0 \end{cases} \tag{29}$$

Several methods exist to identify process parameters from the step response. They use drawing a tangent to a step response or denoting certain points from the step response and developing computational procedures based on them. In many cases, the transfer function with a single time constant and dead time is sufficient. The following equation is a first order process with dead time:

$$G(s) = \frac{K_p e^{-Le}}{1 + Ts} \tag{30}$$

where K_p is the process gain
 L if the dead time
 T the time constant.

Figure 11 shows the step response of the above system and the tangent drawn to it. It also shows determination of the time constant and dead time from the step response using the common Ziegler-Nichols method. Process gain is the ratio between step response in the steady state and step size (=K/M).

2.8.2 Impulse tests

An impulse test resembles the step test. The input variable changes stepwise during a short time for measurement of the response to the output variable. The ideal impulse has zero duration and limitless amplitude. Since this is not possible in practice, one must use a pulse. The mathematical definition of impulse is as follows:

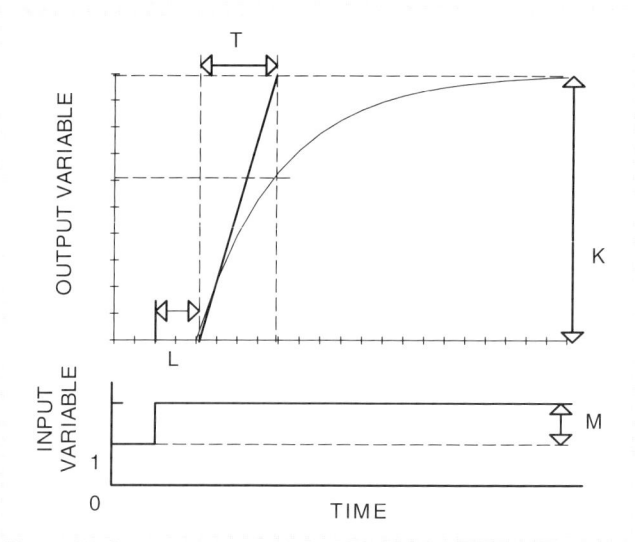

Figure 11. Step response and Ziegler-Nichols method.

$$fe(t) = \begin{cases} 1/\theta & \text{when } t \leq 0 \\ 0 & \text{when } t > 0 \end{cases} \quad (31)$$

Graphically analyzing impulse response is like step responses. Figure 12 shows a procedure for the first order process without dead time. The time constant is the time when the response reaches the following value:

$$\frac{K}{T}e^{-T/T} = \frac{K}{T}e^{-1} = 0.37\frac{K}{T} \quad (32)$$

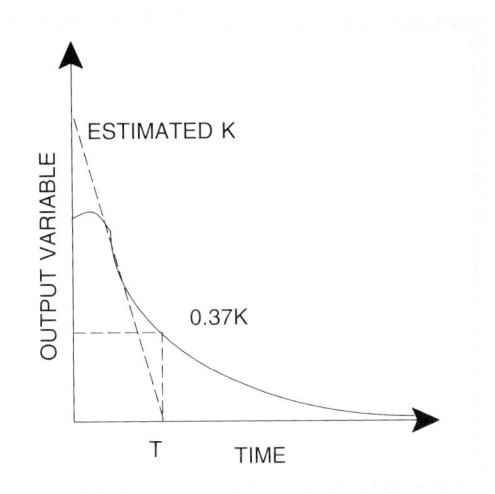

Figure 12. Determination of the model for the first order process from an inpulse response.

In practice, ideal impulses do not apply. Drawing a tangent with maximum slope close to t = 0 determines the realistic value for the time constant.

Transfer functions for the second order processes such as the following:

$$G(s) = \frac{K}{(s/\omega_o)^2 + 2\xi s/\omega_o + 1} \quad (33)$$

use the relationship between the form of the response and the damping ratio, ξ, and the natural frequency, ω.

2.8.3 Random signals

Random signals in process testing are usually binary signals. They can only have two values that change at random intervals. The most usual is a pseudo-random binary sequence (PRBS) that consists of N intervals with a duration T arranged randomly. The sequence repeats one or more times as Fig. 13 shows.

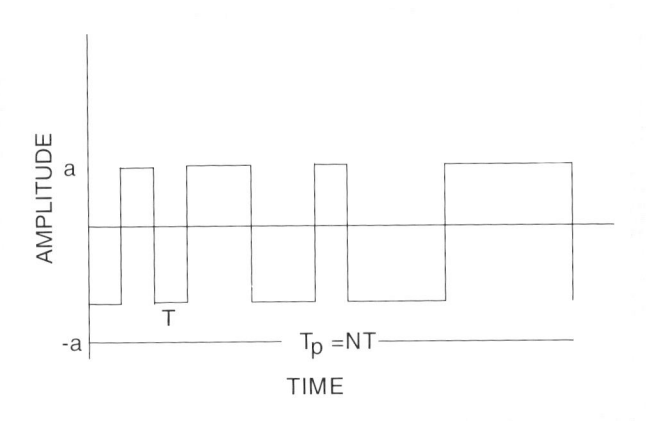

Figure 13. PRBS signal.

The choice of a, T, and N occurs as follows:

- Amplitude a should be as high as the process tolerates. This minimizes the effects of process noise and disturbances.

CHAPTER 10

- Interval T is recommended as $T_i/5$ where T_i is the smallest time constant that has a meaning for testing.

- Test length $T_p = NT$ should be T_{95} at minimum. T_{95} means the time that the response takes to reach 95% of its steady state value. The recommendation is $T_p = 1.5\ T_{95}$.

2.8.4 Other dynamic models

Transfer functions are usually useful in situations where the interest limits the relationship between two variables. For multi variable models, state space presentation is more common. The continuous state equation for linear time invariant system is the following:

$$\frac{dx(t)}{dt} = Ax(t) + Bu(t) \tag{34}$$

$$y(t) = Cx(t) + Du(t) \tag{35}$$

where x are the state variables
 u control variables
 y output variables
 A, B, C, and D are coefficient matrices that are independent of time.

A discrete version of Eq. 34 for a single input and single output system is a stationary moving average model (ARMA):

$$\begin{aligned} y(k) + a_1 y(k-1) + a_2 y(k-2) + \ldots + a_n y(k-n) = \\ b_1 x(k) + b_2 x(k-1) + \ldots + b_m x(k-m) \end{aligned} \tag{36}$$

If the model also includes the dead time, the equation has the following form:

$$\begin{aligned} y(k) + a_1 y(k-1) + a_2 y(k-2) + \ldots + a_n y(k-n) = \\ b_1 x(k-d) + b_2 x(k-1-d) + \ldots + b_m x(k-m-d) \end{aligned} \tag{37}$$

The relationship between different models is available in any control engineering textbook.

2.9 On-line process analysis in the pulp and paper industry

The Finnish Pulp and Paper Research Institute developed a process analysis system [5]. It can analyze the operation of continuous processes. Its primary intentions are to reduce quality variations, improve process efficiency, or both. Figure 14 shows its basic structure. The three levels of this structure have the following functions:

- Process defines the entire process as a graph with two kinds of nodes (process components and quantities) and the physical connections between them.

Process analysis, modeling, and simulation

- A subset of a process with quantities that require simultaneous measurement for modeling purposes is measurements. This incorporates the given data obtained from different sources.
- A model is a subset of measurements for modeling. This uses spectral analysis, statistical process control, and multichannel autoregressive modeling (MAR).

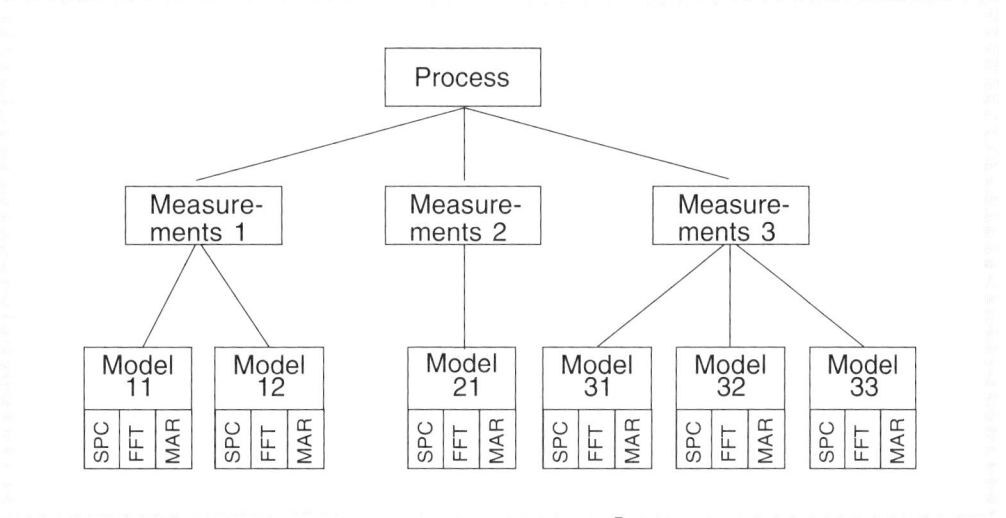

Figure 14. The logical structure of a wet end diagnostic system analyzer[5].

The wet end diagnostic system has two purposes:

- An expert plans the tree structure analysis for a given process, i.e., he defines the measurements, models, or both.
- A nonexpert performs the analysis using the expert's structure of the given situation.

The user information relies on windowing technology that has proven to be a successful approach. The results show as cause-and-effect diagrams with self-documenting features included.

Originally, the wet end diagnostic system only solved wet end problems. Spectral analysis and MAR clarified reasons for basis weight variations and operation of the basis weight valve. This used the data from a 9.1 h separate study of 8 s measuring intervals and three frequency ranges. The results showed that different parts of the process dominate in each range.

On-line applications of the wet end diagnostic system are available[6]. They tell about an installation at a supercalender paper machine producing 220 000 tons per year. The application involves about 80 on-line process measurements such as flows, consistencies, pH-readings, conductivities, quality parameters, etc. The measurements were available at sampling intervals of 10 s, 1 min, and 10 min. The measurements occurred once per second with calculation of averages over sampling intervals. History databases store the last 24 000 readings.

The system benefits fall into the following categories:

- Continuous monitoring of process upsets
- Better availability of process information
- Model-based analysis of operating strategies
- Model-based fault diagnostics and capability analysis
- Decision-making support for process development and investments.

The report[6] concerns examples such as the effect of pH on web breaks, possibilities to decrease the variability of chemical pulp freeness, developing the feedforward control of paper ash content, and modeling the effect of retention time.

Information on another paper machine analyzer is also available[7]. Its development came from the need for a better method to troubleshoot and optimize paper machines. It measures and analyzes paper and print quality using a complete range of high resolution sensors using 0.1 mm resolution and samples of several kilometers length. It measures machine direction and cross direction samples and covers the entire variation of the frequency spectrum from long-term (0.005 Hz) to very short-term (up to 10 000 Hz) variations.

The system has the following applications[7]:

- Cross direction paper quality diagnostics to verify the accuracy of measurements, identify streaks and their sources, and optimize CD actuators
- Machine direction diagnosis to identify the amount of MD variation and causes for it
- Various flock and micro level studies to examine high frequency variations, flock distributions, and formation estimation
- Printability studies for printability index, measuring the amount and distribution of ink in printed paper, blackness, gloss, etc.
- Special applications such as ISO-9000 programs, nip impression measurements, etc.

Another important area of on-line process diagnostics is the condition monitoring for moving parts of the paper machine. Some parts of process variations always exist that control systems cannot compensate This is especially true for rapid variations and transient disturbances. The tools used in condition monitoring usually use spectral analysis such as fast Fourier transformations (FFT) and the principle of synchronized time averaging (STA). Two examples are available in the literature[8, 9].

Process analysis, modeling, and simulation

Figure 15 illustrates STA analysis[9] by showing a press with two rolls having different diameters. Each roll has a flaw exaggerated in the figure. The smaller roll has an indentation, and the larger roll contains a bump. Analysis identifies the effects of the smaller roll.

The caliper (thickness) of the sheet is measured, and the signal shows the thick and thin parts. A rotation phase detector installed on the smaller roll measures the period of its rotations. The caliper signal is split into data strings that always start at the same instance relative to roll position. Adding these data strings amplifies the effects of the smaller roll while averaging other (more random) variations.

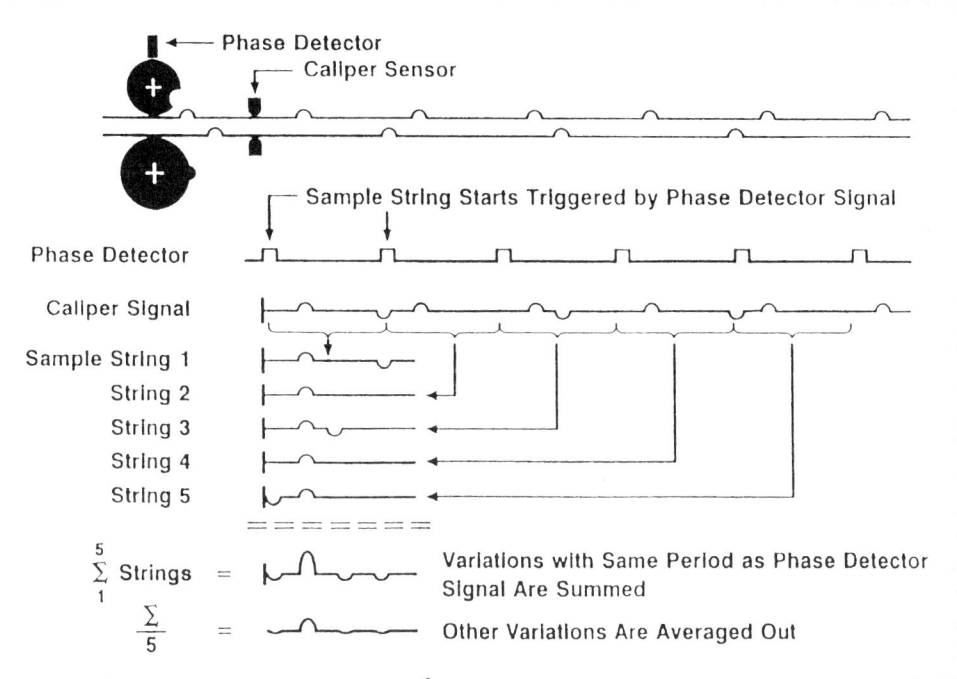

Figure 15. Example of synchronized time averaging[9].

The same reference[9] involves a process monitoring system that includes three modules for the following:

– Analyzing paper quality from on-line sensors

– Monitoring nip roll performances in the press section, size press, and calender

– Monitoring wet end performance of the paper machine.

Paper quality monitoring uses quality measurement signals from scanning measurements or single point sensors. It also reads rotation phase detector signals from monitored equipment such as the fan pump, screens, felts, and rolls. The system simulta-

CHAPTER 10

neously analyzes signals from several scanners from various zones of the machine and automatically separates CD and MD variations. The actual analysis uses STA.

Nip roll performance monitoring measures nip gap variations with high resolution sensor or low frequency accelometers. Detection of felt passing is optical. The contribution of each monitored source to variations in the press section is available to the operator in a display.

Wet end performance is analyzed from pressure and vacuum pulsations and machine element vibration in the stock approach, headbox, and formers. These vibrations are due to various rotating elements such as pumps, screens, rolls, and wires. Several kinds of measurements are useful: fast pressure sensors, rotation phase detectors, and accelometers. The system isolates short-term variations originating from the monitored equipment as Fig. 16 shows[9]. The top window shows the pressure measured at the manifold on the drive. The middle window shows the manifold pressure variations caused by the fourth pressure screen. Almost sinusoidal pulsation at twice the rotation frequency of the screen occurs. This is the primary source for the basis weight variations.

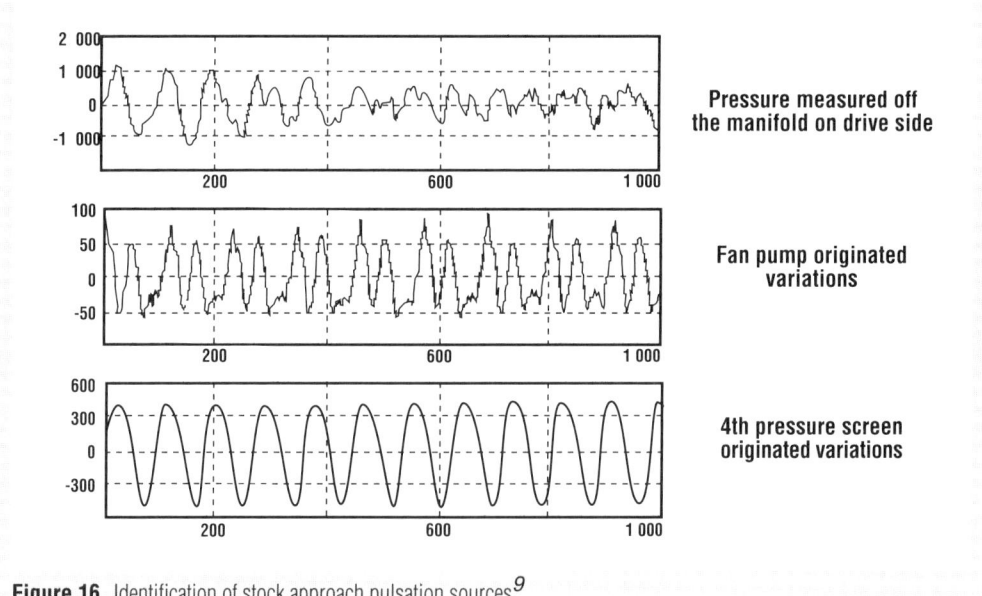

Figure 16. Identification of stock approach pulsation sources[9].

Another investigation[8] introduces an on-line diagnosis system installed at a kraft liner mill in Sweden. After three years, the number of new felts decreased by 40% and the downtime due to web breaks decreased by 40%. Simultaneously, production increased by 3.4%, recycled fibers in the mix increased by 4.3%, and the quality was more uniform.

3 Models and modeling

3.1 Model classification

The following six sections are an update of an earlier publication[10].

Many ways are available to classify models according to different aspects. Development of the model always fulfills a certain purpose that requires defining the model structure and contents as detailed as possible. The purpose and application area define the model type. The actual model formulation depends on model type and available data. Different models also have use in different stages of modeling. Simple models usually have use in the beginning. If necessary, more complicated models are introduced in later stages of the simulation project.

According to the model structure, one can speak about physical and symbolic models. Physical models such as small-scale models, pilot plants, prototypes, and analog models are descriptive and offer versatile ways to study a process. Their costs are usually high. Prototypes and pilot models usually have use at the final stage of the simulation (or design) project after studies with symbolic models (usually mathematical models) have confirmed the feasibility of the system under examination. Mathematical models can be analytical or numerical depending on the principle of solution.

According to time behavior, one can speak about steady-state and dynamic models. Steady-state models have use when time dependent behavior is not significantly important or the system has sufficient time to reach an equilibrium. Such models have permanent use in applications such as flow sheet simulators common in chemical engineering. Dynamic models can be continuous or discrete. They are necessary when time dependent behavior is important such as when time constants play an important role. Continuous systems have differential equations, and time advances as a continuous variable in the simulation. Dynamic models have wide use in various engineering fields such as control system design and training simulators. Figure 17 describes the difference between continuous and discrete dynamic models.

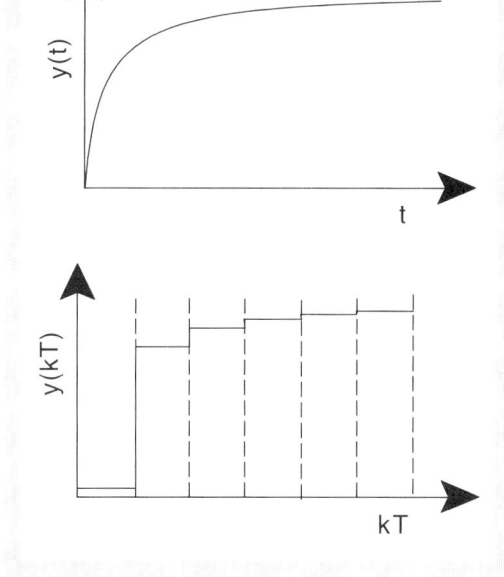

Figure 17. Continuous and discrete dynamic models. Discrete model (lower picture) repeats the model response with discrete intervals.

In modeling tubular reactors and heat exchangers, one uses distributed parameter models to describe the concentration and temperature distributions of the equipment. This leads to partial differential equation (PDE) systems when including time in the simulation. With certain preconditions, one can simplify these models to lumped

CHAPTER 10

parameter models and avoid tedious solution of PDE systems. Lumped parameter models assume that the studied variable has an even distribution and is constant with respect to the space variable. An example is concentration in an ideally mixed reactor. Figure 18 describes the difference between a distributed parameter system and its lumped parameter approximation.

Considering the statistical behavior of a modeled system, one can speak about deterministic models and stochastic models. A classification of models can follow their use. A model can be a communication tool to describe an opinion about a specific problem. Qualitative or even verbal models usually find use in this situation.

Models also explain measurements by giving correlation between some parameters. This is common in all types of research where simple models have use.

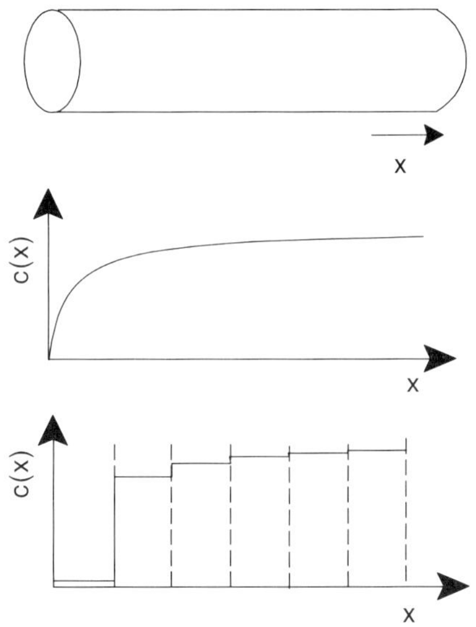

Figure 18. Distributed parameter system such as a tubular reactor and its lumped parameter approximation that assumes constant concentrations in successive sectors of the reactor.

Process and control system analysis employs modeling techniques for testing, tuning, and optimization. This assumes a model that describes the object system in the required conditions. The aim is usually to compare different alternatives to realize the object system. A measure of optimality is therefore necessary.

Training and education use all possible types of models including those developed earlier for other purposes.

Models also fall into classifications according to the modeling principle and to analytical models and experimental models (black box models). The analytical approach starts from the theoretical analytical principles or mechanisms that cover the operation of the physical system. These mechanisms define the method of description used in the modeling and the assumptions. It therefore defines lumped or distributed parameters, steady-state or dynamic models, deterministic or stochastic systems, etc.

Writing material, energy, or momentum balances according to the corresponding conservation laws formulates analytical equations. Assumptions consider the model application and accuracy requirements. A "complete" model is usually so complex that certain simplifications are necessary to solve the model with available resources within a certain time limit. Note that assumptions and simplifications do restrict model availability.

Black box models start from the existing system. Selection of the variables, equations, and parameters base on the problem definition. The screening of variables and the actual experi-mental design use well-known methods that allow one to obtain maximum information on the process with a minimum number of experiments. After the experiments, model parameters are fitted by applying regression analysis methods as an example.

Process analysis, modeling, and simulation

In the black box approach, consideration is necessary for variable selection and experimental conditions. When using regression analysis, the competence region of each variable requires careful consideration. The same is true for the nonlinear effect in the model.

These approaches are certainly not exclusive. The black box approach can numerically solve analytical models, and analytical models are usually verified using some black box techniques as Fig. 19 shows.

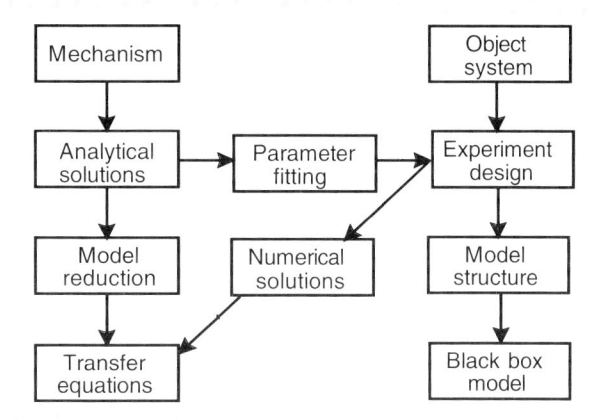

Figure 19. Two modeling principles and their interactions.

3.2 Selection of model type

Selection of model type depends on the problem itself. It is fixed in the early stages of the simulation project. In one project, several types of models may find use as required by the problem. Figure 20 shows that mathematical models usually find use at some point.

In process analysis, two separate cases exist. One is to describe the internal behavior of the process, and the other is to describe the input-output relationships between process control variables. These cases require different types of models. In the first case, detailed models describing the physical and chemical phenomena in the process are necessary. In the second case, simpler input/output models are sufficient.

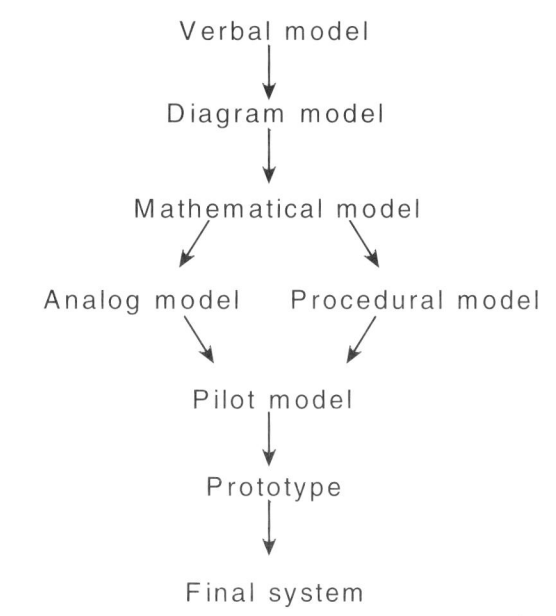

Figure 20. Use of different models in simulation.

Process design and equipment sizing use material and energy balance models and experimental correlation with physical and thermodynamic properties and cost information. Process control and control systems design use simple dynamic models

CHAPTER 10

that can be stochastic, adaptive, or both. Training simulators also use dynamic models exactly like the models used in startup and shutdown simulations and simulation of disturbance situations.

3.3 Model scope

The definition of the simulation scope depends on the problem in question. It must occur at the beginning of the simulation project. It usually means restriction of the simulation according to the level of description, included processes or systems, target accuracy, and model types or solution methods used.

The level of description means how much consideration goes to technical and economical aspects in simulation. In process design and equipment sizing, cost effectiveness is the key phrase. Technical and economical features are therefore part of the models. In the detailed design of some process equipment for a chemical reactor as an example, one sets the economic preconditions in the sizing stage. In the detailed design, primarily the technical aspects receive consideration.

Obviously, models require restriction according to the processes or systems included in the simulation. This means setting balance limits for the entire system under consideration. In flowsheeting, the balance limits of the subsequent units in the simulation model require setting accordingly as Fig. 21 shows.

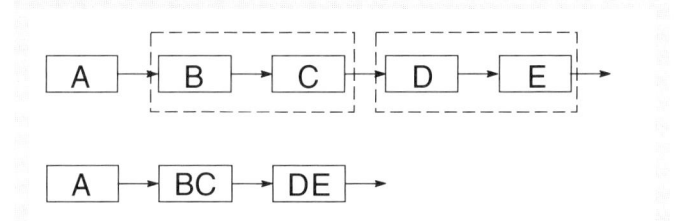

Figure 21. Setting balance limits. In this case, combining units B and C (D and E, respectively) leads to a more concise model of the whole production line than the original model containing five units.

Use of a model largely determines the required accuracy. A model used only to give an understanding of system behavior does not need to give any quantitative predictions. If the model will generate predictions during a certain time span, the model must also give accurate quantitative answers. A model used for control purposes must be a very good representation of the real system, or danger of degenerating process performance through use of an inaccurate model will exist.

3.4 Model construction

Model construction starts from a division of the system to be modeled into two parts: the model and the environment as Fig. 22 shows. In the model, one collects all those variables and aspects that are important for the investigation made. All other char-

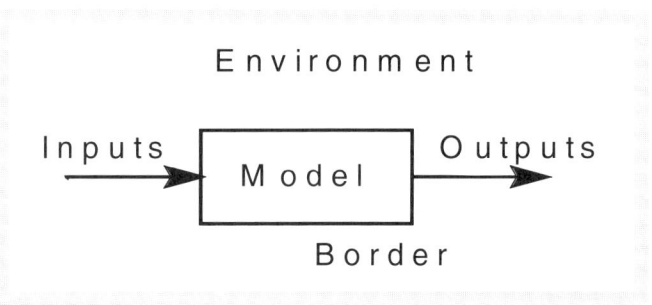

Figure 22. Model and its environment.

acteristics of the system are part of the environment that is a system in the same way. The model and the environment interact, but a common idea in the division is that the influence from the model to the environment is small. One may therefore neglect it. In establishing a simulation model of technical systems, one usually builds a hierarchical structure of models where the model is composed of separate interesting sub-systems. These further have a system of components. The division makes possible building parts of the final model as separate tasks with a final task of integrating the sub-models into one. The sub-models consist of variables and relations between the variables.

3.5 Solution of model equations

The model will ultimately be a set of mathematical equations. Simulating the system then means solution of the model equations. This sometimes requires a reformulation of the problem to a solution algorithm programmed into the computer.

The formulation of a model equation as an algorithm makes possible simulating the system on a digital computer. This again implies that the algorithm has a computer manageable form as a program. The process is primarily facilitated by use of different programming languages and environments. In selecting a specific programming language, consideration of the objective of the simulation study is again important because different languages have different features.

3.6 Model validation and use

The validation of a simulation is the most important phase of any simulation study. In this phase, one is assuring that the model is a true representation of the system in consideration. This means comparing the model and the actual system to ensure that they give the same responses. If this is not the case, the model requires changing by tuning model parameters or by changing the model structure. In the final validation, using independent data, i.e., data not previously used for construction of the model, is important.

To ensure reliable use of the model, all variables must have a certain area of confidence. Note that the model results can be more unreliable near the limits of confidence than in the middle of the confidence region. Extrapolation outside the confidence region is always dangerous.

After building and validating the model, one can use it as a counterfeit real system. This allows experimenting with a system not yet built. One can also undertake a sequence of events that is not possible with a real system because of safety or cost reasons.

When using simulation results, one must know the inherent limitations of their applicability. Very often the simulation results will not have such use. Instead, they might have use as input into a decision process. In such cases, conveying the qualitative implications of the simulation in an easily understandable form to the decision maker is even more important. Model interface and documentation are discussed in Sections 8.3 and 8.4.

Model updating is a critical issue especially with the models used in on-line process control. The need for updating must be clearly specified with the updating procedure. In large off-line models, updating can even be a more tedious job because the process information requires transformation. In the worst case, this will be manual.

CHAPTER 10

Model transferability is sometimes problematic. Parametrization is the least that is necessary when using the developed model in another process. Use of the model structure and its limits and construction require changing. Good documentation is necessary to improve the transferability of the model.

4 Properties of flow sheet simulators

4.1 Definitions

A modular flow sheet simulator describes the process as a set of modules connected by the flows of material and energy between them. The flows constitute numerous components that undergo transformations such as reactions when they pass through the system. Modules are simply material and energy balances with those physical and thermodynamic data and correlations necessary for the calculations. An executive program handles the sequence of module calculations. It also controls the input data reading routines and the printout routines.

Steady-state flow sheet programs have had use since the 1960s in the chemical industry. Their potential applications are for mill engineering design and development of operating strategies for existing mills. Some dynamic flow sheet simulators have also been developed. These programs have use in operator training, development of control systems, analysis of startup and shutdown situations, and operations scheduling.

The first process simulators in the 1960s were very simple and had limited use. The number of different unit process modules was small, data banks for physical and thermodynamic data and correlations were limited, and calculation procedures were simple. Because of the limited properties of computers at that time, their use was also difficult.

Flow sheet simulators of the second generation came into the market at the beginning of the 1970s. They had large libraries for unit modules and physical properties. The numerical methods used were more efficient than in the earlier simulators, and the man to machine interface started to approach the modern level. Simulation results could also connect to cost and investment data.

Third generation code development came in the 1980s. The most important factor was the use of these systems. They integrate simulators in the total data base systems where they can use the same data that is available in all stages of the mill design. The man to machine interface uses the capabilities of work stations and modern personal computers.

The flow sheet simulators use two solution techniques. The modular sequential approach is the most common technique. It computes modules singly in a certain sequence that usually follows the direction of physical flows in the system under study. This requires knowledge about the input flows of the module to be calculated. Calculation of output flows uses the input flows and module parameters. If recycle flows exist, their values require an initial guess in the beginning with iterative improvement during the simulation. The iteration continues until the balance constraints are satisfied and convergence occurs.

The requirements for computer memory are small in this approach because only limited information is necessary for the calculation of each module in the flow sheet. A large number of iterations and long computing times can cause problems especially when considering large processes.

Process analysis, modeling, and simulation

A modular simultaneous approach is another alternative for solving flow-sheeting problems. In this approach, the information contained by the flow sheet and modules is converted into a set of linear equations solved simultaneously.

This method minimizes the number of iterations to provide a rapid solution. The requirements for the computer memory are larger that for the previous case. The system can include constraints for process design that are not possible in the modular sequential approach.

4.2 Structure of flow sheet simulators

Regardless of the solution method, all flow sheet simulators have the following main components as Fig. 23 shows. Note that the extent of each can vary considerably from system to system.

- Data input using interactive methods or different file and data base structures
- Sequencing, i.e., determination of order of calculation
- Executive program
- Printout routines
- Module library
- Routines for design and optimization
- Routines for calculation of physical and thermodynamic properties
- Data reconciliation.

The following text is a brief discussion of these components.

4.3 Example of simulation executives

Simulation executive programs handle the actual operation of a flow sheet program. They manage the input data, store it, call the modules in the preset order of calculation, and direct the printouts through several sub-programs.

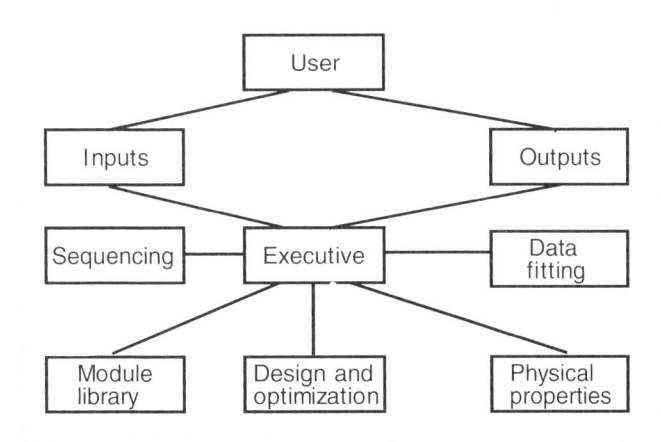

Figure 23. Components of flow sheet simulators.

As an example, consider a process simulator developed at the University of Oulu, Control Engineering Laboratory[11–13].

The original development of the simulator was for calculation of material and energy balances in a kraft mill chemical recovery but later extended for entire mill simulations. The simulation system is a computerized tool for pulp mill balance calculations.

CHAPTER 10

Because of the amount of information handled, material and energy balances require calculation for all process units separately and also for the entire system, since changes in one unit influence the entire mill. The solution is iterative because of the recycle flows involved. The simulation system has had application to double and single line kraft pulp mills. It also includes simple flow dynamics.

The executive programs read the input data, transfer the data between programs, execute the calculations in the correct sequence, display the results, and change the input data if needed as Fig. 24 shows. In many other modular simulation systems, the models use unit operations. This simulation system uses the total material and energy

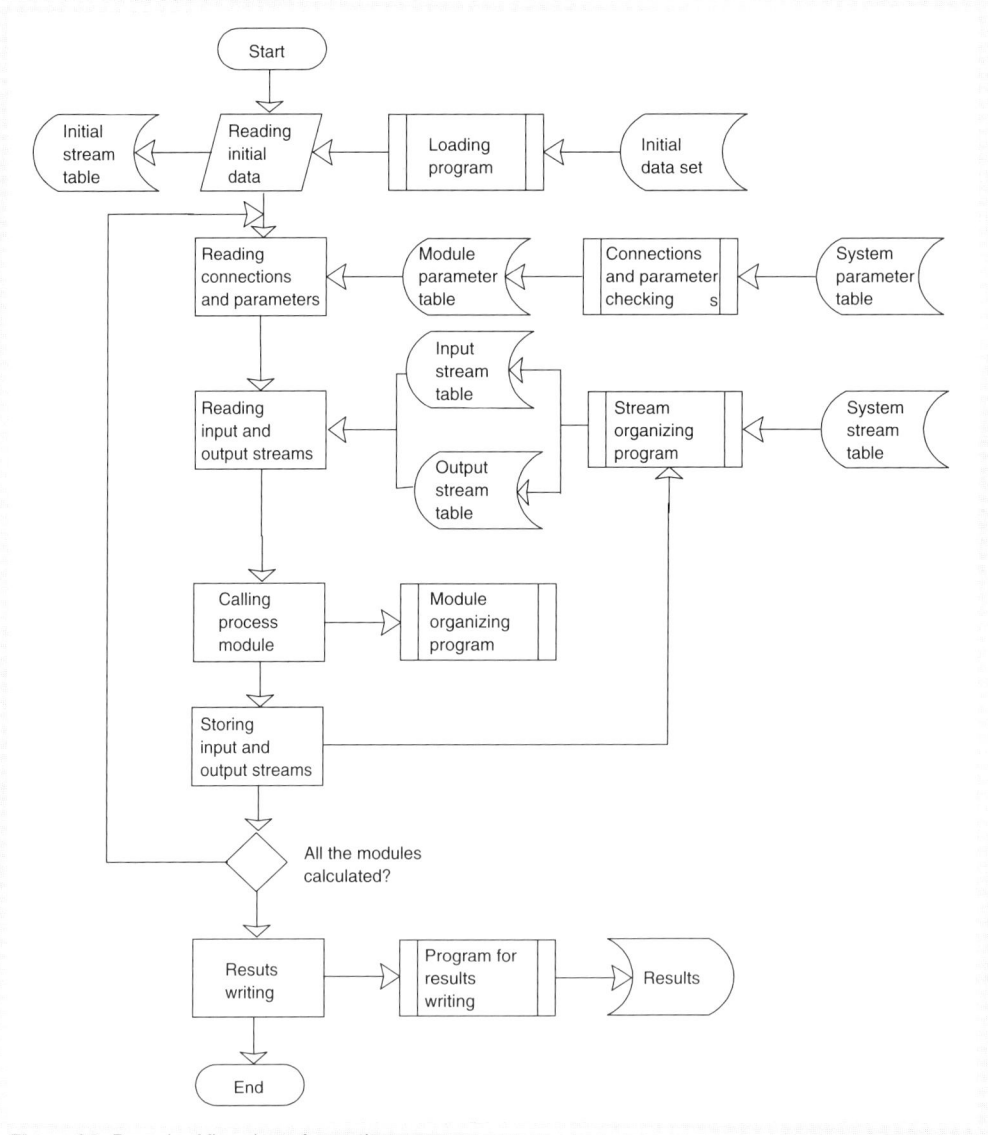

Figure 24. Example of flow sheet of executive programs.

Process analysis, modeling, and simulation

balance models of different process units. These models consider ongoing reactions and losses. An exception is the bleach plant model that uses specific dosage models.

The simulation system can handle eight different types of process streams as Fig. 25 shows. The components of the streams differ from each other. Figure 26 shows the stream vector formulation. The model parameters are input data stored in parameter

```
Fiber flow
Liquor flow
Gas flow
Chip flow
Smelt flow
Makeup flow
Lime flow
Oil flow
```

Figure 25. Stream types for the sample flow sheet system.

N	EN(N)
1	Module number
2	Module type
3	Length of vector
4	Undefined
5	Undefined
6	Number of input flows
7	Stream number
...	
15	Number of output flows
16	Stream number
...	
24	Module parameters
...	
jj	

Figure 27. Stream vector in the sample flow sheet system. First seven parameters define the stream properties. Parameters 8–25 define the flows of different stream components and parameters 26 and forward are reserved for stream specific parameters (for example, causticity and sulfidity in white liquor).

j	Vector component
1	Flow number
2	Undefined
3	Total flow
4	Temperature
5	Pressure
6	Specific heat
7	Density
8	Water flow
9	Flow of component 1
...	
26	Property jj

Figure 26. Stream vector in the sample flow sheet system. First seven parameters define the stream properties. Parameters 8–25 define the flows of different stream components and parameters 26 and forward are reserved for stream specific parameters (for example, causticity and sulfidity in white liquor).

```
Number of processes
Number of components
Calculation order
Total number of flows
Number of feed flows
Stream vectors of feed flows
Number of modules
Module parameter vectors
```

Figure 28. Initial data set in the sample flow sheet system.

277

CHAPTER 10

tables. These parameters correspond to the targets used in the operation of the processes such as kappa number, alkali charge, cooking temperature, dilution factor, strong black liquor solids content, white liquor causticity, etc., as Fig. 27 shows. At the beginning of the calculation, any parameter can change. In addition, input flow rates, temperatures, and concentrations of different components can be respecified as Fig. 28 shows. The simulation system iteratively calculates the steady state material and energy balances.

As examples, Fig. 29 shows a module parameter vector for a washing module, and Fig 30 gives the stream vector for fiber flows.

N	Parameter	Value
1	Module number	7
2	Module type	3
3	Length of parameter vector	27
4	(Free)	
5	(Free)	
6	Number of input flows	2
7	Stream number	7
8		10
9-14		0
15	Number of output flows	5
16	Stream number	12
17	-	5
18	-	11
19	-	108
20	-	9
21-23	-	0
24	Consistency of pulp leaving	12
25	Washing efficiency faetor	12
26	Dilution factor	0
27	Knot percentage	3

Figure 29. Parameter vector for washing module.

1	Number
2	(Free)
3	Total flow
4	Temperature
5	Pressure
6	Specific heat
7	Density
8	Water
9	NaOH
10	Na_2S
11	Na_2CO_3
12	Na_2SO_4
13	$CaCO_3$
14	CaO
15	$Ca(OH)_2$
16	Inert
17	Organic
18	Fiber flow
19	S/Na_2 ratio
20	Organic/Inorganic
21	Na in organic
22	Na in fibers
23	Washing loss
24	Dilution factor
25	
26	Consistency

Figure 30. Fiber flow vector.

Process analysis, modeling, and simulation

4.4 Module library

The modules included in the module library of the flow sheet simulator depend on the problem for which they are designed. Figure 31 shows the module library of our sample simulator.

CONDIG	BATDIG	EVAP	FURN	KILN
WASH1	DWASH	CAUS1	CAUS2	SCREEN
BDENER	CDENER	SPLIT	MIXER	SMDT
CONTL1	PLUG	TANK		

Figure 31. Sample of a module library.

Module names are self-explanatory. Modules PLUG and TANK refer to slow dynamic models for plug flow and storage tanks. The module library also includes some control blocks (CONTL1). This block is a module that can fix some preset value for an output flow by changing certain process parameters as Fig. 32 shows. They act like a feedback controller, but no dynamics are included.

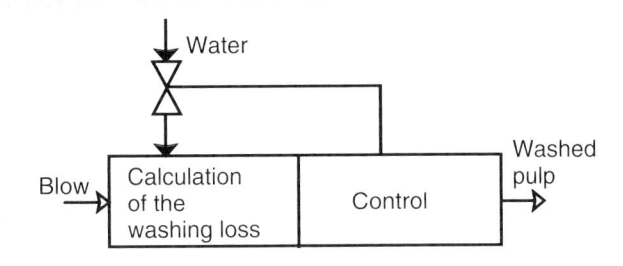

Figure 32. Use of control module in wash plant model.

4.5 Physical and thermodynamic data

To simulate a process, one must know the physical and thermodynamic data connected with the material flows in the system. Some data banks are available that are usually used in flowsheet simulators. In the flowsheet simulator, calculating physical properties continuously as the simulation proceeds must be possible to store the calculated values for later use. Possibilities to enter new data into the system during simulation and produce new data for the user on compounds that lack initial data must also be possible.

Physical and thermodynamic data systems must have the following:

– Wide range of pure component properties (Single constant properties and correlations in those cases where the property changes according to the temperature or pressure must be given. Typically, properties such as molecular weight, boiling point, heat capacity, density, enthalpy, free energy, viscosity, thermal conductivity, solubility parameter, etc., are presented. These values must be easily accessible from the simulator.)

– Models, correlations, and estimation methods (These have use in predicting various properties such as critical constants, heat capacities, vapor pressures, heat of vaporization, etc.)

– Methods to calculate thermodynamic properties for mixtures (Factors such as enthalpy, entropy, free energy, fugacity, activity, heat of solution, etc., require calculation.)

CHAPTER 10

4.6 Design and optimization functions

In modular sequential simulators, the inclusion of design and optimization functions is difficult. The convergence can be slow with large process diagrams, and the use of design oriented models is difficult. One technique is to repeat simulation several times and change process parameters, input flows, or both until the design objective is met. This is very time consuming. Some optimizing routines such as linear programming can also have use to decide which parameter to change and how much. Here the computing time easily becomes a limiting factor. The modular simultaneous approach gives better possibilities for design and optimization with a flow sheet simulator. The design constraints can be added directly into the model, and the solution routine does require changing.

4.7 Dynamic flow sheet simulators

Introduction of the time factor into the flow sheet simulator framework means a considerable increase in the complexity of the system. Several factors contribute to this:

- The dynamics of each module (unit process) require modeling mathematically and identification.

- The resulting set of differential equations requires solution. This may cause computation difficulties especially when including slow and fast dynamics simultaneously. Special procedures are available to overcome this problem, but most computer time is for solving differential equations.

- The steady-state values for all the flows must be given at simulation start-up time. This means that this steady state must be calculated first. It also means that problems do not exist in dynamic part convergence. Dynamic simulation produces more results than steady-state simulation.

5 Flow sheet simulators in the pulp and paper industry

5.1 GEMS

GEMS is a general purpose mass and heat balance simulator originally developed at the University of Idaho. It uses a modular sequential solution structure. It is written especially for the pulp and paper industry and includes a wide variety of process models and a physical property package. In addition to the steady-state modules, it also has two dynamic modules: an ideal, continuous stirred tank and a plug flow delay block. The system has had extensive use for material and energy balance calculations during the 1970s[14-19]. Several different versions are available in the literature.

A set of programs working with a personal computer simulation package makes interactive operations and process optimization possible[20]. The package has four parts. One is a program for running GEMS simulation with screen interaction. It handles inputs and outputs in tabular form. The other is a program for generating a linear programming model. This does not happen interactively, but the input data is generated automatically. For complex situations, this is a time consuming stage. There is also a program for optimization using the linear programming algorithm. Finally, a program estimates model parameters using full simulation models.

Real-time dynamic personal computer GEMS has also had use for training purposes by interfacing it with a distributed control system[21]. A dynamic simulation of the paper machine saveall system running in real time (or faster than real time) has been developed using standard GEMS models. It has also had use in dynamic simulation of a rapid displacement heating (RDH) process where the problem is changes in the process topology (flows stop and change directions)[22, 23].

GEMS has also been integrated with a millwide information system, an expert system shell, and statistical process control into a prototype advisory system for recovery boiler operation[24] and for an expert system for multiple effect evaporators[25]. A recovery boiler model reference is also available[26]. Another reference[27] reports the application of PCGEMS in on-line control and optimization of the refining process. For this purpose, the program executive was rewritten to give 3–4 times faster execution than before.

5.2 MASSBAL

MASSBAL MKII was originally developed by SACDA Inc. in cooperation with Energy Mines and Resources Canada as a generalized simultaneous modular process simulation package for calculating the steady-state heat and mass balances for industrial processes[28]. The system was designed for modeling aqueous processes such as those found in the pulp and paper or mineral industries. It is a successor to various earlier systems developed during the 1970s. It uses several hardware possibilities from personal computers to mainframes and includes earlier program packages in an integrated environment.

Model generation uses the user-friendly multi level keyword based input language. Language design is self-documenting. In a microcomputer environment, an interactive front-end program is available. This allows someone with little or no experience with process simulation to use the system.

Another possibility is to use graphics front-end and configure the simulated system by simply drawing the process flow sheet. It generates the simulation model, executes the simulation, and returns the results on the flow sheet. Fig. 33 shows an example.

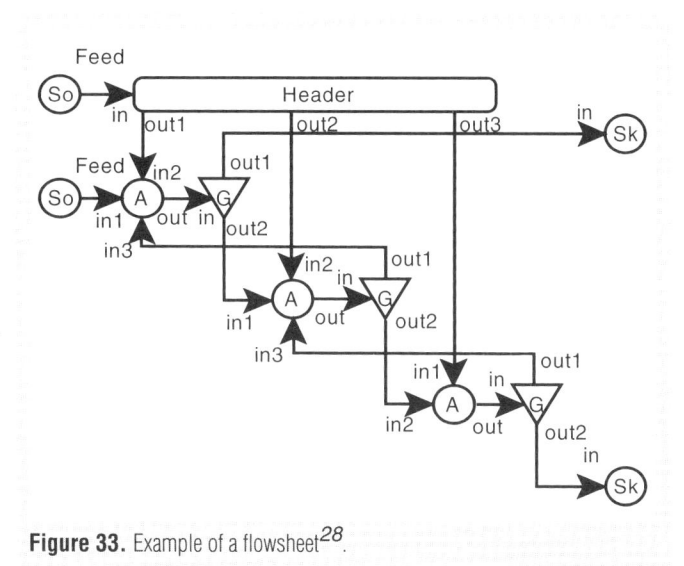

Figure 33. Example of a flowsheet[28].

CHAPTER 10

Still another possibility is to connect the simulator to a real time process distributed control system. This allows doing process optimization, data reconciliation, or predictive calculations.

MASSBAL MKII can handle standard or user specified stream variables. It also includes power streams. Equipment variables are built into the modules. Including economic calculations is also possible. A physical property package includes properties for steam, condensates, calcium and sodium compounds, black liquor, combustion gases, etc. User-supplied correlations or experimental data can change all properties.

The simulation system uses a simultaneous approach for solving mass and energy balances. This results in a flexible system that can handle performance, design, and rating calculations. Solution of all these problems can use the same model but a different configuration. It also supports the data reconciliation features. Optimization uses successive quadratic programming (SQP).

Besides a conventional reporting system, the simulator has a generalized report generator for preparing customized reports or linking simulation results to downstream calculation for engineering drawings, spreadsheet programs, or word processing.

The simulation system has been used for example in kraft liquor cycle simulation in a commercial mill[29]. The simulation considered effects of several key variables such as causticizing efficiency, pulp production rate, green liquor strength, etc.

An operator training system is also available that uses dynamical process models and models for controls and logic. One case[30] reports on a full paper machine application where the training system connects to a distributed control system.

5.3 Other flow sheet simulators

GEMCS (General Engineering and Management Computation System, pronounced gee-max) was originally developed at McMaster University[31]. GEMCS has had use for various chemical engineering applications. Boyle et al.[32] used it to simulate a kraft pulping process. The University of Oulu also developed an extensive simulator for kraft mill studies based on GEMCS[12].

A paper company developed a facility energy model (FEM) that underwent later integration with GEMS resulting in a chemical heat and material process simulation (CHAMPS)[33].

MAPPS is a Fortran-based program using the sequential modular approach for simulation of steady-state behavior of pulp and paper mills[34]. Its original development was for calculating mass and energy flows, but it also has use for equipment design. It can handle up to 12 different stream types including different properties. It has also had use in modeling and design of a fluidized bed dryer.

PAPDYN is a dynamic modular simulation package using a steady-state modular package. An example for washing simulation is available[35].

One pulp simulation system[36] has a built-in computed aided design (CAD) environment. It includes models for unit operations that can easily connect to complete processes and mills. The system calculates material and energy balances for components freely defined by the user. A user interface is built in AutoCAD so the user draws an intelligent process chart using preprogrammed modules or tailored ones. The modules and their connections are automatically generated, and the program will ask and automatically suggest which process parameters should or can be used. The system includes logical checks and easy editing possibilities. Several installations exist in Finnish mills.

One work[37] reports on application SLAM II simulation language for RDH tank farm simulation. Also SIMNON has had use in dynamic modeling and simulation of a recausticizing plant[38].

Some reports have recommended using spreadsheets instead of flow sheet simulators[39]. Operating superintendents or mill process engineers are more likely to use iterative spreadsheets as a starting point for further investigations than flow sheet simulators that seem to be a tool for a research and development laboratory or consultants.

6 Intelligent methods and simulation

Simulation requires many types of expertise. Reaching the appropriate level takes time. First, the application area must be known. Users are increasingly constructing their models from modules available in unit libraries without any help from simulation specialists. Intelligent methods are useful in simulation from modeling to analysis of the results[40]. Nonexpert users especially can take advantage of these methods.

Expert systems help model construction in cases that do not include a considerable amount of uncertainties. As mentioned above, simulation models usually have components provided by the simulation package. Construction can be difficult for nonexpert users, but a conventional expert system can help users select the most suitable modules and specify the parameters correctly. For continuous time simulation, the system checks if the proper integration method was used. Expert systems are available for selecting suitable step sizes and integration algorithms.

Expert systems also help nonexpert users in parameter estimation and in testing and fitting of input data. Some expert systems assist in use of large software packages such as in the selection of suitable methods in statistical analysis packages.

With uncertainties, building sophisticated systems using a conventional expert system is not feasible. A more flexible method for uncertainty processing is therefore necessary. Fuzzy set systems provide a flexible and well-explained way to handle items such as uncertainties in the input as Fig. 34 shows.

Result analysis, optimization, and selection of the key changes are important areas where much information is available from adaptive and intelligent methods. The goal for these systems is to provide advice for users using simulation results. Advice should include instruction about how to change a model and its parameters to reach the goal.

Intelligent methods can assist in searching for errors and running automatic tests. They can also analyze erroneous results and search for probable sources of those errors. In addition, one can construct a higher level qualitative model for use in the validation process. Fuzzy set systems are natural extensions in this area since comparing to qualitative models makes model validation, critical testing, and error analysis more consistent.

CHAPTER 10

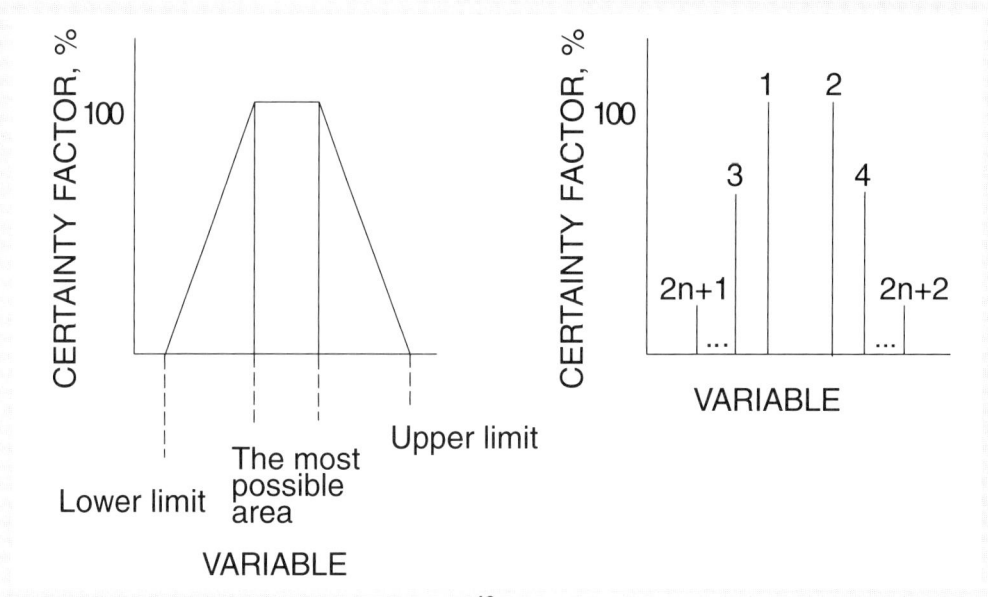

Figure 34. Fuzziness in simulation experimental design[40]. The picture shows the variable range divided into the most possible area where experiments should be concentrated and the area with increasing uncertainty in importance.

When using simulation results, one should be aware of the inherent limitations of their applicability. Simulation results will very often not be useful as such but as input into a decision process. In such cases, conveying the qualitative implications of the simulation in an easily understandable form is even more important. This puts stress on the interface of the model, and interactive interfaces for input information, execution, and reporting are favored.

Simulation runs produce a huge amount of results from which the essential part must be found. Expert systems can filter the results to the users by explaining the results and how they have been achieved. They can also produce new knowledge such as rules, structures, classification, sensitivity, and parameter analysis from simulation data. Expert systems enhance the capabilities of the simulation as a decision support tool for the following:

– Estimating how well the goals are achieved
– Considering possible contradictory constraints
– Monitoring the run to find causal relationships
– Analyzing causal relations found to construct the best causal model
– Defining how sensitive the resulting parameters are to the manipulated parameters.

Process analysis, modeling, and simulation

An example of combining artificial intelligence and simulation is available[41]. The work combined expert systems and simulation in the SIMSMART system that includes dynamic models describing the paper mill processes and the actual control system. The system uses real time process information, but the simulator uses faster than real time data analysis. The simulated data uses first forward chaining to find any unacceptable values of simulated process variables and then backward chaining to find the causes of these malfunctions as Fig. 35 shows.

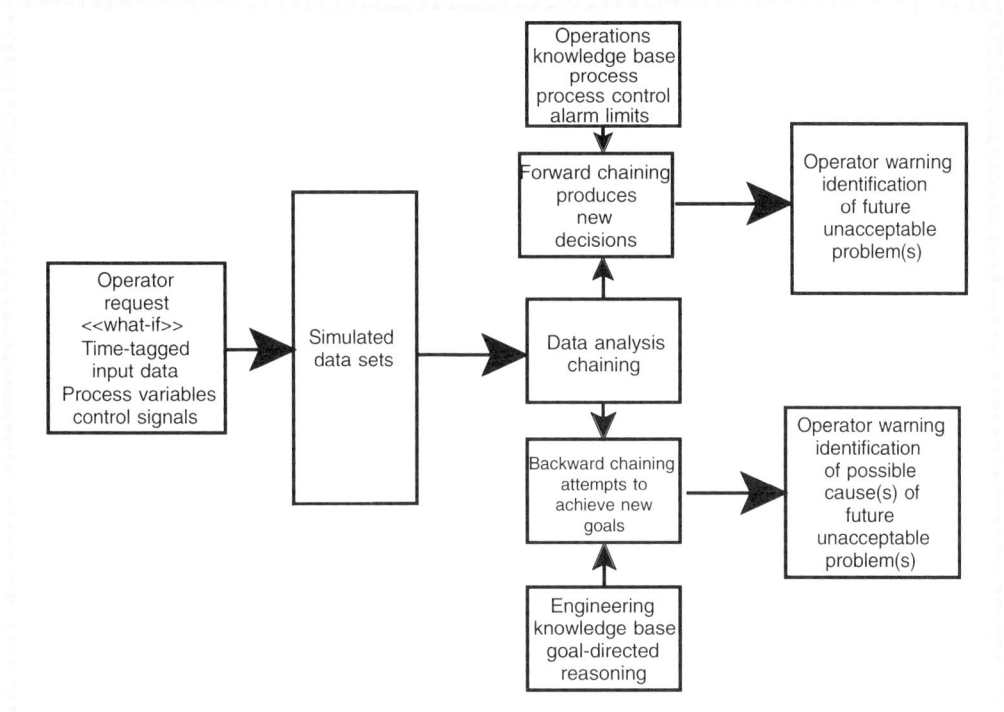

Figure 35. Operation principle of faster than real time data analysis[41].

7 Application viewpoints

Process simulation has two main tasks in the pulp and paper industry:

- To explain the internal behavior of the process, i.e., describe the progress of the process state variables
- To explain the external behavior of the process, i.e., compute the values of process output variables based on the input variables.

These two views are often combined although the arrangements necessary for each case differ considerably. In calculation of the input-output relationships, rather simple models are adequate. In simulation of internal behavior of the process, thorough and detailed process models and numerous preliminary studies are necessary. In planning simulation studies, the purpose of the simulation and the requirements for the results require consideration. The following discussion uses an earlier work[11, 42].

7.1 Process flow-sheeting

In an integrated pulp and paper mill, the number of major material streams can be tens or even hundreds. The number of stream components is also considerably high, because one must consider all the most important active and inactive chemical components.

Despite the large number of streams and variables, a single material and energy balance calculation for a pulp or a paper mill is not an exceedingly difficult task. When studying the effect of variations in different stream components on the entire mill material and energy balance, a simple case by case manual calculation is no more profitable. The use of computer simulation systems gives extensive advantages in comparison with straightforward calculations. After more extensive work in system design and data collection, all kinds of variations in stream components or process connections can be studied by an incremental additional contribution.

The degree of the model sophistication depends on application. To minimize the simulation effort, simple input-output models should be used. Testing the validity of very detailed models is difficult due to the lack of detailed and accurate process information. This increases implementation and updating costs.

Adding dynamics to steady-state simulators means increasing the complexity of the system by having to solve differential equations. This is especially true if separating the fast control dynamics and slow process dynamics is not possible. Inclusion of dynamics means increasing problems in model fitting and verification. The computational load increases as does the volume of results. Dynamic simulators are also necessary in addition to control design for analysis of process dynamics in storage tank sizing, rates of reactions, startups and shutdowns, and grade and rate changes.

7.2 Comparison of process and mill operational alternatives

The comparison of different operational alternatives may require considerable work especially when a detailed explanation is necessary. When a computer simulation system is available, the comparison itself is easy. The predetermination of process models and parameters makes the entire project larger than expected. The total gain is much higher because of more versatile results than with direct calculation or experimental evaluation. With the help of a simulation system, estimating the effect of a change in operational rules is possible for the following:

- Composition of process streams
- Propagation of flow and concentration disturbances
- Process and storage delays
- Operational costs resulting in the optimum control strategy for the mill and processes.

Using simulator training can assure successful start-up of a modernization project[51]. Operators become very familiar and comfortable with the control screens, their use, and the dynamics of the system. Full use of simulation means looking beyond the start-up training requirements to its continued use. It requires modification to remain current with the process and control system. Undertaking a complete simulator-based training project also needs management commitment because it requires development work and maintenance.

8 Problems in modeling and simulation in pulp and paper mills

As already stated, some specific but also rather common problems in pulp and paper mill simulation studies exist [11, 42]. These problems fall into two main groups: fitting of models to the process behavior and the lack of suitable models for certain specific purposes. Because pulp and paper mills consist of many types of sub-processes especially when including the manufacture of byproducts, many problems may arise in addition to the two stated above.

8.1 Adjustment of models

The process stream variables in a detailed simulation study of an integrated pulp and paper mill exceed several thousands. Only some can be determined from the process. The rest require estimation in some other way. The adjustment of model parameters to uncertain process data requires considerable fitting experiments before the real simulation studies can begin. The amount of data for processing is also high.

A typical problem for pulp or paper mill simulation is the adjustment of theoretically derived models to process data. Although in practice the models use material and energy balances meaning the theoretical form of the model should be correct, the model will be correct only at one operating point and incorrect elsewhere.

The most frequent problem in the reconciliation of material and energy balance models is the absence or lack of suitable measurements needed for complete balance checking. The flow rates of the main process streams are normally measured, but several material flows also exist that are not measured at all. Typical examples are the flows of knots and rejects and for sealing waters. Not measuring these gives an inaccurate balance determination.

Concentrations of process streams are not normally measured directly. The most important chemical flows are sampled and analyzed, but this is not the case for all streams. Flow rates and concentrations of several process streams therefore remain undetermined. Representative samples are not always easy to obtain. The measurement of pulp consistency is still a large problem especially for high consistency streams and storage tanks. This means that the fiber production rate is not accurately known until final product output. Simulation can therefore give accurate results only on the theoretical level that cannot be compared with measured production data for short periods. Remember also that the entire mill is only seldom in steady-state, although simulators usually report steady-state solutions.

CHAPTER 10

The lack of suitable balance reconciliation methods is a reason for the high manpower requirement needed for pulp and paper mill simulation. Computerized methods can ease the balance reconciliation[52]. Development of effective methods decreases some problems but does not solve them all.

8.2 Lack of suitable models

Models should fit their purpose. This applies to their scope and complexity. Simulation models should be as complicated as necessary but as simple as possible to minimize the manpower required, assure reliability, and optimize understanding of the results. Many commercial simulation models and model libraries for different pulp and paper processes are available. This fact does not guarantee the suitability of these models for one's own purposes. For example, the models designed for internal process unit simulation should be used with care in the simulation of an entire production line because too many useless equations, variables, and parameters are involved in the solution of a comparatively simple task. This also increases the margin of error for the results.

Another problem in the use of available models is the fact that two identical pulp and paper mills do not exist. This forces the designer to change the available models somewhat for the particular case in question. Complete models are very seldom available, and the lack of suitable models is obvious. At a minimum, the fitting of model parameters is always necessary from process to process and mill to mill.

8.3 Documentation

Good documentation assures continuous use of a simulation system. Using simulation programs after the key individual in system development has left or is busy with other projects must be possible. The level of documentation must correspond to the normal one for the mill and must allow for a nonexpert to start using the simulation system. All the blocks and routines inside the system require documentation with a careful explanation of the input parameters. The ideal system for documentation is to include self-documenting and system support features in the system. These can use structured documentation and hypermedia in the network environment. This allows easy access and updating of documentation.

8.4 Presentation of results

Presentation of results depends heavily on the application. On-line training systems should be able to use displays similar to actual automation systems. This gives the operator the same feeling for the system as during normal operations. Simulators in design should preferably connect with computer aided design systems and use their interface and data base structures. Helping and supporting functions are also necessary. They can take advantage of structured documentation and hypermedia. Simulators in the design of control systems should also connect with the actual design system and use its interface. Stand-alone simulators today usually use different graphical interfaces that allow easy and flexible forms for presentation of results.

References

1. Diamond, W. J., *Practical Experiment Designs for Engineers and Scientists*, Lifetime Learning Publications, Belmont, 1981, pp. 87–125.

2. Anon., "Taguchi Techniques." Available [Online] <http://kernow.curtin.edu.au/www/Taguchi/set7.htm> [Sep. 9, 1997]

3. Isermann, R., *Identifikation dynamischer Systeme*, Springer-Verlag, Berlin, 1988, 344 p.

4. White, K. and Roberts, C., 1994 Control Systems Preprints, SPCI, Stockholm, p. 52.

5. Ritala, R., Paulapuro, H., Penttinen, I., et al., 1990 24th EUCEPA Conference Proceedings, SPCI, Stockholm, p. 45.

6. Pokela, J., Kukkamäki, E., Paulapuro, H., et al., 1994 Control Systems Preprints, SPCI, Stockholm, p. 159.

7. Anon., "Tapio." Available [Online},http:/www.papertek.com/tapio/> [Feb. 2, 1998].

8. Olla, H., Karlsson, H., and Sehlin, E., 1994 Control Systems Preprints, SPCI, Stockholm, p. 166.

9. Ström, U. J. and Pietarila, A., 1990 24th EUCEPA Conference Proceedings, SPCI, Stockholm, p. 65.

10. Wahlström, B. and Leiviskä, K., in1985 IMACS Transactions on Scientific Computation, vol. III (B. Wahlström and K. Leiviskä, Ed.), Elsevier Science Publishers, Amsterdam, p. 3.

11. Jutila, E. and Leiviskä, K., Mathematics and Computers in Simulation XXIII(1):1(1981).

12. Jutila, E., Leiviskä, K., Visuri, M., et al., Tappi 65(1):43(1982).

13. Uronen, P., Leiviskä, K., and Kesti, E., "Benefits and results of computer control in pulp and paper industry," Ministry of Trade and Industry, Series D:76, Helsinki, 1985, pp. 1–8.

14. Corson, S. R.and Edwards, L., Appita 29(5):371(1976).

15. Gunseor, F. D.and Rushton, J. D., Tappi 62(3):63(1979).

16. Venkatesh, V., Corson, S. R., and Edwards, L. L., TAPPI/CPPA 1975 International Mechanical Pulping Conference, TAPPI PRESS, Atlanta, p. 173.

17. Venkatesh, V. and Edwards, L. L., Tappi 60(12):68(1977).

18. Venkatesh, V., Kirkman, A., and Mera, F., Tappi 61(3):87(1978).

19. Xuan, N. N., Venkatesh, V., Gratzl, J. S., et al., Tappi 61(8):53(1978).

20. Boyle, T. J., Tappi J. 70(6):75(1987).

21. Haynes, J. B., Scheldorf, J. J., and Edwards, L. L., Pulp Paper Can. 91(5):45(1990).

22. Scheldorf, J. J. and Edwards, L. L., Tappi J. 76(11):97(1993).

23. Scheldorf, J. J., Edwards, L. L., Lidskog, P., et al., Tappi J. 74(3):109(1991).

24. Smith, D. B., Edwards, L. L., and Damon, R. A., Tappi J. 74(11):93(1991).

25. Brooks, T. R. and Edwards, L. L., Tappi J. 75(11):131(1992).

26. Uloth, V., Richardson, B., Hogikyan, R., et al., Tappi J. 75(11):137(1992).

27. Strand, W. C., Ferritsius, O., and Mokvist, A. V., Tappi J. 74(11):103(1991).

28. Shewchuk, C. F., Pulp Paper Can. 88(5):76(1987).

29. Misra. M. N. and Sowul, L., Pulp Paper Can. 91(8):99(1990).

30. Terrell, J. and McLaughlin, G., Pulp Paper Can. 92(4):59(1991).

31. Johnson, A. I.and Peters, N. P., "GEMCS User's Manual," The University of Western Ontario, 1978.

32. Boyle, T. J., Treiber, S., and Vadnais P., TAPPI 1975 Engineering Conference Proceedings, TAPPI PRESS, Atlanta, p. 185.

33. Thomas, K. V., Tappi 62(2):51(1979).

34. Sell, N. J. and Clay, D. T., Tappi J. 74(6):177(1991).

35. Turner, P. A., Roche, A. A., McDonald, J. D., et al., Pulp Paper Can. 94(9):37(1993).

36. Anon., "Pulpsim article." Available [Online} <http://www.iaf.fi/agco/article.htm> [Feb. 27, 1998].

37. Sezgi, U. S., Kirkman, A. G., Jameel, H., et al., Tappi J. 77(7):213(1994).

38. Wang, L., Tessier, P., and Englezos, P., Tappi J. 77(12):95(1994).

39. Frazier, W. C., Pulp Paper Can. 92(10):29(1991).

40. Juuso, E. K. and Leiviskä, K., Studies in Informatics and Control 6(2):117(1997).

41. Waye, D. and Terroux, A., Pulp Paper Can. 92(1):83(1991).

42. Leiviskä, K., in IMACS Transactions on Scientific Computation –85, Volume III (B. Wahlsröm and K. Leiviskä, Ed.), Elsevier Science Publishers, Amsterdam, 1986, pp.169–173.

43. Shewchuk, C. F., Leaver, E. W., and Schindler, H. E., Pulp Paper Can. 92(9):53(1991).

44. Leiviskä, K., Komokallio, H., Aurasmaa, H., et al., Large Scale Systems 3(1):13(1982).

45. Oldberg, B., TAPPI 1983 Annual Meeting Proceedings, TAPPI PRESS, Atlanta, p. 17.

46. Nettamo, K., Lehtonen, K., and Muukari, P., PIMA, 68(1):56(1986).

47. Sutinen, R., Muukari, P., and Siro, M., Pulp Paper 58(7):93(1984).

48. Mault, K., Tappi J. 67(8):50(1984).

49. Barber, A. L., Murtha, S. A., and Glaser, D. C., TAPPI 1983 Annual Meeting Proceedings, TAPPI PRESS, Atlanta, p. 235.

50. Shewchuk, C. and Leaver, E. W., PIMA 68(1):40(1986).

51. Anderson, H. C., Euhus, L. E., and Parker, D. D., TAPPI J. 80(8):125(1997).

52. Shewchuk, C. F., TAPPI 1983 Annual Meeting Proceedings, TAPPI PRESS, Atlanta, p. 9.

Conversion factors

To convert numerical values found in this book in the RECOMMENDED FORM, divide by the indicated number to obtain the values in CUSTOMARY UNITS. This table is an excerpt from TIS 0800-01 "Units of measurement and conversion factors." The complete document containing additional conversion factors and references to appropriate TAPPI Test Methods is available at no charge from TAPPI, Technology Park/Atlanta, P. O. Box 105113, Atlanta GA 30348-5113 (Telephone: +1 770 209-7303, 1-800-332-8686 in the United States, or 1-800-446-9431 in Canada).

Property	To convert values expressed in RECOMMENDED FORM	Divide by	To obtain values expressed in CUSTOMARY UNITS
Area	square centimeters [cm^2]	6.4516	square inches [in^2]
	square meters [m^2]	0.0929030	square feet [ft^2]
	square meters [m^2]	0.8361274	square yards [yd^2]
Density	kilograms per cubic meter [kg/m^3]	16.01846	pounds per cubic foot [lb/ft^3]
	kilograms per cubic meter [kg/m^3]	1000	grams per cubic centimeter [g/cm^3]
Energy	joules [J]	1.35582	foot pounds-force [ft • lbf]
	joules [J]	9.80665	meter kilogams-force [m • kgf]
	millijoules [mJ]	0.0980665	centimeter grams-force [cm • gf]
	kilojoules [kJ]	1.05506	British thermal units, Int. [Btu]
	megajoules [MJ]	2.68452	horsepower hours [hp • h]
	megajoules [MJ]	3.600	kilowatt hours [kW • h or kWh]
	kilojoules [kJ]	4.1868	kilocalories, Int. Table [kcal]
	joules [J]	1	meter newtons [m • N]
Frequency	hertz [Hz]	1	cycles per second [s^{-1}]
Length	nanometers [nm]	0.1	angstroms [Å]
	micrometers [μm]	1	microns
	millimeters [mm]	0.0254	mils [mil or 0.001 in]
	millimeters [mm]	25.4	inches [in]
	meters [m]	0.3048	feet [ft]
	kilometers [km]	1.609	miles [mi]
Mass	grams [g]	28.3495	ounces [oz]
	kilograms [kg]	0.453592	pounds [lb]
	metric tons (tonne) [t] (= 1000 kg)	0.907185	tons (= 2000 lb)
Mass per unit area	grams per square meter [g/m^2]	3.7597	pounds per ream, 17 x 22 - 500
	grams per square meter [g/m^2]	1.4801	pounds per ream, 25 x 38 - 500
	grams per square meter [$g/m^2 m^2$]	1.4061	pounds per ream, 25 x 40 - 500
	grams per square meter [g/m^2]	4.8824	pounds per 1000 square feet [$lb/1000\ ft^2$]
	grams per square meter [g/m^2]	1.6275	pounds per 3000 square feet [$lb/3000\ ft^2$]
	grams per square meter [g/m^2]	1.6275	pounds per ream, 24 x 36 - 500

Conversion factors

Property	To convert values expressed in RECOMMENDED FORM	Divide by	To obtain values expressed In CUSTOMARY UNITS
Mass per unit volume	grams per liter [g/L]	7.48915	ounces per gallon [oz/gal]
	kilograms per liter [mg/m]	0.119826	pounds per gallon [lz/gal]
	kilograms per cubic meter [kg/m^3]	1	gram per liter [g/L]
	megagrams per cubic meter [Mg/m^3]	27.6799	pounds per cubic inch [lb/in^3]
	kilograms per cubic meter	16.0184	pounds per cubic foot [lb/ft^3]
Power	watts [W]	1.35582	foot pounds-force per second [ft·lbf/s]
	watts [W]	745.700	horsepower [hp] = 550 foot pounds-force per second
	kilowatts [kW]	0.74570	horsepower [hp]
	watts [W]	735.499	metric horsepower
Pressure, stress, force per unit area	kilopascals [kPa]	6.89477	pounds-force per square inch [lbf/in^2 or psi]
	Pascals [Pa]	47.8803	pounds-force per square foot [lbf/ft^2]
	kilopascals [kPa]	2.98898	feet of water (39.2°F) [ft H2O]
	kilopascals [kPa]	0.24884	inches of water (60°F) [in H2O]
	kilopascals [kPa]	3.38638	inches of mercury (32°F) [in Hg]
	kilopascals [kPa]	3.37685	inches of mercury (60°F) [in Hg]
	kilopascals [kPa]	0.133322	millimeters of mercury (0°C) [mm Hg]
	megapascals [Mpa]	0.101325	atmospheres [atm]
	Pascals [Pa]	98.0665	grams-force per square centimeter [gf/cm^2]
	Pascals [Pa]	1	newtons per square meter [N/m^2]
	kilopascals [kPa]	100	bars [bar]
Speed	meters per second [m/s]	0.30480	feet per second [ft/s]
	millimeters per second [mm/s]	5.080	feet per minute [ft/min or fpm]
Viscosity	Pascal seconds [Pa·s]	0.1	poise [P]
	millipascal seconds [mPa·s]	1	centipoise [cP]
Volume, fluid	milliliters [mL]	29.5735	ounces [oz]
	liters [L]	3.785412	gallons [gal]

Index

A
Adaptive control 24–25, 46, 71, 79
Advisory system 102, 111, 281
Autocorrelation 257–259
Autotuning 24

B
Basis weight 53, 64–66, 69, 147, 150,
    152–153, 157, 159–160, 163–169,
    172, 178, 184–185, 187, 190–191,
    228, 265, 268
Brightness 41–43, 53, 56–57, 61, 68, 91,
    93–97, 109, 187, 212, 228

C
Caliper 66–67, 147, 150, 168–172, 187, 267
Causticizing efficiency 116, 282
CD control 147, 150–152, 156, 158,
    165–166, 168–169, 173, 175–176
Closed-loop control 21, 30, 158
Coating weight 69, 173–176
Color 53, 68, 110, 186
Consistency 50–55, 58, 61–63, 71, 91,
    129–138, 148, 161, 164, 178–179,
    213, 221, 252, 289
Control charts 42, 96, 254
Controller tuning 23–26, 94, 287
Correlation analysis 44, 251, 259
Crosscorrelation 258–259
Cusum charts 43–44

D
Data reconciliation 275, 282
Decoupling 28, 153, 160, 162, 175
Diagnostics 31, 96, 111, 182, 266
Digester control .. 13, 25, 73, 81–82, 85–86, 88
Dilution factor 89, 91, 278
Dry line 164–165
Dynamic model 288

E
Efficiency monitoring 226, 229, 232
Expert system 30–32, 102–103, 111, 113, 124,
    218, 221, 225, 228, 230–231, 244,
    281, 283

F
Fiber length 53, 57–59, 61, 134–138, 178
Fiber orientation 66, 162–163, 166–167
Flexible production 182, 194
Formation 63, 65–66, 108, 117–118, 120,
    123, 147, 161, 165, 266
Fouling monitoring 101
Freeness 53, 57–59, 71, 129–130,
    133–139, 141–142, 178, 266
Fuzzy control 47, 124, 136
Fuzzy logic 30, 34–35, 94–95, 106, 109,
    114, 124, 137, 226

G
Gain scheduling 25
Gloss 67–68, 147, 168–169, 171–172, 266
Grade change 182–191, 208, 211, 227, 233

H
H-factor 75–78, 80, 83–86

I
Image analysis 54, 59, 105, 109, 112, 165
Impulse test 262
Intelligent control 136
Internal model control 118, 124

K
Kappa 40, 43–44, 54–57, 73–77, 79–81,
    84–86, 88, 91–95, 212, 255, 259,
    278

L
Linear programming ... 220, 239–240, 280–281
Load allocation 235–240

M

Mathematical model 249
MD control 147, 149–150, 152–155, 176
Millwide control 11, 16–17, 197–200, 206, 208–209, 214, 226–227, 234
Model reference control 25
Moisture 53–54, 64–65, 69–70, 73, 82–83, 118–121, 129–130, 133, 147, 150, 153, 160, 165–166, 168–169, 172, 176, 178, 189–190
Multi variable control 26–27, 118, 124

N

Neural network 37, 47, 91, 112, 125, 137–138, 185, 231–232
Normal distribution 254–256, 260

O

Opacity 66, 68
Open systems 14–15
Open-loop control 21, 189
Optimal state control 28
Optimal state controller 28
Optimization 11, 33, 36, 61, 106–109, 123–124, 129, 142, 152, 166, 168, 191, 197, 216, 218–221, 223, 225, 227, 231, 234–238, 240–241, 243–244, 249–251, 270, 275, 280–283, 287

P

Paper machine control . 13, 147–148, 150, 152
Pareto analysis 41, 96
Performance monitoring 268
Physical model 166
PID-algorithm 22
Porosity 67, 69, 107, 165, 168
Predictive control 75, 125, 185, 190, 194
Production monitoring 227, 229
Profitability monitoring 226, 233

Q

Quality monitoring 188, 211, 226, 268

R

Regression analysis ... 249, 259–261, 270–271
Relative precision index 45

Retention 51–52, 61–63, 78–79, 93, 96–97, 178–179, 211, 250, 259, 266

S

Scheduling 25, 32, 82–84, 206, 213–229, 243–244, 274, 287
Self-tuning control 23, 25–26, 118
Simulation 11, 33, 91, 102, 113, 184, 197, 202, 218–221, 224–225, 230, 235, 241, 249–250, 269, 271–291
Smoothness 67–68, 147, 150, 172
Species change 80–81, 217
Specific energy control 133, 138, 140–141
Statistical process control 40, 47, 113, 259, 265, 281
Steady-state model 288
Steam leveling 82, 84
Step test 261–262
Supervisory control 16, 31, 120, 123

T

t-distribution 255–256

W

Web inspection 180–182
Web tension 167–168

Z

Ziegler-Nichols method 24, 262